CALCULUS OF
VARIATIONS

CALCULUS OF VARIATIONS

With Applications to Physics
and Engineering

•

Robert Weinstock

*Professor of Physics, Oberlin College
(formerly Acting Assistant Professor,
Department of Mathematics,
Stanford University)*

DOVER PUBLICATIONS, INC.

NEW YORK

International Standard Book Number: 0-486-63069-2
Library of Congress Catalog Card Number: 74-75706

Manufactured in the United States of America
Dover Publications, Inc.
180 Varick Street
New York, N.Y. 10014

To Betty

PREFACE

There seems to have been published, up to the present time, no English-language volume in which an elementary introduction to the calculus of variations is followed by extensive application of the subject to problems of physics and theoretical engineering. The present volume is offered as partial fulfillment of the need for such a book. Thus its chief purpose is twofold:

(i) To provide for the senior or first-year graduate student in mathematics, science, or engineering an introduction to the ideas and techniques of the calculus of variations. (The material of the first seven chapters—with selected topics from the later chapters—has been used several times as the subject matter of a 10-week course in the Mathematics Department at Stanford University.)

(ii) To illustrate the application of the calculus of variations in several fields outside the realm of pure mathematics. (By far the greater emphasis is placed upon this second aspect of the book's purpose.)

The range of topics considered may be determined at a glance in the table of contents. Mention here of some of the more significant omissions may be pertinent:

The vague, mechanical "δ method" is avoided throughout. Thus, while no advantage is taken of a sometimes convenient shorthand tactic, there is eliminated a source of confusion which often grips the careful student when confronted with its use.

No attempt is made to treat problems of sufficiency or existence: no consideration is taken of the "second variation" or of the conditions of Legendre, Jacobi, and Weierstrass. Besides being outside the scope of the chief aim of this book, these matters are excellently treated in the volumes of Bolza and Bliss listed in the Bibliography.

Expansion theorems for the eigenfunctions associated with certain boundary-value problems are stated without proof. The proofs, beyond the scope of this volume, can be constructed, in most instances, on the basis of the theory of integral equations.

Space limitations prevent inclusion of such topics as perturbation theory, heat flow, hydrodynamics, torsion and buckling of bars, Schwinger's treatment of atomic scattering, and others. However, the reader who has mastered the essence of the material included should have little difficulty in applying the calculus of variations to most of the subjects which have been squeezed out.

vii

It is hardly necessary to mention the debt I owe to nearly all the works mentioned in the Bibliography; Courant-Hilbert has been especially helpful. In the early stages of the work, comments from my former students Gordon Kent and Peter Szegö were useful. Occasional chats with colleagues in the Stanford Mathematics Department were similarly helpful. I owe a tremendous debt of gratitude to my wife, Elizabeth B. Weinstock, whose keen critical faculty is responsible for several important corrective changes in the text, who worked out nearly all the exercises, who did all the typing of the final draft, and whose complete companionship-in-effort has made the writing of this book a happy experience.

ROBERT WEINSTOCK

PALO ALTO, CALIF.
April, 1952

CONTENTS

CHAPTER 1

INTRODUCTION

The definite integral

$$I = \int_{x_1}^{x_2} f\left(x, y, \frac{dy}{dx}\right) dx \tag{1}$$

is a well-defined quantity—a number—when x_1 and x_2 have definite numerical values, when the integrand f is given as a function of the arguments x, y, (dy/dx), and *when y is given as a function of x*. The "first" problem of the calculus of variations involves comparison of the various values assumed by (1) when different choices of y as a function of x are substituted into the integrand of (1). What is sought, specifically, is the particular function $y = y(x)$ that gives to (1) its *minimum* (or *maximum*) value. Explicit examples of this type of problem are given detailed treatment in Chap. 3. These include the problems of "the shortest distance between two points on a given surface," "the curve of quickest descent between two points," and "the surface of revolution of minimum area."

Generalization of the first problem is effected in many directions. For example, the integrand of (1) may be replaced by a function of several dependent variables, with respect to which a minimum (or maximum) of the definite integral is sought. Further, the functions with respect to which the minimization (or maximization) is carried out may be required to satisfy certain subsidiary conditions. Explicit examples of various aspects of these generalizations are handled in Chaps. 3 and 4. An important special case is the problem of "the maximum area bounded by a closed curve of given perimeter."

Another line along which generalization is pursued is the replacement of (1) by a *multiple* integral whose minimum (or maximum) is sought with respect to one or more functions of the independent variables of integration. Thus, for example, we seek to minimize the double integral

$$\iint_D f\left(x, y, w, \frac{\partial w}{\partial x}, \frac{\partial w}{\partial y}\right) dx\, dy, \tag{2}$$

carried out over a fixed domain D of the xy plane, with respect to functions $w = w(x,y)$. Such problems are dealt with in the opening sections of Chaps. 7 and 9.

1

The techniques of solving the problems of minimizing (or maximizing) (1), (2), and related definite integrals are intimately connected with the problems of maxima and minima that are encountered in the elementary differential calculus.   If, for example, we seek to determine the values for which the function $y = g(x)$ achieves a minimum (or maximum), we form the derivative $(dy/dx) = g'(x)$, set $g'(x) = 0$, and solve for x.   The roots of this equation—the only values of x for which $y = g(x)$ can possibly achieve a minimum[1] (or maximum)—do not, however, necessarily designate the locations of minima (or even of maxima).   The condition $g'(x) = 0$ is merely a *necessary* condition for a minimum (or maximum); conditions of sufficiency involve derivatives of higher order than the first. The vanishing of $g'(x)$ for a given value of x implies merely that the curve representing $y = g(x)$ has a horizontal tangent at that value of x.   A horizontal tangent may imply one of the three circumstances: maximum, minimum, or horizontal inflection; we call any one of the three an *extremum* of $y = g(x)$.

The treatment of many of the problems of the calculus of variations in this volume is analogous to the treatment of maximum and minimum problems through the use of the first derivative only; quite often we merely derive a set of necessary conditions for a minimum (or maximum) and rely upon geometric or physical intuition to establish the applicability of our solution.   In other cases our interest lies only in the attainment of an extremum; in these it is immaterial whether we have a maximum, minimum, or a condition analogous to a horizontal inflection in the elementary case.   The methods involved in establishing the conditions *sufficient* for a minimum (or maximum)—and in proving the *existence* of a minimum (or maximum)—are extremely profound and intricate; such investigations are found elsewhere in the literature.[2]

The chief purpose of the present work is to illustrate the application of the calculus of variations in several fields outside the realm of pure mathematics.   Such applications are found in the chapters following Chap. 4.   By no means can the treatment here of any special field be considered exhaustive in its relationship to the calculus of variations; each of several of the later chapters is amenable to expansion to the length of a volume the size of the present one.

The reader is expected to have as a part of his (or her) permanent knowledge most of the concepts and techniques learned in a first-year calculus course, including a smattering of ordinary differential equations. Furthermore, he (or she) must be familiar with many of the matters

---

[1] It is clear that here "minimum" (or "maximum") refers to *relative* minimum (or *relative* maximum).

[2] For example, see Bliss (1,2), Bolza, and Courant (1) listed in the Bibliography.

encountered in a short course in advanced calculus. Practically all the required results from this latter category are collected in Chap. 2; the corresponding proofs may be found in texts listed in the Bibliography.[1] With one brief exception (11-2), no use is made of the methods of vector analysis. The same statement holds for the use of complex numbers; in the absence of a statement to the contrary, all quantities that appear are to be assumed real.

The wider the reader's knowledge of physics, quite naturally, the fuller will be his (or her) appreciation of several of the results achieved in later chapters. Only the barest acquaintance with the concepts of elementary physics is presupposed, however; the reader to whom the study of physics is *completely* foreign will experience difficulty in following the development at only a very few points.

With respect to purpose the exercises at the end of each chapter may be divided, roughly, into three categories: (i) filling in of details in the development of the text, (ii) illustration of methods and results treated in the text, and (iii) extension of the results achieved in the text. In nearly all cases adequate hints are given; often these hints appear only as final answers.

Study should begin with Chap. 3. The material of Chap. 2 should be referred to only as it is required in the work following.

[1] Goursat, Franklin, and Kellogg.

CHAPTER 2

BACKGROUND PRELIMINARIES

2-1. Piecewise Continuity, Piecewise Differentiability

(a) Let $x \to x_0^-$ denote "x approaches x_0 from the left" and $x \to x_0^+$ denote "x approaches x_0 from the right." In this volume we consider only those functions $f(x)$ for which $\lim_{x \to x_0^-} f(x)$ and $\lim_{x \to x_0^+} f(x)$ both exist for all x_0 *interior* to the interval $(x_1 \leqq x \leqq x_2)$ in which $f(x)$ is defined. At the respective end points we require the existence of $\lim_{x \to x_1^+} f(x)$ and $\lim_{x \to x_2^-} f(x)$. If, for $x_1 < x_0 < x_2$, $\lim_{x \to x_0^-} f(x) = \lim_{x \to x_0^+} f(x) = f(x_0)$, then $f(x)$ is continuous at $x = x_0$; otherwise $f(x)$ exhibits a *jump discontinuity* at $x = x_0$. If $\lim_{x \to x_1^+} f(x) = f(x_1)$, then $f(x)$ is continuous at the left-hand end point $x = x_1$; otherwise $f(x)$ exhibits a jump discontinuity at $x = x_1$. An equivalent statement holds for the right-hand end point $x = x_2$.

A function is said to be *piecewise continuous* in an interval if it possesses at most a finite number of jump discontinuities in the interval.

(b) A function is said to be differentiable at $x = x_0$ if the limit as $x \to x_0$ of the ratio $(\Delta f / \Delta x) = \{[f(x) - f(x_0)]/(x - x_0)\}$ exists. If $\lim_{x \to x_0^-} (\Delta f / \Delta x)$ exists, the function is said to have a left-hand derivative at $x = x_0$; if $\lim_{x \to x_0^+} (\Delta f / \Delta x)$ exists, the function is said to have a right-hand derivative at $x = x_0$.

A function is said to be *piecewise differentiable* in $x_1 \leqq x \leqq x_2$ if it possesses a right- and left-hand derivative at every interior point of the interval and if the two are equal at all but a finite number of points of the interval. Further, the function must possess a right-hand derivative at $x = x_1$ and a left-hand derivative at $x = x_2$. Any point at which the right- and left-hand derivatives are unequal we label "a point of discontinuity of the derivative."

We eliminate consideration of any function whose derivative undergoes infinitely many changes of sign in a finite interval. This elimination precludes, incidentally, the appearance of any function of which the derivative is discontinuous at a point although the right- and left-hand derivatives are equal at the point.

4

2-2.   Partial and Total Differentiation

(a) If $u = f(x,y, \ldots ,z)$, $x = x(r,s, \ldots ,t)$, $y = y(r,s, \ldots ,t), \ldots ,$ $z = z(r,s, \ldots ,t)$, then

$$\frac{\partial u}{\partial r} = \frac{\partial f}{\partial x}\frac{\partial x}{\partial r} + \frac{\partial f}{\partial y}\frac{\partial y}{\partial r} + \cdots + \frac{\partial f}{\partial z}\frac{\partial z}{\partial r}, \tag{1}$$

where r may successively be replaced by $s, \ldots , t$.

(b) If $u = f(x,y, \ldots ,z,t)$, $x = x(t)$, $y = y(t), \ldots , z = z(t)$, then

$$\frac{du}{dt} = \frac{\partial f}{\partial t} + \frac{\partial f}{\partial x}\frac{dx}{dt} + \frac{\partial f}{\partial y}\frac{dy}{dt} + \cdots + \frac{\partial f}{\partial z}\frac{dz}{dt}. \tag{2}$$

(c) The quantity $p(x,y) + q(x,y)y'$—where the prime indicates ordinary differentiation with respect to x—is the derivative (dg/dx) of some function $g(x,y)$ if and only if $(\partial p/\partial y) = (\partial q/\partial x)$. In this event $p = (\partial g/\partial x)$, $q = (\partial g/\partial y)$.

2-3.   Differentiation of an Integral

(a) If

$$I = I(\epsilon) = \int_{x_1(\epsilon)}^{x_2(\epsilon)} f(x,\epsilon)dx,$$

then

$$\frac{dI}{d\epsilon} = I'(\epsilon) = f(x_2,\epsilon)\frac{dx_2}{d\epsilon} - f(x_1,\epsilon)\frac{dx_1}{d\epsilon} + \int_{x_1(\epsilon)}^{x_2(\epsilon)} \frac{\partial f}{\partial \epsilon}dx, \tag{3}$$

provided $(\partial f/\partial \epsilon)$ is a continuous function of ϵ and of x in $x_1 \leqq x \leqq x_2$. In case x_1 and x_2 are strictly constant (independent of ϵ), the right-hand member of (3) reduces to its final term.

(b) If the integrand f of a multiple integral I is a function of a parameter ϵ, as well as of the variables of integration, the derivative $(dI/d\epsilon)$ is computed by replacing f by $(\partial f/\partial \epsilon)$ as integrand function. It is assumed that the region of integration is fixed (independent of ϵ) and that $(\partial f/\partial \epsilon)$ is a continuous function of ϵ and the variables of integration.

2-4. Integration by Parts

We repeatedly employ the rule for integration by parts

$$\int_{x_1}^{x_2} g\frac{df}{dx}dx = gf\Big]_{x_1}^{x_2} - \int_{x_1}^{x_2} f\frac{dg}{dx}dx, \tag{4}$$

in which it is required that f and g be everywhere continuous but *merely piecewise differentiable* in $x_1 \leqq x \leqq x_2$.

2-5. Euler's Theorem on Homogeneous Functions

A function $F(x,y, \ldots ,z,u,v, \ldots ,w)$ is said to be homogeneous, of degree n, in the variables $u, v, \ldots , w$ if, for arbitrary h,

$$F(x,y, \ldots ,z,hu,hv, \ldots ,hw) = h^n F(x,y, \ldots ,z,u,v, \ldots ,w). \quad (5)$$

Any function for which (5) holds satisfies Euler's therorem:

$$u \frac{\partial F}{\partial u} + v \frac{\partial F}{\partial v} + \cdots + w \frac{\partial F}{\partial w} = nF(x,y, \ldots ,z,u,v, \ldots ,w). \quad (6)$$

2-6. Method cf Undetermined Lagrange Multipliers

A necessary condition for a minimum (or maximum) of $F(x,y, \ldots ,z)$ with respect to variables $x, y, \ldots , z$ that satisfy

$$G_i (x,y, \ldots ,z) = C_i \quad (i = 1,2, \ldots ,N), \quad (7)$$

where the C_i are given constants, is

$$\frac{\partial F^*}{\partial x} = \frac{\partial F^*}{\partial y} = \cdots = \frac{\partial F^*}{\partial z} = 0, \quad (8)$$

where $F^* = F + \sum_{i=1}^{N} \lambda_i G_i$. The constants $\lambda_1, \lambda_2, \ldots , \lambda_N$—introduced as undetermined *Lagrange multipliers*—are evaluated, together with the minimizing (or maximizing) values of $x, y, \ldots , z$, by means of the set of equations consisting of (7) and (8).

2-7. The Line Integral

(a) The *line integral* of the function $f(x,y,z)$ from P_1 to P_2 along the finite curve C (assumed to consist of a finite number of smooth arcs) is defined as follows:

We subdivide C into N arcs of lengths $\Delta s_1, \Delta s_2, \ldots , \Delta s_N$. The function $f(x,y,z)$ is evaluated at an arbitrary point (x_k,y_k,z_k) of the kth subdivision and the product $f(x_k,y_k,z_k)\Delta s_k$ is formed, for each $k = 1, 2, \ldots , N$. We form the sum $S_N = \sum_{k=1}^{N} f(x_k,y_k,z_k)\Delta s_k$ and proceed to refine the subdivision in such fashion that N increases without limit and the largest Δs_k approaches zero. If the limit of S_N with respect to this unlimited refinement exists (independently of the specific modes of subdivision), it is by definition

$$\lim S_N = \int_C f(x,y,z)ds \quad (9)$$

—the line integral of f from P_1 to P_2 along C.

Other forms of the line integral are

$$\int_C f(x,y,z)dx, \qquad \int_C f(x,y,z)dy, \qquad \int_C f(x,y,z)dz. \qquad (10)$$

In terms of the definition of (9) these are respectively equal to

$$\int_C \left[f(x,y,z) \frac{dx}{ds} \right] ds, \qquad \int_C \left[f(x,y,z) \frac{dy}{ds} \right] ds, \qquad \int_C \left[f(x,y,z) \frac{dz}{ds} \right] ds.$$

Since, however, the derivatives (dx/ds), (dy/ds), (dz/ds)—computed with respect to the curve C—have algebraic signs that depend upon the direction (along C) assigned to the increase of s, the complete specification of each of (10) requires a statement as to the direction (from P_1 to P_2 or from P_2 to P_1) in which the integration is carried out, *i.e.*, the assignment of the direction in which s is assumed to increase. (Thus, any one of the integrals (10) carried out along C from P_1 to P_2 is the negative of the same integral carried out along C from P_2 to P_1.)

(*b*) To evaluate (9) we introduce the parametric equations $x = x(t)$, $y = y(t)$, $z = z(t)$ of the arc C (where t increases in the direction of increasing s) to form the definite integral

$$\int_{t_1}^{t_2} f(x(t),y(t),z(t)) \sqrt{\left(\frac{dx}{dt}\right)^2 + \left(\frac{dy}{dt}\right)^2 + \left(\frac{dz}{dt}\right)^2} \, dt \qquad (t_1 < t_2), \quad (11)$$

where t_1 and t_2 are the values of t which denote the respective end points of C. (The parameter t is in some cases conveniently chosen to be one of the variables x, y, z, or even s.) The definite integral (11) provides the evaluation of the line integral (9); for the evaluation of the integrals (10), the radical of (11) is replaced respectively by (dx/dt), (dy/dt), (dz/dt).

(*c*) An important example of a line integral is

$$I = \frac{1}{2} \int_C (x \, dy - y \, dx) = \frac{1}{2} \int_{t_1}^{t_2} \left(x \frac{dy}{dt} - y \frac{dx}{dt} \right) dt, \qquad (12)$$

taken counterclockwise about a simple closed curve C in the xy plane. Here the parameter t is chosen so that the point $[x(t),y(t)]$ traverses C once in the counterclockwise sense as t increases from t_1 to t_2. The integral (12) is equal to the *area* enclosed by C.

(*d*) Quite often involved in the integrand of a line integral taken about a simple closed curve C in the xy plane is the *normal derivative* of a function $w(x,y)$. The (outward) normal derivative is defined as

$$\lim_{\Delta n \to 0} \frac{w(x,y) - w(x',y')}{\Delta n} = \frac{\partial w}{\partial n},$$

where (x,y) lies on C, (x',y') lies interior to C on the normal drawn to C at (x,y), and Δn is the distance from (x',y') to (x,y) measured along the normal.

A useful relation is

$$\frac{\partial w}{\partial n} = \frac{\partial w}{\partial x}\frac{dy}{ds} - \frac{\partial w}{\partial y}\frac{dx}{ds},\tag{13}$$

where (dy/ds) and (dx/ds) are computed with respect to C.

2-8. Determinants

(a) The general nth-order determinant

$$\begin{vmatrix} a_{11} & a_{12} & \ldots & a_{1n} \\ a_{21} & a_{22} & \ldots & a_{2n} \\ \ldots & \ldots & \ldots & \ldots \\ a_{n1} & a_{n2} & \ldots & a_{nn} \end{vmatrix}\tag{14}$$

is by definition a linear homogeneous function of the elements a_{k1}, $a_{k2}, \ldots, a_{kn}$ of the kth row, for each $k = 1, 2, \ldots, n$, such that it is identically zero if two rows are identical and has the value 1 when $a_{jk} = 0$ $(k \neq j)$ and $a_{kk} = 1$ $(k = 1,2, \ldots ,n)$. In the special case $n = 2$, the definition provides

$$\begin{vmatrix} a_{11} & a_{12} \\ a_{21} & a_{22} \end{vmatrix} = a_{11}a_{22} - a_{21}a_{12};$$

for $n = 3$,

$$\begin{vmatrix} a_{11} & a_{12} & a_{13} \\ a_{21} & a_{22} & a_{23} \\ a_{31} & a_{32} & a_{33} \end{vmatrix} = a_{11}\begin{vmatrix} a_{22} & a_{23} \\ a_{32} & a_{33} \end{vmatrix} + a_{12}\begin{vmatrix} a_{23} & a_{21} \\ a_{33} & a_{31} \end{vmatrix} + a_{13}\begin{vmatrix} a_{21} & a_{22} \\ a_{31} & a_{32} \end{vmatrix}.$$

(b) A system of n simultaneous linear homogeneous equations

$$\sum_{k=1}^{n} a_{jk}x_k = 0 \qquad (j = 1,2, \ldots ,n)$$

in the n unknowns $x_1, x_2, \ldots, x_n$ has a nontrivial solution—whereby *not all* the x_k are equal to zero—if and only if the determinant (14) of the coefficients vanishes.

(c) The product of two nth-order determinants whose elements are denoted respectively by a_{jk} and b_{jk} $(j,k = 1,2, \ldots ,n$, independently) is the nth-order determinant whose elements are

$$c_{jk} = a_{j1}b_{k1} + a_{j2}b_{k2} + \cdots + a_{jn}b_{kn}$$

$$(j,k = 1,2, \ldots ,n, \text{ independently}).$$

(d) If the elements of (14) are differentiable functions of a variable x, the derivative of (14) with respect to x is the sum of n determinants, the kth of which is formed by replacing each element of the kth row of (14) by its derivative with respect to x, for $k = 1, 2, \ldots, n$.

(e) A set of functions $\phi_1(x,y, \ldots ,z)$, $\phi_2(x,y, \ldots ,z)$, $\ldots$, $\phi_n(x,y, \ldots ,z)$ is said to be *linearly independent* if no relation of the form

$$A_1\phi_1 + A_2\phi_2 + \cdots + A_n\phi_n = 0,$$

where $A_1, A_2, \ldots, A_n$ are constants, holds identically in $x, y, \ldots, z$ unless $A_1 = A_2 = \cdots = A_n = 0$. Otherwise—if such a relation holds in which one or more of the A_k differ from zero—the functions are said to be *linearly dependent*.

If the functions $\phi_1(x)$, $\phi_2(x)$, $\ldots$, $\phi_n(x)$ all satisfy the same nth-order linear homogeneous differential equation, a necessary and sufficient condition that they be linearly dependent is the identical vanishing of their *wronskian*

$$w = \begin{vmatrix} \phi_1(x) & \phi_2(x) & \ldots & \phi_n(x) \\ \phi_1'(x) & \phi_2'(x) & \ldots & \phi_n'(x) \\ \cdots\cdots\cdots\cdots\cdots\cdots\cdots\cdots \\ \phi_1^{(n-1)}(x) & \phi_2^{(n-1)}(x) & \ldots & \phi_n^{(n-1)}(x) \end{vmatrix}, \tag{15}$$

where $\phi^{(k)}(x)$ is the kth derivative of ϕ with respect to x, for $k = 1, 2, \ldots, n-1$. (The prime replaces the superscript 1 in case $k = 1$.)

(f) The functional determinant, or *jacobian*, of $u_1, u_2, \ldots, u_N$ with respect to $x_1, x_2, \ldots, x_N$ is defined as

$$\frac{\partial(u_1,u_2, \ldots ,u_N)}{\partial(x_1,x_2, \ldots ,x_N)} = \begin{vmatrix} \dfrac{\partial u_1}{\partial x_1} & \dfrac{\partial u_2}{\partial x_1} & \ldots & \dfrac{\partial u_N}{\partial x_1} \\ \dfrac{\partial u_1}{\partial x_2} & \dfrac{\partial u_2}{\partial x_2} & \ldots & \dfrac{\partial u_N}{\partial x_2} \\ \cdots\cdots\cdots\cdots\cdots\cdots \\ \dfrac{\partial u_1}{\partial x_N} & \dfrac{\partial u_2}{\partial x_N} & \ldots & \dfrac{\partial u_N}{\partial x_N} \end{vmatrix}.$$

If $u_1, u_2, \ldots, u_N$ are differentiable functions of $y_1, y_2, \ldots, y_N$, and $y_1, y_2, \ldots, y_N$ are differentiable functions of $x_1, x_2, \ldots, x_N$, then

$$\frac{\partial(u_1,u_2, \ldots ,u_N)}{\partial(x_1,x_2, \ldots ,x_N)} = \frac{\partial(u_1,u_2, \ldots ,u_N)}{\partial(y_1,y_2, \ldots ,y_N)} \cdot \frac{\partial(y_1,y_2, \ldots ,y_N)}{\partial(x_1,x_2, \ldots ,x_N)}.$$

The change of coordinate variables $x = x(u,v,w)$, $y = y(u,v,w)$, $z = z(u,v,w)$ is a one-to-one correspondence in any region of space in which the jacobian $[\partial(x,y,z)/\partial(u,v,w)]$ does not vanish. In two dimensions a change of plane coordinate variables $x = x(u,v)$, $y = y(u,v)$ is a one-to-one correspondence in any region of the xy plane in which the jacobian $[\partial(x,y)/\partial(u,v)]$ does not vanish.

A change of variables $x = x(u,v,w)$, $y = y(u,v,w)$, $z = z(u,v,w)$ for the evaluation of a triple integral is carried out according to the rule

$$\iiint\limits_{R} F(x,y,z)dx\,dy\,dz = \iiint\limits_{R'} f(u,v,w)\left|\frac{\partial(x,y,z)}{\partial(u,v,w)}\right|\,du\,dv\,dw, \qquad (16)$$

where f is the function F expressed in terms of u, v, w, and R' is the region R, but described by the variables u, v, w. A formula completely analogous to (16) holds for the transformation of double integrals.

2-9. Formula for Surface Area

If $z = z(x,y)$ is a single-valued continuously differentiable function of x and y, the area of a portion of the surface represented by this function is given by

$$I = \iint\limits_{D}\left[1 + \left(\frac{\partial z}{\partial x}\right)^{2} + \left(\frac{\partial z}{\partial y}\right)^{2}\right]^{\frac{1}{2}} dx\,dy,$$

where the integration is carried out over the domain D of the xy plane onto which the given portion of the surface projects.

2-10. Taylor's Theorem for Functions of Several Variables

If, in some neighborhood of $(x_0,y_0, \ldots ,z_0)$, $F(x,y, \ldots ,z)$ possesses partial derivatives of order N with respect to all combinations of the variables x, y, $\ldots$, z, we have the expansion, valid in that neighborhood,

$$F(x,y, \ldots ,z) = F(x_0,y_0, \ldots ,z_0) + \left(\xi\frac{\partial}{\partial x} + \eta\frac{\partial}{\partial y} + \cdots + \zeta\frac{\partial}{\partial z}\right)F\bigg|_0$$

$$+ \frac{1}{2!}\left(\xi\frac{\partial}{\partial x} + \eta\frac{\partial}{\partial y} + \cdots + \zeta\frac{\partial}{\partial z}\right)^{2}F\bigg|_0 + \cdots$$

$$+ \frac{1}{N!}\left(\xi\frac{\partial}{\partial x} + \eta\frac{\partial}{\partial y} + \cdots + \zeta\frac{\partial}{\partial z}\right)^{N}F\bigg|_{\theta},$$

where $\xi = x - x_0$, $\eta = y - y_0$, $\ldots$, $\zeta = z - z_0$. Each "power" of $[\xi(\partial/\partial x) + \eta(\partial/\partial y) + \cdots + \zeta(\partial/\partial z)]$ is formed according to the laws of algebra, but with the coefficient of $\xi^i\eta^j \ldots \zeta^k$ interpreted as $(\partial^{i+j+\cdots+k}/\partial x^i\partial y^j \ldots \partial z^k)$ multiplied by the proper numerical factor; the subscript "0" implies the evaluation of the derivatives at $x = x_0$, $y = y_0$, $\ldots$, $z = z_0$; and the subscript "θ" implies the evaluation of the Nth-order derivatives at $x = x_0 + \theta\xi$, $y = y_0 + \theta\eta$, $\ldots$, $z = z_0 + \theta\zeta$ $(0 < \theta < 1)$.

2-11. The Surface Integral

(a) The *surface integral* of the function $f(x,y,z)$ over the given finite surface B (assumed to consist of a finite number of smooth portions bounded by curves composed of a finite number of smooth arcs) is defined as follows:

We subdivide B into N portions of area $\Delta S_1, \Delta S_2, \ldots, \Delta S_N$. The function $f(x,y,z)$ is evaluated at an arbitrary point (x_k, y_k, z_k) of the kth subdivision, and we form the product $f(x_k, y_k, z_k)\Delta S_k$ for each $k = 1, 2, \ldots, N$. We form the sum $\Sigma_N = \sum_{k=1}^{N} f(x_k, y_k z_k)\Delta S_k$ and proceed to refine the subdivision in such fashion that N increases without limit and the greatest distance between pairs of points of any subdivision approaches zero. If the limit of Σ_N with respect to this unlimited refinement exists (independently of the specific modes of subdivision), it is by definition

$$\lim \Sigma_N = \iint_B f(x,y,z)dS \tag{17}$$

—the surface integral of f over B.

(b) For the evaluation of (17) one introduces a set of surface coordinates (v,w) such that one and only one pair of values of these variables defines a single point on B through relations of the form $x = x(v,w)$, $y = y(v,w)$, $z = z(v,w)$. With the introduction of these parametric equations, the surface integral (17) is evaluated as the double integral

$$\iint f(x(v,w), y(v,w), z(v,w)) \sqrt{EF - G^2}\, dv\, dw$$

carried out over the values of v and w that completely describe B, where

$$E = \left(\frac{\partial x}{\partial v}\right)^2 + \left(\frac{\partial y}{\partial v}\right)^2 + \left(\frac{\partial z}{\partial v}\right)^2, \qquad F = \left(\frac{\partial x}{\partial w}\right)^2 + \left(\frac{\partial y}{\partial w}\right)^2 + \left(\frac{\partial z}{\partial w}\right)^2,$$
$$G = \frac{\partial x}{\partial v}\frac{\partial x}{\partial w} + \frac{\partial y}{\partial v}\frac{\partial y}{\partial w} + \frac{\partial z}{\partial v}\frac{\partial z}{\partial w}.$$

(In case the curves $v = $ constant meet the curves $w = $ constant at right angles, the quantity G vanishes identically.)

(c) Quite often involved in the integrand of a surface integral carried out over a closed surface B is the normal derivative of a function $U(x,y,z)$. The (outward) *normal derivative* is defined as

$$\lim_{\Delta n \to 0} \frac{U(x,y,z) - U(x',y',z')}{\Delta n} = \frac{\partial U}{\partial n},$$

where (x,y,z) lies on B, (x',y',z') lies interior to B on the normal drawn to B at (x,y,z), and Δn is the distance from (x',y',z') to (x,y,z) measured along the normal.

A useful relation is

$$\frac{\partial U}{\partial n} = \frac{\partial U}{\partial x} \cos (n,x) + \frac{\partial U}{\partial y} \cos (n,y) + \frac{\partial U}{\partial z} \cos (n,z), \qquad (18)$$

where $\cos (n,x)$ is the cosine of the angle between the positive x direction and the normal drawn outward from B at the point at which $(\partial U/\partial n)$ is computed. Cos (n,y) and cos (n,z) have corresponding meanings.

A second useful expression for the (outward) normal derivative is

$$\frac{\partial U}{\partial n} = \pm \frac{\dfrac{\partial U}{\partial x}\dfrac{\partial u}{\partial x} + \dfrac{\partial U}{\partial y}\dfrac{\partial u}{\partial y} + \dfrac{\partial U}{\partial z}\dfrac{\partial u}{\partial z}}{\sqrt{\left(\dfrac{\partial u}{\partial x}\right)^2 + \left(\dfrac{\partial u}{\partial y}\right)^2 + \left(\dfrac{\partial u}{\partial z}\right)^2}}, \qquad (19)$$

where $u(x,y,z) =$ constant is the equation of the surface B. The plus sign is chosen if $(\partial u/\partial n) > 0$—i.e., if $u(x,y,z)$ increases along the normal drawn outward from B; the minus sign is chosen if $(\partial u/\partial n) < 0$.

2-12. Gradient, Laplacian

(a) The *gradient* of the function $\phi(x,y,z)$, denoted by $\nabla\phi$, is defined as the vector whose cartesian components are respectively $(\partial\phi/\partial x)$, $(\partial\phi/\partial y)$, $(\partial\phi/\partial z)$. The magnitude $|\nabla\phi|$ of $\nabla\phi$—the square root of the sum of the squares of the three components—is the normal derivative $(\partial\phi/\partial n)$, where the positive normal direction is perpendicular to the surface $\phi(x,y,z) =$ constant, in the direction of increasing ϕ.

(b) The *scalar product* of two vectors, defined as the product of the respective magnitudes of the vectors multiplied by the cosine of the angle between their directions, is equal to the sum of the products of their respective cartesian components. In particular, if two vectors have the same direction, the scalar product of the two is the product of their magnitudes.

(c) The *laplacian* of a function $\phi(x,y,z)$ is defined as

$$\nabla^2\phi = \frac{\partial^2\phi}{\partial x^2} + \frac{\partial^2\phi}{\partial y^2} + \frac{\partial^2\phi}{\partial z^2}. \qquad (20)$$

If ϕ depends only on x and y, the final term of (20) drops out. In this case $\nabla^2\phi$ is said to denote the *two-dimensional laplacian*.

2-13. Green's Theorem (Two Dimensions)

We consider a domain D of the xy plane bounded by a simple closed curve C that consists of a finite number of smooth arcs. The line integrals which appear are carried out along C in the sense that an observer walking forward along C in the direction of integration constantly has D on his (or her) left.

(a) If $P(x,y)$ and $Q(x,y)$ are everywhere continuous in D and piecewise continuous along C, and if D may be subdivided into a finite number of subdomains in each of which the first partial derivatives of P and Q are continuous, then

$$\iint_D \left(\frac{\partial P}{\partial x} + \frac{\partial Q}{\partial y}\right) dx\, dy = \int_C (P\, dy - Q\, dx). \qquad (21)$$

(b) By writing $P = \eta G$, $Q = \eta F$ in (21), we obtain the two-dimensional analogue of integration by parts

$$\iint_D \left(G\frac{\partial \eta}{\partial x} + F\frac{\partial \eta}{\partial y}\right) dx\, dy = -\iint_D \eta \left(\frac{\partial G}{\partial x} + \frac{\partial F}{\partial y}\right) dx\, dy + \int_C \eta (G\, dy - F\, dx).$$
$$(22)$$

(c) By writing $\eta = \psi$, $G = (\partial \phi/\partial x)$, $F = (\partial \phi/\partial y)$ in (22), we obtain, with the aid of (13) of 2-7(d),

$$\iint_D \psi \nabla^2 \phi\, dx\, dy = -\iint_D \left(\frac{\partial \phi}{\partial x}\frac{\partial \psi}{\partial x} + \frac{\partial \phi}{\partial y}\frac{\partial \psi}{\partial y}\right) dx\, dy + \int_C \psi \frac{\partial \phi}{\partial n}\, ds, \quad (23)$$

with the definition (20) of the (two-dimensional) laplacian.

An important special case of (23) is achieved by setting $\psi = \phi$.

(d) By interchanging ϕ and ψ in (23) and by subtracting the result from (23), we obtain the Green's formula

$$\iint_D (\psi \nabla^2 \phi - \phi \nabla^2 \psi)dx\, dy = \int_C \left(\psi \frac{\partial \phi}{\partial n} - \phi \frac{\partial \psi}{\partial n}\right) ds. \qquad (24)$$

(e) By setting $Q = 0$, $P = [G(\partial \eta/\partial x) - \eta(\partial G/\partial x)]$ in (21), we obtain

$$\iint_D G\frac{\partial^2 \eta}{\partial x^2}\, dx\, dy = \iint_D \eta \frac{\partial^2 G}{\partial x^2}\, dx\, dy + \int_C \left(G\frac{\partial \eta}{\partial x} - \eta \frac{\partial G}{\partial x}\right) dy. \qquad (25)$$

Further, the use of $P = 0$, $Q = [G(\partial \eta/\partial y) - \eta(\partial G/\partial y)]$ in (21) provides

$$\iint_D G\frac{\partial^2 \eta}{\partial y^2}\, dx\, dy = \iint_D \eta \frac{\partial^2 G}{\partial y^2}\, dx\, dy - \int_C \left(G\frac{\partial \eta}{\partial y} - \eta \frac{\partial G}{\partial y}\right) dx. \qquad (26)$$

By setting $P = \frac{1}{2}[G(\partial\eta/\partial y) - \eta(\partial G/\partial y)]$, $Q = \frac{1}{2}[G(\partial\eta/\partial x) - \eta(\partial G/\partial x)]$ in (21), we obtain, finally,

$$\iint_D G \frac{\partial^2 \eta}{\partial x \, \partial y} \, dx \, dy = \iint_D \eta \frac{\partial^2 G}{\partial x \, \partial y} \, dx \, dy + \frac{1}{2} \int_C \left(G \frac{\partial \eta}{\partial y} - \eta \frac{\partial G}{\partial y} \right) dy$$

$$- \frac{1}{2} \int_C \left(G \frac{\partial \eta}{\partial x} - \eta \frac{\partial G}{\partial x} \right) dx. \quad (27)$$

2-14. Green's Theorem (Three Dimensions)

We consider a region R bounded by the surface B which consists of a finite number of smooth sections. (It may happen that B consists of two or more unconnected portions, as in the case of a "hollow" region.)

(a) We let $U(x,y,z)$, $V(x,y,z)$, $W(x,y,z)$ be continuous in R and suppose that B may be subdivided into a finite number of portions on each of which U, V, W are continuous. Further, we assume that R may be subdivided into a finite number of subregions in each of which the first partial derivatives of U, V, W are continuous. Then

$$\iiint_R \left(\frac{\partial U}{\partial x} + \frac{\partial V}{\partial y} + \frac{\partial W}{\partial z} \right) dx \, dy \, dz = \iint_B [U \cos (n,x) + V \cos (n,y)$$

$$+ W \cos (n,z)]dS, \quad (28)$$

where $\cos (n,x)$, $\cos (n,y)$, $\cos (n,z)$ have the meanings assigned in 2-11(c).

(b) By writing $U = \eta F$, $V = \eta G$, $W = \eta H$ in (28), we obtain the three-dimensional analogue of integration by parts

$$\iiint_R \left(F \frac{\partial \eta}{\partial x} + G \frac{\partial \eta}{\partial y} + H \frac{\partial \eta}{\partial z} \right) dx \, dy \, dz = - \iiint_R \eta \left(\frac{\partial F}{\partial x} + \frac{\partial G}{\partial y} + \frac{\partial H}{\partial z} \right) dx \, dy \, dz$$

$$+ \iint_B \eta[F \cos (n,x) + G \cos (n,y) + H \cos (n,z)]dS. \quad (29)$$

(c) By writing $\eta = \psi$, $F = (\partial\phi/\partial x)$, $G = (\partial\phi/\partial y)$, $H = (\partial\phi/\partial z)$ in (29), and with the aid of (18) of 2-11(c), we obtain

$$\iiint_R \psi \nabla^2 \phi \, dx \, dy \, dz = - \iiint_R \left(\frac{\partial\phi}{\partial x} \frac{\partial\psi}{\partial x} + \frac{\partial\phi}{\partial y} \frac{\partial\psi}{\partial y} + \frac{\partial\phi}{\partial z} \frac{\partial\psi}{\partial z} \right) dx \, dy \, dz$$

$$+ \iint_B \psi \frac{\partial\phi}{\partial n} \, dS, \quad (30)$$

with the definition (20) of the laplacian.

An important special case of (30) is obtained by writing $\psi = \phi$.

(d) By setting $\psi = 1$ in (30), we obtain

$$\iiint_R \nabla^2 \phi \, dx \, dy \, dz = \iint_B \frac{\partial \phi}{\partial n} \, dS. \tag{31}$$

(e) In case R includes all of space, and if ψ approaches zero with sufficient rapidity at distances far from the origin of coordinates, (30) becomes, if $\phi = \psi$,

$$\iiint \psi \nabla^2 \psi \, dx \, dy \, dz = - \iiint \left[\left(\frac{\partial \psi}{\partial x} \right)^2 + \left(\frac{\partial \psi}{\partial y} \right)^2 + \left(\frac{\partial \psi}{\partial z} \right)^2 \right] dx \, dy \, dz, \tag{32}$$

where the integrals are carried out over all of space.

CHAPTER 3

INTRODUCTORY PROBLEMS

3-1. A Basic Lemma

(a) In the work of this and succeeding chapters we employ repeatedly one or another form of the following basic lemma:

If x_1 and $x_2(> x_1)$ are fixed constants and $G(x)$ is a particular continuous function for $x_1 \leqq x \leqq x_2$, and if

$$\int_{x_1}^{x_2} \eta(x)G(x)dx = 0 \tag{1}$$

for every choice of the continuously differentiable function $\eta(x)$ for which

$$\eta(x_1) = \eta(x_2) = 0, \tag{2}$$

we conclude that

$$G(x) = 0 \qquad \text{identically in } x_1 \leqq x \leqq x_2. \tag{3}$$

Proof of the foregoing lemma rests upon demonstration of the existence of *at least one* suitable function $\eta(x)$ for which (1) is violated when $G(x)$ is such that (3) does not hold:

We therefore suppose that (3) does not hold—that, namely, there is a particular value x' of x ($x_1 < x' < x_2$) for which $G(x') \neq 0$; for the sake of definiteness, we suppose $G(x') > 0$. Since $G(x)$ is continuous, there must be an interval surrounding x'—say $x_1' \leqq x \leqq x_2'$—in which $G(x) > 0$ everywhere. But (1) cannot then hold for *every* permissible choice of $\eta(x)$. For example, we consider the function defined by

$$\eta(x) = \begin{cases} 0 & \text{for } x_1 \leqq x \leqq x_1', \\ (x - x_1')^2(x - x_2')^2 & \text{for } x_1' \leqq x \leqq x_2', \\ 0 & \text{for } x_2' \leqq x \leqq x_2; \end{cases} \tag{4}$$

for this particular η (which satisfies (2) and is continuously differentiable, clearly) the integral of (1) becomes

$$\int_{x_1}^{x_2} \eta(x)G(x)dx = \int_{x_1'}^{x_2'} (x - x_1')^2(x - x_2')^2 G(x)dx. \tag{5}$$

Since $G(x) > 0$ in $x_1' \leqq x \leqq x_2'$, the right-hand member of (5) is definitely positive—a violation of the hypothesis (1). A similar contradiction is reached if we assume $G(x') < 0$. The lemma is hereby proved.

16

(b) In some applications the basic lemma of (a) is required in a more restrictive form. It is required, for example, that an integral of the form (1) vanish for every continuously *twice*-differentiable $\eta(x)$ for which (2) holds. To prove the necessity of (3) we again suppose $G(x) > 0$ in $x_1' \leqq x \leqq x_2'$, but we choose for $\eta(x)$ the function equal to $(x - x_1')^3 \cdot (x_2' - x)^3$ in $x_1' \leqq x \leqq x_2'$ and zero in the remainder of $x_1 \leqq x \leqq x_2$. The details are left for exercise 1(a) at the end of this chapter.

Similarly, the basic lemma of (a) holds if we require that $\eta(x)$ possess continuous derivatives up to and including any given order [see exercise 1(b)].

(c) If D is a domain of the xy plane, the vanishing of the double integral

$$\iint_D \eta(x,y)G(x,y)dx \, dy \tag{6}$$

for *every* continuously differentiable η that vanishes on the boundary C of D necessitates the identical vanishing of $G(x,y)$, assumed continuous, in D. The proof of this extension of the basic lemma, in essence the same as the proof given in (a) above, is left for end-chapter exercise 1(c). Further, this two-dimensional form of the lemma still holds if we require that $\eta(x,y)$ possess continuous partial derivatives up to and including any given order [see exercise 1(d)].

The extension of the basic lemma to integrals of any given multiplicity is obvious [see exercise 1(e)].

3-2. Statement and Formulation of Several Problems

The problems handled first in this chapter possess an intimate connection which enables us to treat them all as special cases of one general problem whose solution follows in 3-3. For this reason we state briefly and formulate four problems in this section, with the aim of making evident their common character.

(a) We first concern ourselves with the question: What plane curve connecting two given points has the smallest arc length? As a first approach to an answer we fix our attention upon two points (x_1, y_1) and (x_2, y_2) in the xy plane, with $x_1 < x_2$, and a smooth curve of the form

$$y = y(x) \qquad [y(x_1) = y_1, \, y(x_2) = y_2] \tag{7}$$

connecting them. The length of the arc (7) is given by

$$I = \int_{x_1}^{x_2} \sqrt{1 + y'^2} \, dx, \tag{8}$$

where $y' = y'(x)$ denotes the derivative (dy/dx). The problem thus becomes one of choosing the function $y(x)$ in such fashion that the integral (8) has the smallest possible value. In 3-9 below, the restriction (7) that y be a single-valued function of x is removed. This is done by considering arcs in the parametric form $x = x(t)$, $y = y(t)$, where t is the parameter of the curve.

(b) A less trivial problem than the one posed in (a) resides in the question: Given two points on the surface of a sphere, what is the arc, lying on the surface and connecting the two points, which has the shortest possible length? We immediately generalize the problem as follows: Given two points on the surface

$$g(x,y,z) = 0, \tag{9}$$

what is the equation of the arc lying on (9) and connecting these points, which, of all such connecting arcs, has the shortest length?

To formulate the more general problem, we express the equation of the given surface (9) in parametric form, with parameters u and v:

$$x = x(u,v), \qquad y = y(u,v), \qquad z = z(u,v). \tag{10}$$

In terms of the differentials of u and v, the square of the differential of arc length may be written

$$(ds)^2 = (dx)^2 + (dy)^2 + (dz)^2$$
$$= P(u,v)(du)^2 + 2Q(u,v)du\,dv + R(u,v)(dv)^2, \tag{11}$$

where, by direct computation from (10), we have

$$P = \left(\frac{\partial x}{\partial u}\right)^2 + \left(\frac{\partial y}{\partial u}\right)^2 + \left(\frac{\partial z}{\partial u}\right)^2, \qquad R = \left(\frac{\partial x}{\partial v}\right)^2 + \left(\frac{\partial y}{\partial v}\right)^2 + \left(\frac{\partial z}{\partial v}\right)^2, \tag{12}$$

$$Q = \frac{\partial x}{\partial u}\frac{\partial x}{\partial v} + \frac{\partial y}{\partial u}\frac{\partial y}{\partial v} + \frac{\partial z}{\partial u}\frac{\partial z}{\partial v}. \tag{13}$$

(In case the curves $u = $ constant are orthogonal to the curves $v = $ constant on the surface (9), the quantity Q is identically zero.)

If the given fixed points on the surface are (u_1,v_1) and (u_2,v_2), with $u_2 > u_1$, and we limit our consideration to arcs whose equations are expressible in the form

$$v = v(u) \qquad [v(u_1) = v_1, v(u_2) = v_2], \tag{14}$$

the length of the arc is given, according to (11), by

$$I = \int_{u_1}^{u_2} \sqrt{P(u,v) + 2Q(u,v)v' + R(u,v)v'^2}\, du, \tag{15}$$

where $v' = v'(u)$ designates the derivative (dv/du). Our problem, then, is to find the function $v(u)$ that renders the integral (15) a minimum.

Removal of the restriction (14)—namely, that v be a single-valued function of u—is effected in 3-9 by replacing (14) by a parametric representation $u = u(t)$, $v = v(t)$ of the connecting arc on the surface.

(c) In June, 1696, Johann Bernoulli set the following problem before the scholars of his time: "Given two points A and B in a vertical plane, to find for the movable particle M, the path AMB, descending along which by its own gravity, and beginning to be urged from the point A, it may in the shortest time reach the point B." It is tacit in this statement that the particle descends without friction.

Although Newton had earlier considered at least one problem falling within the province of the calculus of variations, the proposal of Bernoulli's brachistochrone problem marked the real beginning of general interest in this subject. (The term "brachistochrone" derives from the Greek *brachistos*, shortest, and *chronos*, time.)

We suppose the points A and B in the xy plane, the y axis directed vertically downward, and the x axis horizontal, with passage from A to B marked by an increase in x. Let the extremizing path have the equation $y = y(x)$. We assume the initial speed v_1 of the particle to be given in the statement of the problem. Let the points A and B have the coordinates (x_1, y_1) and (x_2, y_2), respectively, so that $y(x_1) = y_1$ and $y(x_2) = y_2$.

Since the speed along the curve is given by $v = (ds/dt)$, we have for the total time of descent

$$I = \int_{x=x_1}^{x=x_2} \frac{ds}{v} = \int_{x_1}^{x_2} \frac{\sqrt{1 + y'^2}\, dx}{v}.$$

To compute v as a function of the coordinates we use the fact that a decrease of potential energy is accompanied, in the assumed absence of friction, by an equal increase of kinetic energy; namely, if the particle mass is m and g is the constant acceleration due to gravity, we have

$$\tfrac{1}{2}mv^2 - \tfrac{1}{2}mv_1^2 = mg(y - y_1),$$

whence $v = \sqrt{2g}\,\sqrt{y - y_0}$, where $y_0 = y_1 - (v_1^2/2g)$. (Clearly, $y_1 - y_0$ is the vertical distance through which the particle must descend from rest to achieve the speed v_1.) Thus the time of descent is

$$I = \frac{1}{\sqrt{2g}} \int_{x_1}^{x_2} \frac{\sqrt{1 + y'^2}}{\sqrt{y - y_0}}\, dx. \tag{16}$$

Our problem is that of choosing the function $y(x)$ which renders (16) a minimum.

(d) Given two fixed points (x_1, y_1) and (x_2, y_2), we seek to pass through them the arc $y = y(x)$ whose rotation about the x axis generates a surface of revolution whose area included in $x_1 \leqq x \leqq x_2$ is a minimum. We assume that $y_1 > 0$, $y_2 > 0$ and that $y(x) \geqq 0$ for $x_1 \leqq x \leqq x_2$. That is, we seek to minimize the integral

$$I = 2\pi \int_{x_1}^{x_2} y \sqrt{1 + y'^2}\, dx,$$

which is the area of the surface of revolution, by proper choice of $y = y(x)$ for which $y_1 = y(x_1)$ and $y_2 = y(x_2)$.

3-3. The Euler-Lagrange Equation

(a) A complete solution to each of the problems of the preceding section would carry us beyond the scope of our study. We content ourselves, instead, with deriving the answer to a limited question which runs as follows: Given that there exists a continuous, twice-differentiable function which minimizes the integral connected with any one of the problems of 3-2, what is the differential equation which this function must satisfy? We do not, in the first place, inquire into the existence of the minimum; nor do we take into account the possibility of minimizing functions which do not possess the conditions of smoothness (twice differentiability) that we require. It can be shown,[1] in several cases, that the conditions we impose are not too restrictive; i.e., the answer to our main question is very often the correct one for the same question when the minimizing function is required merely to be continuous and differentiable in sections.

In analytical terms our question is: Given that there exists a twice-differentiable function $y = y(x)$ satisfying the conditions $y(x_1) = y_1$, $y(x_2) = y_2$ which renders the integral

$$I = \int_{x_1}^{x_2} f(x,y,y')dx \tag{17}$$

a minimum, what is the differential equation satisfied by $y(x)$? The constants x_1, y_1, x_2, y_2 are supposed given, and f is a given function of the arguments x, y, y' which is twice differentiable with respect to any, or any combination, of them. Examination shows that each integral of 3-2 above is a special case of (17).

(b) We denote the function that minimizes (17) by $y(x)$ and proceed to form the one-parameter family of "comparison" functions $Y(x)$, defined by

$$Y(x) = y(x) + \epsilon\eta(x), \tag{18}$$

where $\eta(x)$ is an arbitrary differentiable function for which

$$\eta(x_1) = \eta(x_2) = 0 \tag{19}$$

[1] See Bliss (1,2).

and ϵ is the parameter of the family. Thus, for each function $\eta(x)$, we have a single one-parameter family of the form (18); with $\eta(x)$ given, each value of ϵ designates a single member of that one-parameter family. The condition (19) ensures that $Y(x_1) = y(x_1) = y_1$ and $Y(x_2) = y(x_2) = y_2$; that is, all the comparison functions possess the required end-point values of the functions with respect to which the minimization is carried out. By suitable choice of $\eta(x)$ and ϵ it is possible to represent any differentiable function having the required end-point values by an expression of the form (18). The essential importance of the form (18) lies in the fact that no matter which family $Y(x)$ we happen to deal with—no matter, that is, which function $\eta(x)$ is chosen—the minimizing function $y(x)$ is a member of that family for the choice of parameter value $\epsilon = 0$.

Geometrically, the discussion of the preceding paragraph deals with one-parameter families of curves $y = Y(x)$ connecting the points (x_1,y_1) and (x_2,y_2). The minimizing arc $y = y(x)$ is a member of each family for $\epsilon = 0$. The vertical deviation of any curve $y = Y(x)$ from the actual minimizing arc is given by $\epsilon\eta(x)$. (See Fig. 3-1.) For any permissible choice of $\eta(x)$ it is possible to choose a range of values of ϵ—say $-\epsilon_0 < \epsilon < \epsilon_0$—which renders the product $|\epsilon\eta(x)|$ arbitrarily small for all x between x_1 and x_2. The region of the plane covered by the curves $y = Y(x)$ for which $|y(x) - Y(x)|$ is

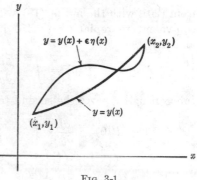

FIG. 3-1.

below any assigned positive number is said to constitute a "neighborhood" of the minimizing arc $y = y(x)$.

Replacing y and y' in (17) respectively by $Y(x)$ and $Y'(x)$, we form the integral

$$I(\epsilon) = \int_{x_1}^{x_2} f(x,Y,Y')dx, \tag{20}$$

where, for a given function $\eta(x)$, this integral is clearly a function of the parameter ϵ. The argument Y' is given, through (18), by

$$Y' = Y'(x) = y'(x) + \epsilon\eta'(x). \tag{21}$$

We thus see, with the aid of (18), that the setting of ϵ equal to zero is equivalent to replacing Y and Y' respectively by y and y'. Thus the integral (20) is a minimum with respect to ϵ for the value $\epsilon = 0$, according to our designation that $y(x)$ is the actual minimizing function. This fact holds *no matter what is the choice of* $\eta(x)$.

The problem at hand is reduced in this way to an ordinary minimum problem of the differential calculus with respect to the single variable ϵ. But, unlike most ordinary minimum problems, we know in advance the value of the variable for which the minimum is achieved—namely, $\epsilon = 0$. Thus we know that the necessary condition for a minimum, the vanishing of the first derivative of I with respect to ϵ, must hold for $\epsilon = 0$; that is,

$$I'(0) = 0. \tag{22}$$

Using the rule given in 2-3(a) for the derivative of an integral with respect to a parameter, we obtain

$$\frac{dI}{d\epsilon} = I'(\epsilon) = \int_{x_1}^{x_2} \left(\frac{\partial f}{\partial Y} \frac{\partial Y}{\partial \epsilon} + \frac{\partial f}{\partial Y'} \frac{\partial Y'}{\partial \epsilon} \right) dx = \int_{x_1}^{x_2} \left(\frac{\partial f}{\partial Y} \eta + \frac{\partial f}{\partial Y'} \eta' \right) dx \tag{23}$$

from (20), with the aid of (18) and (21). Since setting ϵ equal to zero is equivalent to replacing (Y, Y') by (y, y'), we have, according to (22) and (23),

$$I'(0) = \int_{x_1}^{x_2} \left(\frac{\partial f}{\partial y} \eta + \frac{\partial f}{\partial y'} \eta' \right) dx = 0.$$

Integrating by parts the second term of this integral, we obtain[1]

$$I'(0) = \frac{\partial f}{\partial y'} \eta \Bigg]_{x_1}^{x_2} + \int_{x_1}^{x_2} \left[\frac{\partial f}{\partial y} - \frac{d}{dx} \left(\frac{\partial f}{\partial y'} \right) \right] \eta \, dx$$

$$= \int_{x_1}^{x_2} \left[\frac{\partial f}{\partial y} - \frac{d}{dx} \left(\frac{\partial f}{\partial y'} \right) \right] \eta \, dx = 0, \tag{24}$$

because of the restriction (19). Since (24) must hold for all η, we may use the basic lemma of 3-1(a) to conclude that

$$\frac{\partial f}{\partial y} - \frac{d}{dx} \left(\frac{\partial f}{\partial y'} \right) = 0. \tag{25}$$

This equation—the so-called *Euler-Lagrange differential equation*—is in general of second order. Its solution for any given problem of the type

---

[1] We use the designation $\Big]_{x_1}^{x_2}$ throughout this work as follows:

$$G(x) \Bigg]_{x_1}^{x_2} = G(x_2) - G(x_1).$$

This must be distinguished from the notation

$$G(x) \Big|_{x_0} = G(x_0).$$

enunciated in 3-2 supplies the twice-differentiable minimizing function of the integral of the problem, provided the minimum exists.

(c) It is to be noted in the procedure of (b) above that the condition $I'(0) = 0$ is not a sufficient condition for a minimum of $I(\epsilon)$ for $\epsilon = 0$. In fact the relation $I'(0) = 0$ might even indicate a maximum of $I(\epsilon)$ for $\epsilon = 0$. In all the problems considered in 3-2, however, it is simple to convince oneself that no maximum exists for the integrals involved. The distance along a smooth arc connecting fixed points can be made as large as we please; such is the case also for the time of descent down a curve and for the area of the surface of revolution generated by a smooth arc between fixed points. Yet the relation $I'(0) = 0$ may also indicate the existence of what corresponds in the ordinary differential calculus to a horizontal inflection at $\epsilon = 0$. That is, we may have a situation in which—for at least *some* choice of the function $\eta(x)$ introduced in (18)— the difference $[I(\epsilon) - I(0)]$ may change sign as ϵ passes through zero, although the curve of $I(\epsilon)$ plotted as a function of ϵ possesses a horizontal tangent at $\epsilon = 0$ for *every* choice of $\eta(x)$. As pointed out in (a) above, detailed investigation as to which of the three situations—minimum, maximum, or "stationary" value—prevails is in general beyond the scope of our study. There are, however, a few specific cases in which it is demonstrable in an elementary fashion that $I'(0) = 0$ definitely implies a minimum for $\epsilon = 0$, and these are handled in conjunction with specific problems or are left for the exercises.

There are many problems that arise in our study wherein we have no concern as to whether the condition $I'(0) = 0$ implies a maximum, minimum, or stationary value of $I(\epsilon)$ at $\epsilon = 0$. We therefore find it useful to apply the term *extremum* to the value $I(0)$ for all three situations. The function $y(x)$ which renders the integral I an extremum is accordingly called the *extremizing function*. Thus a function $y(x)$ which satisfies the Euler-Lagrange equation (25) and the imposed end-point conditions is by definition the extremizing function for the integral upon whose integrand f the equation (25) is generated.

Even in those cases for which an extremum is an actual minimum, it is not necessarily an absolute minimum. It is recalled from the ordinary differential calculus that a minimum characterized by the vanishing of the first derivative is merely a *relative* minimum with respect to values of the independent variable in a neighborhood of the value for which the first derivative vanishes. Thus, if $I(0)$ is a minimum achieved through the function $y(x)$ which renders $I'(0) = 0$, it must be a minimum only relative to values of ϵ in the neighborhood of zero. In terms of the neighborhood of the arc $y = y(x)$ defined in (b) above, the function $y(x)$ minimizes the integral I if and only if there exists a neighborhood of

$y = y(x)$ such that every arc $y = Y(x) \not\equiv y(x)$ satisfying the required end-point conditions and lying entirely within the neighborhood, renders the value of I larger than $I(0)$. Thus, even though $I(0)$ is a minimum for which $I'(0) = 0$, there may be functions $Y(x)$—for which $y = Y(x)$ lies outside the neighborhood described—which render the integral I even smaller than $I(0)$. Although this fact plays no role in our study, it is mentioned here to point out a limitation of the theory as here developed. This limitation is of particular significance in the minimum-surface-of-revolution problem posed in 3-2(d) and considered further in 3-7(b) below.

3-4. First Integrals of the Euler-Lagrange Equation. A Degenerate Case

(a) A particularly simple Euler-Lagrange equation results if the integrand function f is explicitly independent of the dependent variable y. For then we have that $(\partial f/\partial y)$ vanishes identically, and (25) of 3-3(b) becomes

$$\frac{d}{dx}\left(\frac{\partial f}{\partial y'}\right) = 0,$$

or

$$\frac{\partial f}{\partial y'} = C_1, \tag{26}$$

an arbitrary constant. Thus the quest for the extremizing function is reduced to the solution of an equation involving y' and x only, a *first-order* differential equation.

If, further, f is explicitly independent of the independent variable x, as well as being independent of y, the partial derivative $(\partial f/\partial y')$ is a function of y' alone, so that the solution of (26) is simply $y' = C_2$, where the constant C_2 is some function of C_1. Thus the extremizing functions for cases in which the integrand f depends explicitly on y' alone are necessarily linear functions of x. This fact immediately affords the solution of the shortest-distance-in-the-plane problem of 3-2(a)—a straight line!

(b) We have the readily verifiable identity

$$\frac{d}{dx}\left(y'\frac{\partial f}{\partial y'} - f\right) = y'\frac{d}{dx}\left(\frac{\partial f}{\partial y'}\right) - \frac{\partial f}{\partial x} - \frac{\partial f}{\partial y}y'$$

$$= -y'\left[\frac{\partial f}{\partial y} - \frac{d}{dx}\left(\frac{\partial f}{\partial y'}\right)\right] - \frac{\partial f}{\partial x}, \tag{27}$$

which suggests an obvious first integral of the Euler-Lagrange equation in the special case that f is explicitly independent of the independent variable x. For since $(\partial f/\partial x) = 0$ in this event, we see that the Euler-

Lagrange equation (25) implies the vanishing of the first member of (27)—namely,

$$\frac{d}{dx}\left(y'\,\frac{\partial f}{\partial y'} - f\right) = 0,$$

or

$$y'\,\frac{\partial f}{\partial y'} - f = C_1, \tag{28}$$

an arbitrary constant (not necessarily the same as C_1 in (a) above). Thus the extremizing function may be obtained as the solution of a first-order differential equation involving y and y' only.

(c) If the integrand function f of

$$I = \int_{x_1}^{x_2} f(x,y,y')dx$$

is explicitly the total derivative with respect to x of some function of x and y, then surely the integral I is independent of the particular choice of the function $y(x)$, so long as the prescribed end-point values $y(x_1) = y_1$ and $y(x_2) = y_2$ are achieved. For if $f = (dg/dx)$, I is equal to the difference of the prescribed values of $g(x,y)$ at the end points—namely, $[g(x_2,y_2) - g(x_1,y_1)]$. It is interesting to see what form the Euler-Lagrange equation assumes in this event.

We have in this case

$$f = \frac{dg}{dx} = \frac{\partial g}{\partial x} + \frac{\partial g}{\partial y}\,y',$$

so that the Euler-Lagrange equation (25) reads

$$\frac{\partial^2 g}{\partial y\,\partial x} + \frac{\partial^2 g}{\partial y^2}\,y' - \frac{d}{dx}\left(\frac{\partial g}{\partial y}\right) = 0.$$

But evaluation of the total derivative of $(\partial g/\partial y)$ with respect to x shows that this last equation is *identically* satisfied since

$$\frac{\partial^2 g}{\partial y\,\partial x} = \frac{\partial^2 g}{\partial x\,\partial y}.$$

This result suggests the question: What is the most general case in which the Euler-Lagrange equation is identically satisfied? To discover the answer we expand (25) of 3-3(b) as

$$\frac{\partial f}{\partial y} - \frac{\partial^2 f}{\partial x\,\partial y'} - \frac{\partial^2 f}{\partial y\,\partial y'}\,y' - \frac{\partial^2 f}{\partial y'^2}\,y'' = 0. \tag{29}$$

Since the first three terms on the left contain at the highest the first derivative of y, the identical satisfaction of (29) requires the coefficient

$(\partial^2 f/\partial y'^2)$ of y'' to vanish identically. But this is equivalent to stating that f must be a linear function of y'; that is,

$$f = p(x,y) + q(x,y)y'. \tag{30}$$

Forming the Euler-Lagrange equation (25) for this particular f, we have

$$\frac{\partial p}{\partial y} + \frac{\partial q}{\partial y} y' - \frac{dq}{dx} = \frac{\partial p}{\partial y} - \frac{\partial q}{\partial x} = 0$$

for all x and y. But this, according to 2-2(c), is precisely the condition that (30) be the total derivative (dg/dx) of some function $g(x,y)$.

Thus we have that a necessary and sufficient condition for an Euler-Lagrange equation to be identically satisfied is that the integrand function be explicitly the derivative (dg/dx) of some function $g(x,y)$. Implicit in this result lies another fact of some significance: A necessary and sufficient condition that the addition of a term to the integrand of a given integral leave unaltered the corresponding Euler-Lagrange equation is that the additional term be the derivative (dg/dx) of some function $g(x,y)$. This follows from the first result because of the linearity of the Euler-Lagrange equation with respect to the integrand function f.

3-5. Geodesics

(a) We return to the problem posed above in 3-2(b) for the arc of minimum length connecting two points on a given surface. Such an arc is termed a *geodesic* for the surface. The special case for the plane, presented in 3-2(a), is solved in 3-4(a) above. According to (15) of 3-2(b) the integrand function for the problem is

$$f = \sqrt{P + 2Qv' + Rv'^2}, \tag{31}$$

where P, Q, and R are three given functions of the surface coordinates u and v; it is assumed that the minimizing arc has the form $v = v(u)$. According to (25) of 3-3(b)—with u and v here playing the respective roles of x and y—v must satisfy the Euler-Lagrange equation which reads, through (31),

$$\frac{\frac{\partial P}{\partial v} + 2v' \frac{\partial Q}{\partial v} + v'^2 \frac{\partial R}{\partial v}}{2\sqrt{P + 2Qv' + Rv'^2}} - \frac{d}{du}\left(\frac{Q + Rv'}{\sqrt{P + 2Qv' + Rv'^2}}\right) = 0.$$

In the special case where P, Q, and R are explicitly functions of u alone, this last result becomes

$$\frac{Q + Rv'}{\sqrt{P + 2Qv' + Rv'^2}} = C_1, \tag{32}$$

an arbitrary constant. In the event $Q = 0$, which is the case if the surface curves $u = $ constant are orthogonal to the curves $v = $ constant, v can be expressed directly in terms of u as an integral—namely,

$$v = C_1 \int \frac{\sqrt{P}\, du}{\sqrt{R^2 - C_1^2 R}}, \tag{33}$$

as we obtain from (32) with the facts that $v' = (dv/du)$ and that P and R are given functions of u alone. The constant of integration in (33) and C_1 are determined so that (33) passes through the given fixed end points.

(b) Still supposing that $Q = 0$, but now that P and R are explicit functions of v alone, we are in a position to use the result of 3-4(b), which is applicable in the case where f is explicitly independent of the independent variable. From (28) of 3-4(b) and (31) of 3-5(a)—with $Q = 0$—we obtain

$$\frac{Rv'^2}{\sqrt{P + Rv'^2}} - \sqrt{P + Rv'^2} = C_1,$$

whence, since $v' = (dv/du)$,

$$u = C_1 \int \frac{\sqrt{R}\, dv}{\sqrt{P^2 - C_1^2 P}}. \tag{34}$$

(c) As a particular case we consider the geodesic connecting two points on a sphere. The parameters u, v most convenient for describing position on the sphere surface are the colatitude v and the longitude u, with

$$x = a \sin v \cos u, \qquad y = a \sin v \sin u, \qquad z = a \cos v, \tag{35}$$

where a is the radius of the sphere. It is directly verifiable that v is the angle between the positive z axis and the line drawn from the sphere center to the designated point, that u is the angle between the xz plane $(x > 0)$ and the half plane bounded by the z axis and containing the designated point, and that $x^2 + y^2 + z^2 = a^2$.

From (12) and (13) of 3-2(b) the parametric equations (35) of the sphere give

$$P = a^2 \sin^2 v, \qquad R = a^2, \qquad Q = 0.$$

We are thus able to use the result (34) of (b) above to obtain

$$u = C_1 \int \frac{dv}{\sqrt{a^2 \sin^4 v - C_1^2 \sin^2 v}} = \int \frac{\csc^2 v\, dv}{\sqrt{[(a/C_1)^2 - 1] - \cot^2 v}}$$

$$= -\sin^{-1} \frac{\cot v}{\sqrt{(a/C_1)^2 - 1}} + C_2,$$

whence it follows that

$$(\sin C_2)a \sin v \cos u - (\cos C_2)a \sin v \sin u - \frac{a \cos v}{\sqrt{(a/C_1)^2 - 1}} = 0.$$

With the use of (35) we see that the sphere geodesic lies on the plane

$$x \sin C_2 - y \cos C_2 - \frac{z}{\sqrt{(a/C_1)^2 - 1}} = 0,$$

which passes through the center of the sphere. Hence the familiar result: The shortest arc connecting two points on the surface of a sphere is the intersection of the sphere with the plane containing the given points and the center of the sphere—the so-called great-circle arc.

(d) To obtain the geodesic on a general surface of revolution we consider the surface

$$y^2 + z^2 = [g(x)]^2 \tag{36}$$

generated by revolving the curve $y = g(x)$, with $g \geqq 0$, about the x axis. A convenient parametric representation of this surface is

$$x = u, \qquad y = g(u) \cos v, \qquad z = g(u) \sin v, \tag{37}$$

which is readily verified to satisfy (36). From (12) and (13) of 3-2(b) equations (37) give directly

$$P = 1 + [g'(u)]^2, \qquad R = [g(u)]^2, \qquad Q = 0.$$

The result (33) of (a) above is therefore applicable; from it we obtain

$$v = C_1 \int \frac{\sqrt{1 + [g'(u)]^2}\, du}{g(u)\, \sqrt{[g(u)]^2 - C_1^2}}.$$

In 4-5(c) the general geodesic problem is again considered, but from a point of view somewhat different from that taken in the present chapter.

3-6. The Brachistochrone

(a) With the results of 3-4 we are in a position to solve the brachistochrone problem formulated in 3-2(c). The integrand

$$f = \frac{\sqrt{1 + y'^2}}{\sqrt{y - y_0}} \tag{38}$$

of the integral (16) of 3-2(c) giving the time of descent is explicitly independent of the independent variable x. We may therefore write down immediately a first integral of the Euler-Lagrange equation—namely,

$y'(\partial f/\partial y') - f = C_1$—according to the result of 3-4(b). From (38) we thus obtain

$$\frac{y'^2}{\sqrt{(y - y_0)(1 + y'^2)}} - \frac{\sqrt{1 + y'^2}}{\sqrt{y - y_0}} = C_1.$$

Solving this last result for $y' = (dy/dx)$, we integrate both sides of the resulting expression to obtain, on writing for the arbitrary constant $C_1 = (2a)^{-\frac{1}{2}}$,

$$x = \int \frac{\sqrt{y - y_0}\, dy}{\sqrt{2a - (y - y_0)}}. \tag{39}$$

To evaluate this integral we substitute

$$y - y_0 = 2a \sin^2 \frac{\theta}{2}; \tag{40}$$

with this (39) becomes

$$x = 2a \int \sin^2 \frac{\theta}{2}\, d\theta = a(\theta - \sin \theta) + x_0, \tag{41}$$

where x_0 is the constant of integration.

(b) Rewriting (40) and combining it with (41), we have

$$x = x_0 + a(\theta - \sin \theta), \qquad y = y_0 + a(1 - \cos \theta) \tag{42}$$

for the parametric equations of the required curve of most rapid descent. These are recognized to be the equations of the cycloid generated by the motion of a fixed point on the circumference of a circle of radius a which rolls on the positive side of the given line $y = y_0$. It can be shown[1] that by adjustment of the arbitrary constants a and x_0 it is always possible to construct one and only one cycloid (42) of which one arch contains the two points between which the brachistochrone is required to extend. Moreover, this arc renders the time of descent an absolute minimum as compared with all other connecting arcs. (The constant y_0 is not arbitrary, but is given, according to 3-2(c), by

$$y_0 = y_1 - \frac{v_1^2}{2g}, \tag{43}$$

where y_1 is the ordinate of the starting point (x_1,y_1), v_1 the prescribed initial speed, and g the constant acceleration due to gravity.)

The techniques we employ here to solve the brachistochrone problem were not available to Johann Bernoulli in 1696. The method which is essentially the one devised by Bernoulli to solve the problem is developed below in Chap. 5.

[1] See Bliss (1), p. 55.

3-7. Minimum Surface of Revolution

(a) In the problem of "the minimum surface of revolution" given in 3-2(d) the integrand function

$$f = y \sqrt{1 + y'^2} \tag{44}$$

is explicitly independent of x, so that we may use (28) of 3-4(b) to obtain a first integral of the Euler-Lagrange equation. With (44) equation (28) becomes

$$\frac{yy'^2}{\sqrt{1 + y'^2}} - y \sqrt{1 + y'^2} = C_1,$$

from which we obtain directly

$$x = -C_1 \int \frac{dy}{\sqrt{y^2 - C_1^2}} = -C_1 \cosh^{-1} \frac{y}{C_1} + C_2;$$

or

$$y = b \cosh \frac{x - a}{b}, \tag{45}$$

where we write $C_1 = -b$, $C_2 = a$; b is positive.

The curve represented by (45) is called a *catenary*; the corresponding surface of revolution about the x axis is called a *catenoid* of revolution. In the problem at hand we are required to adjust the arbitrary constants a and b so that the catenary passes through the given end points (x_1,y_1) and (x_2,y_2). The possibility of fulfilling this requirement is discussed directly below:

(b) We choose a one-parameter family of catenaries from among (45) characterized by the fact that every member passes through the left-hand end point (x_1,y_1). Thus for this one-parameter family the constants a and b are related by the condition

$$y_1 = b \cosh \frac{x_1 - a}{b}, \tag{46}$$

which we obtain by substituting (x_1,y_1) for (x,y) in (45). Our problem is to discover which, if any, of this litter of catenaries passes through the second end point (x_2,y_2). In Fig. 3-2 there are plotted several of the curves (45) through (x_1,y_1). With this figure as reference we make a few assertions without proof:

Every member of the family defined by (45) *and* (46) is tangent to the dotted curve OE, the envelope of the family No member of the family passes through any point B that is separated from (x_1,y_1) by the envelope.   One and only one member passes through any given point F on the envelope, which is its point of tangency.   Through any point G, not separated from (x_1,y_1) by the envelope, there pass exactly two members of the family.

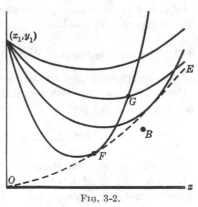

The assertions of the above paragraph reflect, as do those that follow, on the limitations of the theory as developed in our study.   If, for example, the point (x_2,y_2) is at B, no member of (45) fits the required endpoint conditions and we conclude that the minimum surface area is not generated by any curve of the form $y = y(x)$, where $y(x)$ is twice differentiable.   It can be shown,[1] in this

Fig. 3-2.

event, that the minimum area is generated by the broken line whose three segments are

$$x = x_1\ (0 \leqq y \leqq y_1),\ y = 0\ (x_1 \leqq x \leqq x_2),\ x = x_2\ (0 \leqq y \leqq y_2).$$

This is the so-called *Goldschmidt discontinuous solution*.

But even when there is a *unique* catenary connecting the given end points—when (x_2,y_2) lies on the envelope OE—it turns out that this catenary does not render the surface area a minimum.   Once again, the minimum is afforded by the Goldschmidt solution.   The catenary does not even provide a relative minimum in the sense of 3-3(c).

In the case in which two catenaries of the family (45) fit the required end-point conditions—if (x_2,y_2) is not separated from (x_1,y_1) by the envelope OE—a relative minimum of the surface area is supplied by the *upper* catenary of the pair, but no minimum area is generated by the catenary whose point of tangency with OE lies in the interval $(x_1 < x < x_2)$. Although the criterion cannot be stated in simple terms, we may assert further that, if (x_2,y_2) is sufficiently far above (or to the left of) the envelope, the upper catenary generates a surface area that is an absolute minimum.   Otherwise it provides merely a relative minimum, and the absolute minimum is supplied by the Goldschmidt solution.   In *every* case the Goldschmidt solution—by which the surface area generated is clearly $\pi(y_1^2 + y_2^2)$—affords a relative minimum.

(c) It is obvious that the information given in (b) above is by no means supplied by the limited theory here developed.   Nevertheless, since the

---

[1] See Bliss (1) for a detailed discussion of this problem, with proofs carrying beyond the scope of the present study.

principal aim of this work is to study the role of the calculus of variations as a branch of applied mathematics, we must be content, quite often, to bypass such sophisticated problems as those of existence, singular solutions, and the like.

3-8. Several Dependent Variables

(a) We now proceed to derive the differential equations that must be satisfied by the twice-differentiable functions $x(t)$, $y(t)$, $\ldots$, $z(t)$ that extremize the integral

$$I = \int_{t_1}^{t_2} f(x,y, \ldots ,z,\dot{x},\dot{y}, \ldots ,\dot{z},t)dt \tag{47}$$

with respect to those functions x, y, $\ldots$, z which achieve prescribed values at the fixed limits of integration t_1 and t_2, where $t_1 < t_2$. The superior dot indicates ordinary differentiation with respect to the independent variable t.

We denote the set of actual extremizing functions by $x(t)$, $y(t)$, $\ldots$, $z(t)$ and proceed to form the one-parameter family of comparison functions

$$X(t) = x(t) + \epsilon\xi(t), \qquad Y(t) = y(t) + \epsilon\eta(t), \qquad \ldots ,$$
$$Z(t) = z(t) + \epsilon\zeta(t), \quad (48)$$

where ξ, η, $\ldots$, ζ are arbitrary differentiable functions for which

$$\xi(t_1) = \xi(t_2) = \eta(t_1) = \eta(t_2) = \cdots = \zeta(t_1) = \zeta(t_2) = 0 \tag{49}$$

and ϵ is the parameter of the family. The condition (49) assures us that every member of each comparison family satisfies the required prescribed end-point conditions. We see, moreover, that no matter what the choice of ξ, η, $\ldots$, ζ, the set of extremizing functions $x(t)$, $y(t)$, $\ldots$, $z(t)$ is a member of each comparison family for the parameter value $\epsilon = 0$. Thus if we form the integral

$$I(\epsilon) = \int_{t_1}^{t_2} f(X,Y, \ldots ,Z,\dot{X},\dot{Y}, \ldots ,\dot{Z},t)dt \tag{50}$$

by replacing x, y, $\ldots$, z, etc., in (47) by X, Y, $\ldots$, Z, etc., respectively, we have that $I(0)$ is the extremum value sought. We therefore conclude that

$$I'(0) = 0. \tag{51}$$

It follows from (48) that

$$\dot{X} = \dot{x} + \epsilon\dot{\xi}, \qquad \dot{Y} = \dot{y} + \epsilon\dot{\eta}, \qquad \ldots , \qquad \dot{Z} = \dot{z} + \epsilon\dot{\zeta}. \tag{52}$$

Using the rule given in 2-3(a), we form the derivative $(dI/d\epsilon)$ of the integral (50):

$$I'(\epsilon) = \int_{t_1}^{t_2} \left(\frac{\partial f}{\partial X} \xi + \frac{\partial f}{\partial \dot{X}} \dot{\xi} + \frac{\partial f}{\partial Y} \eta + \frac{\partial f}{\partial \dot{Y}} \dot{\eta} + \cdots + \frac{\partial f}{\partial Z} \zeta + \frac{\partial f}{\partial \dot{Z}} \dot{\zeta} \right) dt,$$
(53)

where we use (48) and (52) to derive the sequence of substitutions $(\partial X/\partial \epsilon) = \xi, \ldots, (\partial \dot{Z}/\partial \epsilon) = \dot{\zeta}$. It is clear from (48) and (52) that setting $\epsilon = 0$ is equivalent to replacing $X, Y, \ldots, Z, \dot{X}, \dot{Y}, \ldots, \dot{Z}$ by $x, y, \ldots, z, \dot{x}, \dot{y}, \ldots, \dot{z}$, respectively. Thus, because of (51), we obtain from (53), on setting $\epsilon = 0$,

$$I'(0) = \int_{t_1}^{t_2} \left(\frac{\partial f}{\partial x} \xi + \frac{\partial f}{\partial \dot{x}} \dot{\xi} + \frac{\partial f}{\partial y} \eta + \frac{\partial f}{\partial \dot{y}} \dot{\eta} + \cdots + \frac{\partial f}{\partial z} \zeta + \frac{\partial f}{\partial \dot{z}} \dot{\zeta} \right) dt = 0.$$
(54)

This last relation must hold for all choices of the functions $\xi(t)$, $\eta(t), \ldots, \zeta(t)$. In particular, it holds for the special choice in which $\eta, \ldots, \zeta$ are identically zero, but for which $\xi(t)$ is still arbitrary, consistent with (49). With this selection of $\xi, \eta, \ldots, \zeta$ we integrate by parts the second term of the second member of (54) to obtain, since $\xi(t_1) = \xi(t_2) = 0$,

$$\int_{t_1}^{t_2} \left[\frac{\partial f}{\partial x} - \frac{d}{dt} \left(\frac{\partial f}{\partial \dot{x}} \right) \right] \xi \, dt = 0.$$
(55)

Since (55) holds for all ξ, we conclude by applying the basic lemma of 3-1(a) that

$$\frac{\partial f}{\partial x} - \frac{d}{dt} \left(\frac{\partial f}{\partial \dot{x}} \right) = 0.$$
(56)

Through similar treatment of the successive pairs of terms of the second member of (54) we derive like equations, with x replaced by $y, \ldots, z$. Joining these equations with (56), we have

$$\frac{\partial f}{\partial x} - \frac{d}{dt} \left(\frac{\partial f}{\partial \dot{x}} \right) = 0, \qquad \frac{\partial f}{\partial y} - \frac{d}{dt} \left(\frac{\partial f}{\partial \dot{y}} \right) = 0, \qquad \ldots, \qquad \frac{\partial f}{\partial z} - \frac{d}{dt} \left(\frac{\partial f}{\partial \dot{z}} \right) = 0$$
(57)

for the system of simultaneous Euler-Lagrange equations which must be satisfied by the functions $x(t), y(t), \ldots, z(t)$ which render the integral (47) an extremum.

(b) The readily verified identity

$$\frac{d}{dt}\left(\frac{\partial f}{\partial \dot{x}}\dot{x} + \frac{\partial f}{\partial \dot{y}}\dot{y} + \cdots + \frac{\partial f}{\partial \dot{z}}\dot{z} - f\right)$$

$$= -\dot{x}\left[\frac{\partial f}{\partial x} - \frac{d}{dt}\left(\frac{\partial f}{\partial \dot{x}}\right)\right] - \dot{y}\left[\frac{\partial f}{\partial y} - \frac{d}{dt}\left(\frac{\partial f}{\partial \dot{y}}\right)\right]$$

$$- \cdots - \dot{z}\left[\frac{\partial f}{\partial z} - \frac{d}{dt}\left(\frac{\partial f}{\partial \dot{z}}\right)\right] - \frac{\partial f}{\partial t} \quad (58)$$

suggests an important first integral of the system (57) in the special case in which the integrand function f is explicitly independent of the independent variable t. For in this case we have $(\partial f/\partial t) = 0$; with (57) this implies that the right-hand member of (58) vanishes. Thus we have a first integral

$$\frac{\partial f}{\partial \dot{x}}\dot{x} + \frac{\partial f}{\partial \dot{y}}\dot{y} + \cdots + \frac{\partial f}{\partial \dot{z}}\dot{z} - f = C_1 \quad (59)$$

of the system (57) whenever $(\partial f/\partial t) = 0$; C_1 is an arbitrary constant.

Other first integrals may be obtained directly from (57) in case f is explicitly independent of any of the dependent variables $x, y, \ldots, z$. If, for example, $(\partial f/\partial x) = 0$, the first of (57) implies directly that $(\partial f/\partial \dot{x}) = $ constant, etc.

3-9. Parametric Representation

(a) The results of 3-8 are directly applicable to problems of the type introduced in 3-2 and solved in 3-5 to 3-7 when these are generalized so as to include parametric relationships $x = x(t)$, $y = y(t)$ between the variables x and y, rather than have the solutions restricted to relationships in which one of the variables is a single-valued function of the other, as, for example, $y = y(x)$. In some problems the requirement of single-valuedness is excessively restrictive; for it turns out that the Euler-Lagrange equation—derived under the assumption that the extremizing function is single-valued—may have for the solution which satisfies the given end-point conditions a relationship in which the dependent variable is *not* a single-valued function of the independent variable. One cannot, without further justification, accept such a solution as valid.

We proceed to show, however, that the extremizing relationship between a pair of variables x and y is the same, whether the solution is derived under the assumption that y is a single-valued function of x or that a more general parametric representation is required to express the relationship between x and y. We do this by showing that the solution of the Euler-Lagrange equation derived on the basis of the assumption of the single-valuedness of y as a function of x satisfies also the system of

Euler-Lagrange equations derived on the basis of the parametric relationship between x and y.

Under the assumption that y is a single-valued function of x, the integral to be extremized is given as

$$I = \int_{x_1}^{x_2} f(x,y,y')dx, \tag{60}$$

where y is required to have the values y_1 and y_2 at $x = x_1$ and $x = x_2$. If instead we use the parametric representation $x = x(t)$, $y = y(t)$, where $x(t_j) = x_j$, $y(t_j) = y_j$ for $j = 1,2$, the integral (60) is transformed through the relationships

$$y' = \frac{dy}{dx} = \frac{\dot{y}}{\dot{x}}, \qquad dx = \dot{x}\, dt, \tag{61}$$

where the superior dot indicates differentiation with respect to t:

$$I = \int_{t_1}^{t_2} f\left(x,y,\frac{\dot{y}}{\dot{x}}\right) \dot{x}\, dt. \tag{62}$$

The Euler-Lagrange equation corresponding to (60) is, according to 3-3(b),

$$\frac{\partial f}{\partial y} - \frac{d}{dx}\left(\frac{\partial f}{\partial y'}\right) = 0. \tag{63}$$

According to 3-8(a) the system of Euler-Lagrange equations associated with (62) is, if we write

$$g(x,y,\dot{x},\dot{y}) = f(x,y,y')\dot{x} \qquad \left(y' = \frac{\dot{y}}{\dot{x}}\right), \tag{64}$$

$$\frac{\partial g}{\partial x} - \frac{d}{dt}\left(\frac{\partial g}{\partial \dot{x}}\right) = 0, \qquad \frac{\partial g}{\partial y} - \frac{d}{dt}\left(\frac{\partial g}{\partial \dot{y}}\right) = 0. \tag{65}$$

From (64) we obtain

$$\frac{\partial g}{\partial x} = \frac{\partial f}{\partial x}\dot{x}, \qquad \frac{\partial g}{\partial \dot{x}} = f - \dot{x}\frac{\partial f}{\partial y'}\frac{\dot{y}}{\dot{x}^2} = f - y'\frac{\partial f}{\partial y'}. \tag{66}$$

With the aid of the second relation of (61) and the identity (27) of 3-4(b) we thus have

$$\frac{d}{dt}\left(\frac{\partial g}{\partial \dot{x}}\right) = \dot{x}\frac{d}{dx}\left(f - y'\frac{\partial f}{\partial y'}\right) = \dot{x}\left\{y'\left[\frac{\partial f}{\partial y} - \frac{d}{dx}\left(\frac{\partial f}{\partial y'}\right)\right] + \frac{\partial f}{\partial x}\right\}. \tag{67}$$

We further obtain from (64)

$$\frac{\partial g}{\partial y} = \frac{\partial f}{\partial y}\dot{x}, \qquad \frac{\partial g}{\partial \dot{y}} = \dot{x}\frac{\partial f}{\partial y'}\frac{1}{\dot{x}} = \frac{\partial f}{\partial y'}, \tag{68}$$

whence, according to the second of (61), we have

$$\frac{d}{dt}\left(\frac{\partial g}{\partial \dot{y}}\right) = \dot{x}\,\frac{d}{dx}\left(\frac{\partial f}{\partial y'}\right).$$

Combining this last result with the first of (68), and (67) with the first of (66), we obtain the pair of equations

$$\frac{\partial g}{\partial x} - \frac{d}{dt}\left(\frac{\partial g}{\partial \dot{x}}\right) = -\dot{y}\left[\frac{\partial f}{\partial y} - \frac{d}{dx}\left(\frac{\partial f}{\partial y'}\right)\right],$$

$$\frac{\partial g}{\partial y} - \frac{d}{dt}\left(\frac{\partial g}{\partial \dot{y}}\right) = \dot{x}\left[\frac{\partial f}{\partial y} - \frac{d}{dx}\left(\frac{\partial f}{\partial y'}\right)\right].$$

From this result we conclude that any relationship, single-valued or not, that satisfies the Euler-Lagrange equation (63)—derived on the basis of an assumed single-valued solution $y = y(x)$—satisfies also the system (65), whose derivation requires no assumption of single-valuedness of y as a function of x. We are therefore justified in accepting as a valid extremizing relationship any solution of the single Euler-Lagrange equation (63); the single-valuedness assumption employed in its derivation is shown to be unessential.

(b) Underlying the result demonstrated in (a) above is the assumed possibility of representing the quantity to be extremized in the two different forms (60) and (62). It is of course only to those problems in which the dual representation is possible that the result is applicable. The problems enumerated in 3-2 are all of this class, as we may readily verify. There are, however, other types of problems to which the result is not applicable, but in these there arises no question of the sort that leads to the investigation carried out in (a).

3-10.   Undetermined End Points

Two simple generalizations of the brachistochrone problem, for example, indicate a necessity for extending the theory of this chapter to include hitherto neglected problems involving undetermined end points:

(i) What is the arc of quickest descent from a fixed point to a given vertical line?

(ii) What is the arc of quickest descent from a fixed point to a given curve?

The first of these questions involves the extremization of an integral whose limits of integration are prescribed, but the extremization must be carried out with respect to functions not prescribed at the upper limit. The general case for such problems is handled in (a) below. In the second of the above problems, we are given neither the upper limit of the integral to be extremized nor the upper-limit end-point value of the functions with respect to which extremization is effected; we are, however, provided with a relationship between these two undetermined quantities—namely, the equation of the curve (supposed to lie in a vertical plane) on which the descent is to end. The general case is treated in (c) below.

(a) We seek to extremize the integral

$$I = \int_{x_1}^{x_2} f(x,y,y')dx,$$

with x_1 and x_2 given, with respect to functions that attain the value y_1 for $x = x_1$, but for which no value is prescribed at $x = x_2$. In the manner of 3-3(b) we suppose that the twice-differentiable $y(x)$ is the actual extremizing function and set up the one-parameter family

$$Y(x) = y(x) + \epsilon\eta(x) \tag{69}$$

of comparison functions. The differentiable function $\eta(x)$ is arbitrary to within the condition

$$\eta(x_1) = 0, \tag{70}$$

and ϵ is the parameter of the family defined by η; for the value $\epsilon = 0$, $y(x)$ is a member of every family (69). Thus the integral

$$I(\epsilon) = \int_{x_1}^{x_2} f(x,Y,Y')dx \tag{71}$$

—where, according to (69),

$$Y' = y' + \epsilon\eta' \tag{72}$$

—is an extremum for $\epsilon = 0$. Thus we have $I'(0) = 0$ for all choices of η.

After differentiating (71) with respect to ϵ according to (3) of 2-3(a), we set ϵ equal to zero—equivalent, by (69) and (72), to replacing (Y,Y') by (y,y')—to obtain

$$I'(0) = \int_{x_1}^{x_2} \left(\frac{\partial f}{\partial y} \eta + \frac{\partial f}{\partial y'} \eta' \right) dx = 0, \tag{73}$$

with the use of $(\partial Y/\partial\epsilon) = \eta$ and $(\partial Y'/\partial\epsilon) = \eta'$. Integrating by parts the second term of (73), we get for (73) with the aid of (70)

$$I'(0) = \frac{\partial f}{\partial y'}\bigg|_{x_2} \eta_2 + \int_{x_1}^{x_2} \left[\frac{\partial f}{\partial y} - \frac{d}{dx}\left(\frac{\partial f}{\partial y'} \right) \right] \eta\, dx = 0, \tag{74}$$

where η_2 is written for $\eta(x_2)$.

Since (74) must hold for all choices of $\eta(x)$ consistent with (70), it must in particular hold for those η for which $\eta(x_2) = \eta_2 = 0$. For such $\eta(x)$

the second member of (74) reduces to the integral alone; by application of the basic lemma of 3-1(a) we conclude

$$\frac{\partial f}{\partial y} - \frac{d}{dx}\left(\frac{\partial f}{\partial y'}\right) = 0. \tag{75}$$

With this result, and for general $\eta(x)$ once again, the second member of (74) reduces to its first term. Now, by choosing $\eta(x_2) = \eta_2 = 1$, the vanishing for all η of the term remaining requires fulfillment of the end-point condition

$$\frac{\partial f}{\partial y'}\bigg|_{x_2} = 0. \tag{76}$$

From (75) we see that the differential equation is determined by the integrand function f, and not at all by the end-point conditions; for (75) is precisely the Euler-Lagrange equation derived in 3-3(b) under the condition of fixed end points. The two constants of integration obtained in the solution of (75), a second-order equation, are determined by the end-point conditions $y(x_1) = y_1$ and (76)—provided, of course, a solution of the problem exists.

The case in which the left-hand end point is free is left for the end-chapter exercises. The result is the condition (76) applied at $x = x_1$—namely,

$$\frac{\partial f}{\partial y'}\bigg|_{x_1} = 0. \tag{77}$$

(b) Application of the result (76) to the problem of the curve of quickest descent from a fixed point to the vertical line $x = x_2$ gives, for the integrand function (38) of 3-6(a),

$$\frac{\partial f}{\partial y'} = \frac{y'}{\sqrt{(y - y_0)(1 + y'^2)}} = 0 \qquad \text{for } x = x_2.$$

That is, the tangent to the cycloid giving the quickest descent must be horizontal at the intersection with the line $x = x_2$. The construction is always possible.

It is to be pointed out that the result (77) for the case in which the left-hand end point is free is not applicable to the problem of quickest descent from a vertical line to a fixed point. For (77), as well as (76), is derived under the assumption that the integrand function f is not explicitly a function of the value of y at the free end point; but as we see from (38) and (43) of 3-6, the brachistochrone integrand function f depends explicitly upon y_1. The problem of most rapid descent from a given curve to a fixed point is handled separately in 3-11 below.

(c) We seek to extremize the integral

$$I = \int_{x_1}^{x_2} f(x,y,y')dx \tag{78}$$

with respect to functions which attain the value y_1 for $x = x_1$ and which satisfy the given relation

$$g(x,y) = 0 \tag{79}$$

at the upper limit of integration, as yet undetermined. To this end we set up the one-parameter family of comparison functions

$$Y(x) = y(x) + \epsilon\eta(x), \tag{80}$$

where $y(x)$ is the actual extremizing function, and $\eta(x)$ is arbitrary to within differentiability and

$$\eta(x_1) = 0, \tag{81}$$

which ensures that all the comparison arcs pass through the prescribed point (x_1,y_1). The point of intersection of a comparison arc $y = Y(x)$ with the given curve (79) is denoted by (X_2,Y_2). For the special case $\epsilon = 0$—the actual extremizing arc—(X_2,Y_2) is denoted by (x_2,y_2). We thus have, with (80),

$$g(X_2,Y_2) = 0, \qquad Y_2 = y(X_2) + \epsilon\eta(X_2). \tag{82}$$

Since these relations hold for all ϵ, we have that the total derivative of $g(X_2,Y_2)$ with respect to ϵ must vanish. From (82) we therefore obtain, on noting that X_2 is a function of ϵ alone for any given $\eta(x)$,

$$0 - \frac{\partial g}{\partial X_2}\frac{dX_2}{d\epsilon} + \frac{\partial g}{\partial Y_2}\frac{dY_2}{d\epsilon}$$
$$= \frac{\partial g}{\partial X_2}\frac{dX_2}{d\epsilon} + \frac{\partial g}{\partial Y_2}\left[y'(X_2)\frac{dX_2}{d\epsilon} + \eta(X_2) + \epsilon\eta'(X_2)\frac{dX_2}{d\epsilon}\right]. \tag{83}$$

We set $\epsilon = 0$, whence X_2, Y_2 and $(dX_2/d\epsilon)$ become respectively x_2, y_2 and $(dX_2/d\epsilon)_0$; solving (83), with $\epsilon = 0$, for the latter quantity, we obtain

$$\left(\frac{dX_2}{d\epsilon}\right)_0 = -\frac{\eta_2(\partial g/\partial y_2)}{(\partial g/\partial x_2) + (\partial g/\partial y_2)y_2'}, \tag{84}$$

where we write $\eta_2 = \eta(x_2)$ and $y_2' = y'(x_2)$. The result (84) is employed directly below.

We form the integral

$$I(\epsilon) = \int_{x_1}^{X_2} f(x,Y,Y')dx$$

by replacing x_2, y, y' in (78) by X_2, Y, Y', respectively, where

$$Y' = y'(x) + \epsilon\eta'(x),$$

according to (80). Thus since X_2, Y, Y' reduce to the respective extremizing quantities x_2, y, y' for $\epsilon = 0$, we have that $I(0)$ is an extremum, so that $I'(0) = 0$. Using the rule (3) given in 2-3(a), we form the derivative

$$I'(\epsilon) = \frac{dX_2}{d\epsilon}f\Big|_{X_2} + \int_{x_1}^{X_2}\left(\frac{\partial f}{\partial Y}\eta + \frac{\partial f}{\partial Y'}\eta'\right)dx,$$

since $(\partial Y/\partial\epsilon) = \eta$ and $(\partial Y'/\partial\epsilon) = \eta'$. Setting $\epsilon = 0$ and then integrating by parts the second term of the integral on the right, we obtain, with the use of (81),

$$I'(0) = \left(\frac{dX_2}{d\epsilon}\right)_0 f\Big|_{x_2} + \frac{\partial f}{\partial y'}\Big|_{x_2}\eta_2 + \int_{x_1}^{x_2}\left[\frac{\partial f}{\partial y} - \frac{d}{dx}\left(\frac{\partial f}{\partial y'}\right)\right]\eta\,dx = 0.$$

With the aid of (84) this becomes

$$\left[\frac{\partial f}{\partial y'}\Big|_{x_2} - \frac{\dfrac{\partial g}{\partial y_2}f|_{x_2}}{\left(\dfrac{\partial g}{\partial x_2}\right) + y_2'\left(\dfrac{\partial g}{\partial y_2}\right)}\right]\eta_2 + \int_{x_1}^{x_2}\left[\frac{\partial f}{\partial y} - \frac{d}{dx}\left(\frac{\partial f}{\partial y'}\right)\right]\eta\,dx = 0,$$

for all choices of η consistent with (81). Repeating the line of argument carried out in (a) above with application of the basic lemma of 3-1(a), we conclude that $y = y(x)$ satisfies the Euler-Lagrange equation (75) and, in addition to the left-hand end-point requirement $y(x_1) = y_1$, the right-hand end-point condition

$$\frac{\partial f}{\partial y'}\Big|_{x_2} - \frac{\dfrac{\partial g}{\partial y_2}f|_{x_2}}{\left(\dfrac{\partial g}{\partial x_2}\right) + y_2'\left(\dfrac{\partial g}{\partial y_2}\right)} = 0. \tag{85}$$

A similar result for $x = x_1$ is to be obtained if the left-hand end point is required to lie on a given curve $h(x,y) = 0$. In this case $(\partial g/\partial x_2)$ and $(\partial g/\partial y_2)$ are replaced by $(\partial h/\partial x_1)$ and $(\partial h/\partial y_1)$, respectively, in the end-point condition.

(d) For the curve of quickest descent from a fixed point to a given curve $g(x,y) = 0$ application of the end-point condition (85) to the integrand (38) for the brachistochrone gives the result

$$y_2'\left[\left(\frac{\partial g}{\partial x_2}\right)\Big/\left(\frac{\partial g}{\partial y_2}\right)\right] = 1.$$

Since the slope of $g(x,y) = 0$ at (x_2,y_2) is given by the negative of the coefficient of y_2', and since y_2' is the slope of the extremizing curve at (x_2,y_2), this relation implies the orthogonality of the two curves at the point of intersection.

3-11. Brachistochrone from a Given Curve to a Fixed Point

The solution to the problem of the arc of most rapid descent from a given curve

$$h(x,y) = 0 \tag{86}$$

to a fixed point (x_2,y_2) is furnished by the function $y(x)$ which extremizes the integral

$$I = \int_{x_1}^{x_2} f(y_1,y,y')dx \tag{87}$$

with respect to functions which attain the value y_2 for $x = x_2$ and which satisfy the given relation (86) at the lower limit of integration, as yet undetermined. The function f is given by

$$f = \frac{\sqrt{1 + y'^2}}{\sqrt{y - y_0}}, \qquad y_0 = y_1 - \frac{v_1^2}{2g}, \tag{88}$$

according to (38) and (43) of 3-6, where v_1 and g are known constants.

To solve this problem we set up the one-parameter family of comparison functions

$$Y(x) = y(x) + \epsilon\eta(x), \tag{89}$$

where $y(x)$ is the actual extremizing function, and $\eta(x)$ is arbitrary to within differentiability and

$$\eta(x_2) = 0, \tag{90}$$

so that every comparison arc passes through (x_2,y_2). The point of intersection of a comparison arc $y = Y(x)$ with the given curve (86) is denoted by (X_1,Y_1), and is denoted by (x_1,y_1) when $\epsilon = 0$—that is, for the actual extremizing arc $y = y(x)$. We thus have, with (89),

$$h(X_1,Y_1) = 0, \qquad Y_1 = y(X_1) + \epsilon\eta(X_1), \tag{91}$$

for all ϵ. Taking the total derivative of $h(X_1,Y_1) = 0$, we obtain from (91), in the manner used to reach (84) of 3-10(c),

$$\left(\frac{dX_1}{d\epsilon}\right)_0 = -\frac{\eta_1(\partial h/\partial y_1)}{(\partial h/\partial x_1) + (\partial h/\partial y_1)y_1'}, \qquad \left(\frac{dY_1}{d\epsilon}\right)_0 = \frac{\eta_1(\partial h/\partial x_1)}{(\partial h/\partial x_1) + (\partial h/\partial y_1)y_1'}, \tag{92}$$

where we write $\eta_1 = \eta(x_1)$, $y_1' = y'(x_1)$, and $(dX_1/d\epsilon)_0$, $(dY_1/d\epsilon)_0$ for $(dX_1/d\epsilon)$, $(dY_1/d\epsilon)$ when $\epsilon = 0$.

The special dependence of the integrand function (88) on y and y_1 provides another formula useful directly below—namely,

$$\frac{\partial f}{\partial y_1} = -\frac{\partial f}{\partial y}. \tag{93}$$

This may be verified directly from (88) or be recognized as an immediate consequence of the fact that y and y_1 appear in f in the combination $(y - y_1)$ only.

We set up the integral

$$I(\epsilon) = \int_{X_1}^{x_2} f(Y_1, Y, Y') dx \tag{94}$$

by replacing x_1, y_1, y, y' in (87) by X_1, Y_1, Y, Y', respectively, where $Y' = y'(x) + \epsilon\eta'(x)$, according to (89). Since X_1, Y_1, Y, Y' reduce to the corresponding extremizing quantities for $\epsilon = 0$, we have that (94) is an extremum for $\epsilon = 0$, so that $I'(0) = 0$. Using the rule (3) of 2-3(a), we differentiate (94) with respect to ϵ, then set $\epsilon = 0$, to obtain

$$I'(0) = 0 = -\left(\frac{dX_1}{d\epsilon}\right)_0 f\Big|_{x_1} + \int_{x_1}^{x_2}\left[\frac{\partial f}{\partial y_1}\left(\frac{dY_1}{d\epsilon}\right)_0 + \frac{\partial f}{\partial y}\eta + \frac{\partial f}{\partial y'}\eta'\right] dx$$

$$= -\left(\frac{dX_1}{d\epsilon}\right)_0 f\Big|_{x_1} - \frac{\partial f}{\partial y'}\Big|_{x_1}\eta_1 + \left(\frac{dY_1}{d\epsilon}\right)_0 \int_{x_1}^{x_2}\frac{\partial f}{\partial y_1} dx$$

$$+ \int_{x_1}^{x_2}\left[\frac{\partial f}{\partial y} - \frac{d}{dx}\left(\frac{\partial f}{\partial y'}\right)\right]\eta\, dx,$$

in which the final form is arrived at through integration by parts and subsequent use of (90). With the help of both equations of (92) we are led to the result

$$\eta_1\left[\frac{f\Big|_{x_1}\dfrac{\partial h}{\partial y_1}}{\dfrac{\partial h}{\partial x_1} + \dfrac{\partial h}{\partial y_1}y_1'} - \frac{\partial f}{\partial y'}\Big|_{x_1} + \frac{\dfrac{\partial h}{\partial x_1}}{\dfrac{\partial h}{\partial x_1} + \dfrac{\partial h}{\partial y_1}y_1'}\int_{x_1}^{x_2}\frac{\partial f}{\partial y_1} dx\right]$$

$$+ \int_{x_1}^{x_2}\left[\frac{\partial f}{\partial y} - \frac{d}{dx}\left(\frac{\partial f}{\partial y'}\right)\right]\eta\, dx = 0. \tag{95}$$

Since (95) holds for all $\eta(x)$ which satisfy (90), it holds in particular for those $\eta(x)$ which also satisfy $\eta_1 = \eta(x_1) = 0$. For such $\eta(x)$ the left-hand member of (95) reduces to the second integral alone, and we may apply the basic lemma of 3-1(a) to conclude

$$\frac{\partial f}{\partial y} - \frac{d}{dx}\left(\frac{\partial f}{\partial y'}\right) = 0. \tag{96}$$

With this result, and for arbitrary η once again, (95) reduces to the term proportional to η_1; thus by choosing $\eta(x)$ so that $\eta_1 = 1$, we obtain the condition

$$\frac{f\Big|_{x_1} \dfrac{\partial h}{\partial y_1}}{\dfrac{\partial h}{\partial x_1} + \dfrac{\partial h}{\partial y_1} y_1'} - \frac{\partial f}{\partial y'}\Big|_{x_1} + \frac{\dfrac{\partial h}{\partial x_1}}{\dfrac{\partial h}{\partial x_1} + \dfrac{\partial h}{\partial y_1} y_1'} \int_{x_1}^{x_2} \frac{\partial f}{\partial y_1}\, dx = 0. \tag{97}$$

Because of (93) and (96) the integral of (97) is readily evaluated as

$$\int_{x_1}^{x_2} \frac{\partial f}{\partial y_1}\, dx = - \int_{x_1}^{x_2} \frac{\partial f}{\partial y}\, dx = - \int_{x_1}^{x_2} \frac{d}{dx}\left(\frac{\partial f}{\partial y'}\right) dx = \frac{\partial f}{\partial y'}\Big|_{x_1} - \frac{\partial f}{\partial y'}\Big|_{x_2}$$

With the use of this result, (97) becomes, at length,

$$\frac{\partial f}{\partial y'}\Big|_{x_2} \frac{\partial h}{\partial x_1} + \left(\frac{\partial f}{\partial y'} y' - f\right)\Big|_{x_1} \frac{\partial h}{\partial y_1} - 0. \tag{98}$$

But since f is explicitly independent of x, we may use the result (28) of 3-4(b), whereby (96) implies

$$\left(\frac{\partial f}{\partial y'} y' - f\right)\Big|_{x_1} = \left(\frac{\partial f}{\partial y'} y' - f\right)\Big|_{x_2}. \tag{99}$$

With the aid of (99) and (88) equation (98) finally reads

$$-\left[\left(\frac{\partial h}{\partial x_1}\right)\Big/\left(\frac{\partial h}{\partial y_1}\right)\right] = -\frac{1}{y_2'}, \tag{100}$$

where $y_2' = y'(x_2)$ is the slope of the extremizing curve at the *right-hand* end point (x_2, y_2). Since the left-hand member of (100) is the slope of $h(x,y) = 0$ at the left-hand end point (x_1, y_1), this final result expresses the interesting fact that the tangent to the brachistochrone at the right-hand end point is perpendicular to the tangent to the given curve $h(x,y) = 0$ at the left-hand end point! (The brachistochrone is again a cycloid, inasmuch as the extremizing function must satisfy the same differential equation (96) as in the fixed-end-point case handled above in 3-6.)

EXERCISES

1. (a) Carry through the details of the proof of the basic lemma of 3-1(a) in which $\eta(x)$ is assumed to be *twice* continuously differentiable in $x_1 \leqq x \leqq x_2$.

(b) Extend the proof of the basic lemma to the case in which $\eta(x)$ is required to possess a continuous derivative of kth order in addition to satisfying (2). HINT: If $G(x) > 0$ in $x_1' \leqq x \leqq x_2'$, let $\eta = [(x - x_1')(x_2' - x)]^{k+1}$ in this subinterval, with $\eta = 0$ in the remainder of $x_1 \leqq x \leqq x_2$.

(c) Prove that $G(x,y) = 0$ identically in D is necessary for the vanishing of (6) for all continuously differentiable $\eta(x,y)$ that vanish on C, provided $G(x,y)$ is continuous in D. HINT: If $G(x,y) > 0$ in $x_1' \leqq x \leqq x_2'$, $y_1' \leqq y \leqq y_2'$, let $\eta = 0$ everywhere in D outside this rectangle; in the rectangle let $\eta = [(x - x_1')(x - x_2')(y - y_1')(y - y_2')]^2$.

(d) Extend part (c) to the case in which η is required to have continuous partial derivatives of all orders up to and including the kth. HINT: See part (b) above.

(e) Extend the basic lemma to integrals of multiplicity m, with η required to satisfy differentiability conditions of the type required in part (d).

2. (a) Regarding the left-hand member of the obvious inequality

$$\int_{x_1}^{x_2} [y(x) + th(x)]^2 \, dx \geqq 0$$

as a quadratic function of t, where t is arbitrary, prove the (Schwarz's) inequality

$$\int_{x_1}^{x_2} h^2 \, dx \int_{x_1}^{x_2} g^2 \, dx \geqq \left\{ \int_{x_1}^{x_2} gh \, dx \right\}^2, \tag{101}$$

where the equality sign holds if and only if $g(x) = Ah(x)$, where A is some constant.

(b) Given that $y(x_1) = y_1$, $y(x_2) = y_2$ and that $p(x)$ is a known function, use (101) to prove that the absolute minimum of

$$I = \int_{x_1}^{x_2} p^2 y'^2 \, dx \tag{102}$$

is

$$\frac{(y_2 - y_1)^2}{\displaystyle\int_{x_1}^{x_2} (dx/p^2)}, \tag{103}$$

and that this minimum is attained if and only if

$$y' = \frac{A}{p^2}, \tag{104}$$

where A is an arbitrary constant. HINT: $\displaystyle\int_{x_1}^{x_2} y' \, dx = y_2 - y_1$.

(c) Show that (104) is a first integral of the Euler-Lagrange equation associated with the integral (102). Thus it is shown that the extremum of (102) is an absolute minimum. Verify that (103) is the value of (102) when (104) is substituted.

3. (a) Show that, if y satisfies the Euler-Lagrange equation associated with the integral

$$I = \int_{x_1}^{x_2} (p^2 y'^2 + q^2 y^2) dx, \tag{105}$$

where $p(x)$ and $q(x)$ are known functions, I has the value $(p^2 yy') \Big]_{x_1}^{x_2}$.

(b) Show that, if y satisfies the Euler-Lagrange equation associated with (105), and if $z(x)$ is an arbitrary differentiable function for which

$$z(x_1) = z(x_2) = 0, \tag{106}$$

then

$$\int_{x_1}^{x_2} (p^2 y' z' + q^2 yz) dx = 0.$$

Hence show that by replacing y in (105) by the function $(y + z)$, where the condition (106) ensures that y and $(y + z)$ have the same end-point values, the value of I is increased by the *nonnegative* amount

$$\int_{x_1}^{x_2} (p^2z'^2 + q^2z^2)dx.$$

Thus it is shown that the extremizing function y renders (105) an absolute minimum with respect to differentiable functions assuming the required end-point values y_1 and y_2.

4. (a) Given

$$I = \int_{x_1}^{x_2} f(y,y')dx, \tag{107}$$

reverse the roles of dependent and independent variables in order to rewrite (107) in the form

$$I = \int_{y_1}^{y_2} g(y,x')dy, \tag{108}$$

where $x' = (dx/dy)$.

(b) Write down the obvious first integral of the Euler-Lagrange equation associated with (108), according to 3-4(a). Rewrite this first integral in terms of f and with x once again as independent variable, in order to achieve the result (28) of 3-4(b).

This method of deriving (28) should meet the objection often raised against pulling the identity (27) out of thin air. HINT: This derivation is implicit in the work of 3-9(a), up to and including the second of equations (66).

5. Show that the family of geodesics on the paraboloid of revolution

$$x = u, \; y = \sqrt{u} \cos v, \; z = \sqrt{u} \sin v$$

has the form

$$u - C^2 = u(1 + 4C^2) \sin^2 \{v - 2C \log k[2\sqrt{u - C^2} + \sqrt{4u + 1}]\}, \tag{109}$$

where C and k are arbitrary constants. Although v is in general not a single-valued function of u here, the validity of (109) rests upon the result of 3-9(a).

6. (a) Prove that any geodesic on one nappe of the right circular cone

$$x^2 = b^2(y^2 + z^2) \tag{110}$$

has the following property: If the nappe is cut from the vertex along a generator and the surface of the cone is made to lie flat on a plane surface, the geodesic becomes a straight line. HINT: Show first that, if the cone is described in terms of the parameters r, θ in the form

$$x = \frac{br}{\sqrt{1 + b^2}}, \quad y = \frac{r \cos (\theta \sqrt{1 + b^2})}{\sqrt{1 + b^2}}, \quad z = \frac{r \sin (\theta \sqrt{1 + b^2})}{\sqrt{1 + b^2}},$$

which satisfies (110), the variables r and θ represent ordinary polar coordinates on the *flattened* surface of the cone, with the origin at the vertex. Identify the geodesic $r = r(\theta)$ as the equation of a straight line in polar coordinates.

(b) Prove the analogous property for geodesics on a right circular cylinder.

(c) Prove the same for an arbitrary cylindrical surface.

7. (a) Derive the differential equation satisfied by the four-times-differentiable function $y(x)$ which extremizes the integral

$$I = \int_{x_1}^{x_2} f(x,y,y',y'')dx$$

under the condition that both y and y' are prescribed at x_1 and x_2.

(b) Show that, if neither y nor y' is prescribed at either end point, the conditions

$$\frac{\partial f}{\partial y''} = 0, \qquad \frac{\partial f}{\partial y'} - \frac{d}{dx}\left(\frac{\partial f}{\partial y''}\right) = 0$$

must be met at $x = x_1$ and $x = x_2$.

(c) Generalize the result of (a) by supposing

$$f = f(x,y,y', \ldots ,y^{(n)}),$$

where $y^{(n)}$ designates the nth derivative of $y(x)$.

8. (a) Show that, if we define

$$I(\epsilon) = \int_{1}^{x_2} f(x,Y')dx$$

as in 3-3(b), we have

$$I''(0) = \int_{1}^{x_2} \frac{\partial^2 f}{\partial y'^2} \eta^2\, dx.$$

Thus we conclude that, if $(\partial f/\partial y) = 0$ and

$$\frac{\partial^2 f}{\partial y'^2} \geqq 0 \qquad \text{for } x_1 \leqq x \leqq x_2, \tag{111}$$

the extremum $I(0)$ is surely a minimum if a minimizing function of the single-valued form $y = y(x)$, with y twice differentiable, exists. (The inequality (111) is meant to exclude the identical vanishing of $(\partial^2 f/\partial y'^2)$ in the interval.)

(b) Show that the function f for the geodesic on a surface of revolution in 3-5(d) satisfies the conditions of (a).

9. Show that, if $y_1 = y_2$, no catenary of the family (45) passes through (x_1,y_1) and (x_2,y_2) if

$$-\frac{p}{\cosh p} < \frac{x_2 - x_1}{2y_1}, \tag{112}$$

where p is the *negative* root of

$$p \sinh p - \cosh p = 0. \tag{113}$$

HINT: First show that $a = \frac{1}{2}(x_1 + x_2)$ when $y_1 = y_2$ and let $p = [(x_1 - a)/b]$. Next show that equality of the two members of (112) is required for a catenary to pass through both end points. Hence, if the *maximum* of the left-hand member of (112) is less than the right-hand member, there is no p for which equality can obtain. Show that the maximizing p is the negative root of (113).

Approximately, (112) reads $(x_2 - x_1) > 1.32y_1$.

10. Use Euler's theorem on homogeneous functions (2-5) to prove that the first integral (59) reduces to an identity, with $C_1 = 0$, when f is homogeneous in $\dot{x}, \dot{y}, \ldots , \dot{z}$ of the first degree.

11. (a) Show, in 3-10(a), that if neither end-point value is prescribed, both (76) and (77) must be fulfilled.

(b) Generalize the result of (a) to the case of several dependent variables.

12. Derive the result of 3-10(a) as a special case of the result of 3-10(c) by choosing $g(x,y) = x - x_2$. NOTE: The notation $(\partial g/\partial x_2)$ is an abbreviation for $(\partial g/\partial x)$ evaluated at $x = x_2$, $y = y_2$—and similarly for $(\partial g/\partial y_2)$.

13. (a) Show that, if

$$f = p(x,y) \sqrt{1 + y'^2},$$

the condition (85) requires the orthogonality of the extremizing arc and the given curve $g(x,y) = 0$, for all $p(x,y)$.

(b) From (a) it follows, with the result of 3-4(a), that the arc which extremizes the length from a fixed point to a given plane curve is a normal drawn from the fixed point to the curve. Using the methods of elementary differential calculus, demonstrate the role played, in determining whether the normal represents a minimum or maximum distance, by the position of the center of curvature of the given curve at the point to which the normal is drawn. In what case is the distance neither a minimum nor maximum?

14. (a) A brachistochrone is required to be constructed from a given curve $h(x,y) = 0$ to a second given curve $g(x,y) = 0$. What relationship must the two given curves bear to one another at the respective points of intersection with the brachistochrone?

(b) A brachistochrone extends from the line $y = x + 4$ to the parabola $y^2 = x$. Show that the point of intersection of the brachistochrone with the parabola is $(\frac{1}{4}, \frac{1}{2})$

CHAPTER 4

ISOPERIMETRIC PROBLEMS

We consider in this chapter a class of problems in which the functions eligible for the extremization of a given definite integral are required to conform with certain restrictions that are added to the usual continuity requirements and possible end-point conditions. In the case of greatest importance for application and extension in chapters following, the additional restrictions reside in the prescription of the values of certain auxiliary definite integrals. We call problems in which such conditions are involved *isoperimetric*, after the best known problem of the class—that of finding the closed curve of given perimeter for which the area is a maximum. Further, we briefly treat cases in which the additional restrictions are expressed through ordinary finite equations or through differential equations.

4-1. The Simple Isoperimetric Problem

(*a*) We seek to derive the differential equation which must be satisfied by the function which renders the integral

$$I = \int_{x_1}^{x_2} f(x,y,y')dx \tag{1}$$

an extremum with respect to continuously differentiable functions $y = y(x)$ for which the second integral

$$J = \int_{x_1}^{x_2} g(x,y,y')dx \tag{2}$$

possesses a given prescribed value, and with $y(x_1) = y_1$, $y(x_2) = y_2$ both prescribed. The given functions f and g are twice differentiable with respect to their arguments.

In essence we follow the procedure of 3-3(*b*) by letting $y(x)$ denote the actual extremizing function and introducing a family $Y(x)$ of "comparison" functions with respect to which we carry out the extremization. We cannot, however, express $Y(x)$ as merely a *one*-parameter family of functions because any change of value of the *single* parameter would in general alter the value of J, whose constancy must be maintained as prescribed. For this reason we introduce the *two*-parameter family

$$Y(x) = y(x) + \epsilon_1\eta_1(x) + \epsilon_2\eta_2(x), \tag{3}$$

in which $\eta_1(x)$ and $\eta_2(x)$ are arbitrary differentiable functions for which

$$\eta_1(x_1) = \eta_1(x_2) = 0 = \eta_2(x_1) = \eta_2(x_2). \tag{4}$$

The condition (4) ensures that

$$Y(x_1) = y(x_1) = y_1 \quad \text{and} \quad Y(x_2) = y(x_2) = y_2,$$

as prescribed, for all values of the parameters ϵ_1 and ϵ_2.

We replace y by $Y(x)$, given by (3), in both (1) and (2) so as to form respectively

$$I(\epsilon_1,\epsilon_2) = \int_{x_1}^{x_2} f(x,Y,Y')dx \tag{5}$$

and

$$J(\epsilon_1,\epsilon_2) = \int_{x_1}^{x_2} g(x,Y,Y')dx. \tag{6}$$

Clearly, the parameters ϵ_1 and ϵ_2 are not independent; because J is to be maintained at a constant value, it is clear from (6) that there is a functional relation between them namely,

$$J(\epsilon_1,\epsilon_2) = \text{constant} \quad \text{(prescribed)}. \tag{7}$$

Since $y(x)$ is assumed to be the actual extremizing function, we have, because of (3), that (5) is an extremum with respect to values of ϵ_1, ϵ_2 which satisfy (7), when $\epsilon_1 = \epsilon_2 = 0$—for arbitrary choice of the functions η_1 and η_2 consistent with (4). (It should be noted that the definition of $y(x)$ implies that (7) is satisfied for $\epsilon_1 = \epsilon_2 = 0$.)

(b) The procedure of (a) above reduces our simple isoperimetric problem to the elementary task of determining the conditions which must be fulfilled in order that the ordinary function $I(\epsilon_1,\epsilon_2)$ of two variables ϵ_1, ϵ_2 be an extremum under the restriction (7). To solve this problem we use the method of Lagrange multipliers described in 2-6. We thus introduce the function of ϵ_1, ϵ_2

$$I^* = I(\epsilon_1,\epsilon_2) + \lambda J(\epsilon_1,\epsilon_2) = \int_{x_1}^{x_2} f^*(x,Y,Y')dx, \tag{8}$$

where, according to (1) and (2),

$$f^* = f + \lambda g. \tag{9}$$

The constant λ is the undetermined multiplier whose value remains to be determined by the conditions of each individual problem to which the method is applied. Thus, according to 2-6 and (a) above, we must have

$$\frac{\partial I^*}{\partial \epsilon_1} = \frac{\partial I^*}{\partial \epsilon_2} = 0 \quad \text{when } \epsilon_1 = \epsilon_2 = 0. \tag{10}$$

From (8), with the aid of (3), it follows that

$$\frac{\partial I^*}{\partial \epsilon_j} = \int_{x_1}^{x_2} \left\{ \frac{\partial f^*}{\partial Y} \frac{\partial Y}{\partial \epsilon_j} + \frac{\partial f^*}{\partial Y'} \frac{\partial Y'}{\partial \epsilon_j} \right\} dx = \int_{x_1}^{x_2} \left\{ \frac{\partial f^*}{\partial Y} \eta_j + \frac{\partial f^*}{\partial Y'} \eta_j' \right\} dx$$

$$(j = 1,2). \quad (11)$$

Setting $\epsilon_1 = \epsilon_2 = 0$, so that, according to (3), (Y,Y') is replaced by (y,y'), we thus have that

$$\left. \frac{\partial I^*}{\partial \epsilon_j} \right|_0 = \int_{x_1}^{x_2} \left\{ \frac{\partial f^*}{\partial y} \eta_j + \frac{\partial f^*}{\partial y'} \eta_j' \right\} dx = 0 \qquad (j = 1,2), \qquad (12)$$

because of (10). (The symbol $|_0$ indicates the setting of $\epsilon_1 = \epsilon_2 = 0$.) Integrating by parts the second term of the integrand of (12), we obtain, with the aid of (4),

$$\int_{x_1}^{x_2} \eta_j \left\{ \frac{\partial f^*}{\partial y} - \frac{d}{dx} \left(\frac{\partial f^*}{\partial y'} \right) \right\} dx = 0 \qquad (j = 1,2). \qquad (13)$$

Because of the arbitrary character of the functions $\eta_1(x)$ and $\eta_2(x)$ the two relations embodied in (13) are essentially one. At any rate we apply the basic lemma of 3-1(a) to either and so obtain the differential equation

$$\frac{\partial f^*}{\partial y} - \frac{d}{dx} \left(\frac{\partial f^*}{\partial y'} \right) = 0 \qquad (14)$$

as the Euler-Lagrange equation which must be satisfied by the function $y(x)$ which extremizes (1) under the restriction that (2) be maintained at a prescribed value.

Solution of the second-order equation (14) yields a function $y(x)$ that involves three undetermined quantities: two constants of integration and, because of (9), the Lagrange multiplier λ. If the solution of a given isoperimetric problem of the type under discussion exists, these quantities are fixed by fitting $y = y(x)$ to the required end-point conditions $y(x_1) = y_1$ and $y(x_2) = y_2$ and by giving to the integral J of (2) its prescribed value.

4-2. Direct Extensions

Since the methods embodied in the paragraphs following are essentially identical with those employed in 4-1 and in various sections of Chap. 3, each result is stated with only a bare outline of the mode of derivation. Many of the details are called for in exercises at the end of this chapter.

(a) In a somewhat more general isoperimetric problem than that which is treated in the preceding section, we seek to extremize the integral

$$I = \int_{x_1}^{x_2'} f(x,y,y')dx \qquad (15)$$

with respect to continuously differentiable functions $y(x)$ for which the
N integrals

$$J_k = \int_{x_1}^{x_2} g_k(x,y,y')dx \qquad (k = 1,2, \ldots ,N) \tag{16}$$

possess given prescribed values, and with $y(x_1) = y_1$, $y(x_2) = y_2$ both
prescribed. By introducing an $(N + 1)$-parameter family of comparison functions $Y(x)$ and subsequent application of the method of Lagrange
multipliers as in 4-1(b), we should directly reach the conclusion that the
extremizing function $y(x)$ must satisfy the Euler-Lagrange equation (14)
of 4-1(b), where here

$$f^* = f + \sum_{k=1}^{N} \lambda_k g_k; \tag{17}$$

the constants $\lambda_1, \lambda_2, \ldots , \lambda_N$ are the undetermined multipliers whose
values we may ascertain through the specific conditions imposed in any
given problem. The details are left for exercise 3(a) at the end of this
chapter.

(b) A combination of the argument carried out in 4-1 with that of
3-10(a) yields the conditions which must be satisfied by the extremizing
function when one or both of the end-point values is left unspecified in
an isoperimetric problem. If neither end-point value is prescribed for
the functions eligible for the extremization of (15), with the condition
(16), we must have

$$\frac{\partial f^*}{\partial y'} = 0 \qquad \text{for } x = x_1 \text{ and } x = x_2; \tag{18}$$

f^* is given by (17). If one, but not the other, end-point value of y is
prescribed, the actual extremizing function is such that (18) is satisfied
at the *other* end point.

(c) Application of the analysis of 3-10(c) to the isoperimetric problem
in which one end point of every arc $y = y(x)$ eligible for the extremization
is required to lie on the curve $h(x,y) = 0$ yields the condition

$$\frac{\partial f^*}{\partial y'} - \frac{(\partial h/\partial y)f^*}{(\partial h/\partial x) + y'(\partial h/\partial y)} = 0 \tag{19}$$

that must be satisfied by the extremizing function at the end point in
question. Again, f^* is given by (17) of (a) above.

(d) We may combine the argument of (a) above with that of 3-8(a)
to arrive at the system of differential equations which must be satisfied

by the functions which extremize the integral[1]

$$I = \int_{t_1}^{t_2} f(x,y, \ \ldots \ ,z,\dot{x},\dot{y}, \ \ldots \ ,\dot{z},t)dt \tag{20}$$

with respect to continuously differentiable functions $x(t)$, $y(t)$, $\ldots$, $z(t)$ for which the N integrals

$$J_k = \int_{t_1}^{t_2} g_k(x,y, \ \ldots \ ,z,\dot{x},\dot{y}, \ \ldots \ ,\dot{z},t)dt \qquad (k = 1,2, \ \ldots \ ,N) \tag{21}$$

possess given prescribed values; it is further supposed that the eligible functions x, y, $\ldots$, z achieve prescribed values at $t = t_1$ and $t = t_2$. The required equations are

$$\frac{\partial f^*}{\partial x} - \frac{d}{dt}\left(\frac{\partial f^*}{\partial \dot{x}}\right) = 0, \qquad \frac{\partial f^*}{\partial y} - \frac{d}{dt}\left(\frac{\partial f^*}{\partial \dot{y}}\right) = 0, \qquad \ldots \ ,$$

$$\frac{\partial f^*}{\partial z} - \frac{d}{dt}\left(\frac{\partial f^*}{\partial \dot{z}}\right) = 0, \tag{22}$$

where f^* is given by the expression (17), but in which f, g_1, g_2, $\ldots$, g_N are the functions which appear in the integrands of (20) and (21).

(e) The methods and results of 3-4(a,b) and 3-8(b) are directly applicable to isoperimetric problems:

(i) If f^* is explicitly independent of the dependent variable y, a first integral of the Euler-Lagrange equation (14) is

$$\frac{\partial f^*}{\partial y'} = C_1, \tag{23}$$

where C_1 is an arbitrary constant.

(ii) If f^* is explicitly independent of the independent variable x, a first integral of the Euler-Lagrange equation (14) is

$$y' \frac{\partial f^*}{\partial y'} - f^* = C_1, \tag{24}$$

with C_1 an arbitrary constant (not necessarily the same as C_1 in (23)).

(iii) If f^* is explicitly independent of the independent variable t, a first integral of the system of equations (22) is

$$\frac{\partial f^*}{\partial \dot{x}} \dot{x} + \frac{\partial f^*}{\partial \dot{y}} \dot{y} + \cdots + \frac{\partial f^*}{\partial \dot{z}} \dot{z} - f^* = C_1, \tag{25}$$

where C_1 is an arbitrary constant.

---

[1] As in 3-8, the superior dot denotes differentiation with respect to the independent variable t.

(*f*) The argument and results of 3-9 are directly applicable to the iso-perimetric problems of (*a*), (*b*), (*c*) above (and therefore to the special case of 4-1); that is, although the derivation of (14) is based upon the tacit assumption that the extremizing function $y(x)$ is a single-valued function of x, we may accept as valid any solution of (14)—satisfying the pertinent set of end-point conditions—in which the single-valuedness requirement is violated. (The remarks of 3-9(*b*) must of course be taken into account.)

4-3. Problem of the Maximum Enclosed Area

(*a*) The *original* isoperimetric problem may be stated as follows: We consider the aggregate P of all closed non-self-intersecting plane curves for which the total length has the given value L. Of these we seek one for which the enclosed area is the greatest.

With the means at our disposal it is necessary that we make the restrictive assumption that the parametric representation

$$x = x(t), \qquad y = y(t) \tag{26}$$

of any member of the aggregate P is such that the functions (26) are continuously differentiable with respect to t. But without loss of generality we may suppose that the representation (26) describes any given curve of P in the counterclockwise sense as t increases from t_1 to t_2; since the curve is closed, we have $x(t_1) = x(t_2) = x_0$ and $y(t_1) = y(t_2) = y_0$. It is no essential restriction to suppose that t_1, t_2, x_0, y_0 have respectively the same values for every member of P.

According to 2-7(*c*) the area enclosed by a given member of P as described by (26) is the integral

$$I = \tfrac{1}{2} \int_{t_1}^{t_2} (x\dot{y} - y\dot{x})dt, \tag{27}$$

where $\dot{x} = (dx/dt)$, $\dot{y} = (dy/dt)$. The total length of the curve, given by

$$J = \int_{t_1}^{t_2} \sqrt{\dot{x}^2 + \dot{y}^2}\, dt, \tag{28}$$

has the same value L for every member of P. We seek the particular functions for which (27) is an extremum (maximum, in the present case) with respect to functions $x(t)$, $y(t)$ which bestow upon (28) the given value L and for which $x(t_1) = x(t_2) = x_0$, $y(t_1) = y(t_2) = y_0$.

From 4-2(*d*) we have that the maximizing functions must satisfy the system of equations (22)—namely,

$$\frac{\partial f^*}{\partial x} - \frac{d}{dt}\left(\frac{\partial f^*}{\partial \dot{x}}\right) = 0, \qquad \frac{\partial f^*}{\partial y} - \frac{d}{dt}\left(\frac{\partial f^*}{\partial \dot{y}}\right) = 0 \tag{29}$$

—where, according to (17), (20), (21), together with (27) and (28),

$$f^* = \tfrac{1}{2}(x\dot{y} - y\dot{x}) + \lambda \sqrt{\dot{x}^2 + \dot{y}^2}. \tag{30}$$

Direct substitution of (30) into (29) yields

$$\frac{1}{2}\dot{y} - \frac{d}{dt}\left(-\frac{1}{2}y + \frac{\lambda\dot{x}}{\sqrt{\dot{x}^2 + \dot{y}^2}}\right) = 0,$$

$$-\frac{1}{2}\dot{x} - \frac{d}{dt}\left(\frac{1}{2}x + \frac{\lambda\dot{y}}{\sqrt{\dot{x}^2 + \dot{y}^2}}\right) = 0,$$

from which we obtain, by direct integration with respect to t,

$$y - \frac{\lambda\dot{x}}{\sqrt{\dot{x}^2 + \dot{y}^2}} = C_1, \qquad x + \frac{\lambda\dot{y}}{\sqrt{\dot{x}^2 + \dot{y}^2}} = C_2, \tag{31}$$

with C_1 and C_2 arbitrary constants.

With the introduction[1] of $(dy/dx) = (\dot{y}/\dot{x})$ both of (31) may be integrated in a conventional manner. We achieve the same result, however, by solving for $(y - C_1)$ and $(x - C_2)$, then squaring and adding the equations obtained:

$$(x - C_2)^2 + (y - C_1)^2 = \lambda^2. \tag{32}$$

Thus we have the well-known result that the closed curve of given perimeter for which the enclosed area is a maximum is a circle.[2] Since the location of the circle is immaterial, the constants of integration C_2 and C_1—the coordinates of the center, according to (32)—remain arbitrary. Also, since λ^2 is the square of the radius, we have $\lambda^2 = (L/2\pi)^2$, where L is the given perimeter.

(b) A problem closely related to the original isoperimetric problem is the following: We consider the aggregate P' of all non-self-intersecting plane arcs for which the total length has the given value L' and whose end points lie on the x axis. Of these we seek one for which the area enclosed by it and the x axis is the greatest.

For the sake of simplicity we let $y = y(x)$ represent any member of the aggregate[3] P' and make the restrictive assumption that $y(x)$ is continuously differentiable. Without loss of generality we may suppose that the left-hand end point is fixed at the given point $(x_1,0)$; the right-hand end point $(x_2,0)$ is unspecified.

---

[1] See end-chapter exercise 5.

[2] In view of the restriction that $x(t)$ and $y(t)$ be continuously differentiable for all functions eligible for the maximization, our result, strictly, should read " . . . that the closed *smooth* curve. . . ."

[3] The choice of this type of representation is justified in 4-2(*f*).

The area enclosed by any member of P' and the x axis is given by

$$I = \int_{x_1}^{x_2} y \, dx; \tag{33}$$

the total length, equal to the fixed value L' for every member of P', is given by

$$J = \int_{x_1}^{x_2} \sqrt{1 + y'^2} \, dx. \tag{34}$$

We seek the equation of the particular arc for which (33) is an extremum (a maximum) with respect to arcs $y = y(x)$ whose left-hand end points coincide at $(x_1, 0)$, whose right-hand end points lie on the curve $y = 0$, and which give to (34) the prescribed value L'.

To this end we apply the Euler-Lagrange equation (14) of 4-1(b) to the integrand

$$f^* = y + \lambda \sqrt{1 + y'^2} \tag{35}$$

derived from (33) and (34). We thus obtain

$$1 - \lambda \frac{d}{dx} \left(\frac{y'}{\sqrt{1 + y'^2}} \right) = 0,$$

whence, by direct integration,

$$\frac{\lambda y'}{\sqrt{1 + y'^2}} = x - C_1.$$

From this it follows that

$$dy = \frac{\pm (x - C_1) dx}{\sqrt{\lambda^2 - (x - C_1)^2}}$$

and therefore that

$$y = \mp \sqrt{\lambda^2 - (x - C_1)^2} + C_2,$$

or

$$(x - C_1)^2 + (y - C_2)^2 = \lambda^2. \tag{36}$$

To derive the condition which must be satisfied at $x = x_2$ we apply the end-point relation (19) of 4-2(c) to the integrand (35), with $y = 0$ as the curve $h(x,y) = 0$. With this we obtain

$$\frac{\lambda y'}{\sqrt{1 + y'^2}} - \frac{y + \lambda \sqrt{1 + y'^2}}{y'} = 0 \qquad \text{at } x = x_2;$$

or since $y = 0$ at $x = x_2$ and since $\lambda = 0$ is ruled out by (36), we have

$$\frac{1}{y' \sqrt{1 + y'^2}} = 0 \qquad \text{at } x = x_2. \tag{37}$$

The fulfillment of (37), clearly, is possible only if the maximizing arc possesses a *vertical* tangent at $x = x_2$. This fact, combined with (36), directly implies that the required arc is a semicircle of radius (L'/π).

4-4. Shape of a Hanging Rope

We may apply the result of 4-1(b) to the problem of determining the shape of a perfectly flexible rope of uniform density that hangs at rest with its end points fixed. The basis for this application resides in the physical principle which states that a mechanical system in stable equilibrium is characterized by a minimum of potential energy consistent with its constraints.

With the trivial assumption that the rope hangs in a vertical plane we let $y = y(x)$ be a representative member of the aggregate P of all possible configurations (in the vertical plane determined by the fixed end points of the rope) that may be assumed by the rope, consistent with the facts that its end points are fixed and its total length has the given value L. The coordinate x is measured horizontally in the vertical plane, and y is the *upward* distance from a fixed horizontal reference plane. (According to 4-2(f) the designation $y = y(x)$ is not excessively restrictive in the event y is not a single-valued function of x in the equilibrium configuration.) Thus, if σ denotes the constant mass per unit length of the rope, the potential energy (relative to $y = 0$) of an element of length ds at (x,y) is given by $gy\sigma\,ds$, where g is the constant acceleration due to gravity. Accordingly, the total potential energy of the rope in the arbitrary configuration $y = y(x)$ is given by

$$I = \sigma g \int_0^L y\,ds = \sigma g \int_{x_1}^{x_2} y\,\sqrt{1 + y'^2}\,dx, \tag{38}$$

where (x_1,y_1), (x_2,y_2) are the respective fixed end points of the rope $(x_1 < x_2)$.

According to the minimum-energy principle the equilibrium configuration is supplied by the particular relation $y = y(x)$ for which (38) is a minimum with respect to functions $y(x)$ for which $y(x_1) = y_1$, $y(x_2) = y_2$, and for which the total arc length

$$J = \int_{x_1}^{x_2} \sqrt{1 + y'^2}\,dx \tag{39}$$

has the prescribed value L. We may therefore apply the Euler-Lagrange equation (14) of 4-1(b) to the integrand function

$$f^* = \sigma g y\,\sqrt{1 + y'^2} + \lambda\,\sqrt{1 + y'^2} \tag{40}$$

formed from (38) and (39). Since f^* is explicitly independent of the independent variable x, however, we may use 4-2(e,ii) and so substitute (40) into (24) of that section:

$$(\sigma gy + \lambda) \left(\frac{y'^2}{\sqrt{1 + y'^2}} - \sqrt{1 + y'^2} \right) = C_1,$$

whence[1]

$$y = -\frac{\lambda}{\sigma g} - \frac{C_1}{\sigma g} \cosh \frac{\sigma g(x - a)}{C_1} \tag{41}$$

where a is an arbitrary constant of integration.

Thus, according to (41), the shape of a hanging rope is that of a catenary with vertical axis. By specifying that the catenary pass through (x_1,y_1) and (x_2,y_2) and that the arc included between these points have the length L we may assign values to the constants C_1, a, λ which appear in (41). The construction is always possible (although the actual computation of C_1, a, λ may involve serious numerical difficulties).

Because a rope (or chain) hangs in the shape of a hyperbolic cosine this curve has been given the name catenary. (The Latin for chain is *catena*.)

4-5. Restrictions Imposed through Finite or Differential Equations

(a) To the problem of 3-8, the extremization of a given integral with respect to several integrand functions, we add a set of restrictions which must be satisfied by the functions eligible for the extremization. These restrictions consist of a set of finite or differential equations or a combination of both, with the total number of equations less than the number of integrand functions. Specifically, we proceed to derive the system of differential equations which must be satisfied by the set of functions which extremize the integral

$$I = \int_{t_1}^{t_2} f(x,y, \ldots ,z,\dot{x},\dot{y}, \ldots ,\dot{z},t)dt \tag{42}$$

with respect to the k continuously differentiable functions x, y, $\ldots$, z which achieve prescribed values at $t = t_1$ and $t = t_2$ and which satisfy the N given (consistent and independent) equations

$$G_j(x,y, \ldots ,z,\dot{x},\dot{y}, \ldots ,\dot{z},t) = 0 \qquad (j = 1,2, \ldots ,N < k). \tag{43}$$

(If a given G_j is explicitly independent of the derivatives $\dot{x}$, $\dot{y}$, $\ldots$ $\dot{z}$, the corresponding equation $G_j = 0$ is a *finite*, rather than a differential, equation.)

---

[1] The details are left for the reader; compare 3-7(a).

As in 3-8(a), we denote the actual extremizing functions by $x(t)$, $y(t)$, . . . , $z(t)$ and introduce the one-parameter family of comparison functions

$$X(t) = x(t) + \epsilon\xi_1(t), \qquad Y(t) = y(t) + \epsilon\xi_2(t), \qquad \cdots,$$
$$Z(t) = z(t) + \epsilon\xi_k(t), \quad (44)$$

where $\xi_1, \xi_2, \ldots, \xi_k$ are differentiable functions for which

$$\xi_i(t_1) = \xi_i(t_2) = 0 \qquad (i = 1,2, \ldots ,k), \qquad (45)$$

and which are otherwise arbitrary *to within consistency with the set of constraints*—formed by replacing $(x,y, \ldots ,z)$ by the comparison functions $(X,Y, \ldots ,Z)$ in (43)—

$$G_j(X,Y, \ldots ,Z,\dot{X},\dot{Y}, \ldots ,\dot{Z},t) = 0 \qquad (j = 1,2, \ldots ,N). \quad (46)$$

We replace, further, $(x,y, \ldots ,z)$ by $(X,Y, \ldots ,Z)$ in the integrand of (42) and so form the integral

$$I(\epsilon) = \int_{t_1}^{t_2} f(X,Y, \ldots ,Z,\dot{X},\dot{Y}, \ldots ,\dot{Z},t)dt. \qquad (47)$$

Because of the designation of $x(t)$, $y(t)$, . . . , $z(t)$ as the actual extremizing functions, it follows from (44) that $I(\epsilon)$ is an extremum for $\epsilon = 0$; that is,

$$I'(0) = 0, \qquad (48)$$

for any permissible choice of $\xi_1, \xi_2, \ldots, \xi_k$.

In the manner of 3-8(a) we form the derivative $I'(\epsilon)$ of (47) with the aid of (44)—from which we derive $(\partial X/\partial\epsilon) = \xi_1$, $(\partial\dot{X}/\partial\epsilon) = \dot{\xi}_1$, etc.— to obtain

$$I'(\epsilon) = \int_{t_1}^{t_2} \left\{ \frac{\partial f}{\partial X}\xi_1 + \frac{\partial f}{\partial\dot{X}}\dot{\xi}_1 + \frac{\partial f}{\partial Y}\xi_2 + \frac{\partial f}{\partial\dot{Y}}\dot{\xi}_2 + \cdots \right.$$
$$\left. + \frac{\partial f}{\partial Z}\xi_k + \frac{\partial f}{\partial\dot{Z}}\dot{\xi}_k \right\} dt.$$

Setting $\epsilon = 0$—that is, replacing $(X,Y, \ldots ,Z)$ by $(x,y, \ldots ,z)$, according to (44)—we obtain, with (48),

$$I'(0) = \int_{t_1}^{t_2} \left\{ \frac{\partial f}{\partial x}\xi_1 + \frac{\partial f}{\partial\dot{x}}\dot{\xi}_1 + \frac{\partial f}{\partial y}\xi_2 + \frac{\partial f}{\partial\dot{y}}\dot{\xi}_2 + \cdots \right.$$
$$\left. + \frac{\partial f}{\partial z}\xi_k + \frac{\partial f}{\partial\dot{z}}\dot{\xi}_k \right\} dt = 0. \quad (49)$$

We cannot, however, continue from this point as in 3-8(a) because of the mutual dependence of the functions $\xi_1, \xi_2, \ldots, \xi_k$ as embodied in (46). To obtain an explicit expression of this dependence we note that the N equations (46) are satisfied identically[1] for all ϵ, so that each may be differentiated with respect to ϵ as follows:

$$\frac{\partial G_j}{\partial X} \xi_1 + \frac{\partial G_j}{\partial \dot{X}} \dot{\xi}_1 + \frac{\partial G_j}{\partial Y} \xi_2 + \frac{\partial G_j}{\partial \dot{Y}} \dot{\xi}_2 + \cdots + \frac{\partial G_j}{\partial Z} \xi_k + \frac{\partial G_j}{\partial \dot{Z}} \dot{\xi}_k = 0$$

$$(j = 1,2, \ldots ,N).$$

(Here we again make use of (44) to evaluate $(\partial X/\partial \epsilon)$, etc.)   In particular for $\epsilon = 0$ we have

$$\frac{\partial G_j}{\partial x} \xi_1 + \frac{\partial G_j}{\partial \dot{x}} \dot{\xi}_1 + \frac{\partial G_j}{\partial y} \xi_2 + \frac{\partial G_j}{\partial \dot{y}} \dot{\xi}_2 + \cdots + \frac{\partial G_j}{\partial z} \xi_k + \frac{\partial G_j}{\partial \dot{z}} \dot{\xi}_k = 0$$

$$(j = 1,2, \ldots ,N), \quad (50)$$

since setting $\epsilon = 0$ means replacing $(X,Y, \ldots ,Z)$ by $(x,y, \ldots ,z)$, according to (44).

Multiplying the jth equation of the system (50) by the unspecified function $\mu_j(t)$, for all $j = 1, 2, \ldots , N$, we add the left-hand members (all equal to zero for any choices of the μ_j) to the integrand of (49) and so obtain

$$\int_{t_1}^{t_2} \left\{ \left[\frac{\partial f}{\partial x} + \sum_{j=1}^{N} u_j \frac{\partial G_j}{\partial x} \right] \xi_1 + \left[\frac{\partial f}{\partial \dot{x}} + \sum_{j=1}^{N} \mu_j \frac{\partial G_j}{\partial \dot{x}} \right] \dot{\xi}_1 + \cdots \right.$$

$$\left. + \left[\frac{\partial f}{\partial z} + \sum_{j=1}^{N} \mu_j \frac{\partial G_j}{\partial z} \right] \xi_k + \left[\frac{\partial f}{\partial \dot{z}} + \sum_{j=1}^{N} \mu_j \frac{\partial G_j}{\partial \dot{z}} \right] \dot{\xi}_k \right\} dt$$

$$= \int_{t_1}^{t_2} \left\{ \frac{\partial F}{\partial x} \xi_1 + \frac{\partial F}{\partial \dot{x}} \dot{\xi}_1 + \frac{\partial F}{\partial y} \xi_2 + \frac{\partial F}{\partial \dot{y}} \dot{\xi}_2 + \cdots + \frac{\partial F}{\partial z} \xi_k + \frac{\partial F}{\partial \dot{z}} \dot{\xi}_k \right\} dt$$

$$= 0, \quad (51)$$

where we define

$$F = f + \sum_{j=1}^{N} \mu_j(t) G_j. \quad (52)$$

Integrating by parts the second, fourth, $\ldots$, $2k$th terms of (51), we get, with the aid of (45),

$$\int_{t_1}^{t_2} \left\{ \left[\frac{\partial F}{\partial x} - \frac{d}{dt} \left(\frac{\partial F}{\partial \dot{x}} \right) \right] \xi_1 + \left[\frac{\partial F}{\partial y} - \frac{d}{dt} \left(\frac{\partial F}{\partial \dot{y}} \right) \right] \xi_2 + \cdots \right.$$

$$\left. + \left[\frac{\partial F}{\partial z} - \frac{d}{dt} \left(\frac{\partial F}{\partial \dot{z}} \right) \right] \xi_k \right\} dt = 0. \quad (53)$$

Because of the set (50) of N equations among them, we cannot regard the k functions $\xi_1, \xi_2, \ldots , \xi_k$ as being free for arbitrary choice.   In

[1] It is sufficient that they be satisfied identically in ϵ only for a neighborhood about $\epsilon = 0$.

fact there is some subset of $N(< k)$ of these functions whose assignment is restricted by the assignment of the remaining $(k - N)$. For the sake of definiteness we suppose that $\xi_1, \xi_2, \ldots, \xi_N$ are the functions of the set whose dependence upon the choices of the *arbitrary* (to within (45)) $\xi_{N+1}, \xi_{N+2}, \ldots, \xi_k$ is governed by (50).[1] At this point we assign the unspecified functions $\mu_1(t), \mu_2(t), \ldots, \mu_N(t)$ to be any set of N functions which make vanish (for all t between t_1 and t_2) the coefficients of $\xi_1, \xi_2, \ldots, \xi_N$ in the integrand of (53). That is, if we let $u_1, u_2, \ldots, u_N$ denote the first N functions of the list $x, y, \ldots, z$, the functions $\mu_j(t)$ are chosen so as to satisfy

$$\frac{\partial F}{\partial u_i} - \frac{d}{dt}\left(\frac{\partial F}{\partial \dot{u}_i}\right) = 0 \qquad (i = 1, 2, \ldots, N). \tag{54}$$

With the choice of all the μ_j so fixed (53) reads as follows:

$$\int_{t_1}^{t_2} \left\{ \left[\frac{\partial F}{\partial u_{N+1}} - \frac{d}{dt}\left(\frac{\partial F}{\partial \dot{u}_{N+1}}\right) \right] \xi_{N+1} + \cdots \right.$$
$$\left. + \left[\frac{\partial F}{\partial u_k} - \frac{d}{dt}\left(\frac{\partial F}{\partial \dot{u}_k}\right) \right] \xi_k \right\} dt = 0, \tag{55}$$

where $u_{N+1}, u_{N+2}, \ldots, u_k = z$ denote the final $(k - N)$ functions of the list $x, y, \ldots, z$. Since the functions $\xi_{N+1}, \xi_{N+2}, \ldots, \xi_k$ are, to within (45), completely arbitrary, we may employ the device used in 3-8(a) to conclude, on the basis of the lemma of 3-1(a), that each of the coefficients of $\xi_{N+1}, \xi_{N+2}, \ldots, \xi_k$ in the integrand of (55) must vanish individually. We have, that is,

$$\frac{\partial F}{\partial u_i} - \frac{d}{dt}\left(\frac{\partial F}{\partial \dot{u}_i}\right) = 0 \qquad (i = N + 1, N + 2, \ldots, k). \tag{56}$$

Thus, on combining (56) with (54) and noting that $u_1, u_2, \ldots, u_k$ constitute the complete list $x, y, \ldots, z$, we reach the conclusion that the k extremizing functions $x(t), y(t), \ldots, z(t)$ satisfy the system of k Euler-Lagrange differential equations

$$\frac{\partial F}{\partial x} - \frac{d}{dt}\left(\frac{\partial F}{\partial \dot{x}}\right) = 0, \qquad \frac{\partial F}{\partial y} - \frac{d}{dt}\left(\frac{\partial F}{\partial \dot{y}}\right) = 0, \qquad \ldots,$$
$$\frac{\partial F}{\partial z} - \frac{d}{dt}\left(\frac{\partial F}{\partial \dot{z}}\right) = 0, \tag{57}$$

where F is given by (52).

---

[1] Although actually some other subset of N functions of the set $\xi_1, \xi_2, \ldots, \xi_k$ may constitute the dependent set, the above choice of enumeration can always be achieved by a proper permutation of the letters $x, y, \ldots, z$. The final result is in no way dependent upon our specific choice.

We must, actually, consider the system of $(N + k)$ equations, consisting of the combination of (43) and (57), as being required for the determination of the $(N + k)$ unknown functions $x, y, \ldots , z, \mu_1,$ $\mu_2, \ldots , \mu_N$.

(b) As a first application of the result of (a) above we consider (with certain obvious changes in notation) the problem presented in exercise 7(a), Chap. 3; that is, we seek the differential equation satisfied by the function which extremizes

$$I = \int_{t_1}^{t_2} f(y,\dot{y},\ddot{y},t)dt \tag{58}$$

with respect to twice-differentiable functions $y(t)$ for which y and $\dot{y}$ are prescribed at $t = t_1$ and $t = t_2$. (By $\ddot{y}$ is meant, of course, the *second* derivative of y with respect to t.) To bring this problem within the scope of (a) we rewrite (58) as

$$I = \int_{t_1}^{t_2} f(y,z,\dot{z},t)dt \tag{59}$$

and accordingly affix the condition—which plays the role of (43) with $N = 1$—

$$z - \dot{y} = 0. \tag{60}$$

(Thus the second function z is prescribed at $t = t_1$ and $t = t_2$.)

In accordance with (52) we employ (59) and (60) to form

$$F = f(y,z,\dot{z},t) + \mu(t)(z - \dot{y}). \tag{61}$$

With (61) the system of differential equations (57) here reads

$$\frac{\partial f}{\partial y} + \frac{d\mu}{dt} = 0, \qquad \frac{\partial f}{\partial z} + \mu - \frac{d}{dt}\left(\frac{\partial f}{\partial \dot{z}}\right) = 0.$$

Eliminating the function μ between the two equations and then eliminating z by means of (60), we obtain the single differential equation

$$\frac{\partial f}{\partial y} - \frac{d}{dt}\left(\frac{\partial f}{\partial \dot{y}}\right) + \frac{d^2}{dt^2}\left(\frac{\partial f}{\partial \ddot{y}}\right) = 0 \tag{62}$$

which must be satisfied by the function $y(t)$ which extremizes (58).

(c) The special case of (a) above in which the equations (43) are all *finite* equations is directly applicable to the geodesic problem considered in 3-5. The distance between two given points in space, as measured along the smooth arc $x = x(t)$, $y = y(t)$, $z = z(t)$ connecting them, is given by the integral

$$I = \int_{t_1}^{t_2} \sqrt{\dot{x}^2 + \dot{y}^2 + \dot{z}^2} \, dt, \tag{63}$$

where t_1 and t_2 are the values of t which respectively designate the given points. If the arc is required to lie in the surface

$$G(x,y,z) = 0, \tag{64}$$

we may thus state the general geodesic problem as follows: We seek the functions which extremize the integral (63) with respect to continuously differentiable functions x, y, z which satisfy (64) and which are prescribed at $t = t_1$ and $t = t_2$.

To solve this problem under the jurisdiction of (a) we use (63) and (64) to form, according to (52), the function

$$F = \sqrt{\dot{x}^2 + \dot{y}^2 + \dot{z}^2} + \mu(t)G(x,y,z). \tag{65}$$

With (65) the Euler-Lagrange equations (57) read

$$\mu \frac{\partial G}{\partial x} - \frac{d}{dt}\left(\frac{\dot{x}}{f}\right) = 0, \qquad \mu \frac{\partial G}{\partial y} - \frac{d}{dt}\left(\frac{\dot{y}}{f}\right) = 0, \qquad \mu \frac{\partial G}{\partial z} - \frac{d}{dt}\left(\frac{\dot{z}}{f}\right) = 0, \tag{66}$$

where, for sake of brevity, we write

$$f = \sqrt{\dot{x}^2 + \dot{y}^2 + \dot{z}^2} = \frac{ds}{dt}. \tag{67}$$

[The final form of (67) is employed in exercise 10 at the end of this chapter.] The function $\mu(t)$ is eliminated from (66) to give the pair of equations

$$\frac{\dfrac{d}{dt}\left(\dfrac{\dot{x}}{f}\right)}{\dfrac{\partial G}{\partial x}} = \frac{\dfrac{d}{dt}\left(\dfrac{\dot{y}}{f}\right)}{\dfrac{\partial G}{\partial y}} = \frac{\dfrac{d}{dt}\left(\dfrac{\dot{z}}{f}\right)}{\dfrac{\partial G}{\partial z}}, \tag{68}$$

which, together with the equation of the given surface (64), determine the equations of the required geodesic arc.

(d) Application of (68) to the problem of the geodesic on the sphere (solved by other means in 3-5(c)) involves writing

$$x^2 + y^2 + z^2 - a^2 = 0 \tag{69}$$

for (64); a is the radius of the given sphere. Thus we have

$$(\partial G/\partial x) = 2x, \qquad (\partial G/\partial y) = 2y, \qquad (\partial G/\partial z) = 2z,$$

so that equations (68) read, in slightly expanded form,

$$\frac{f\ddot{x} - \dot{x}\dot{f}}{2xf^2} = \frac{f\ddot{y} - \dot{y}\dot{f}}{2yf^2} = \frac{f\ddot{z} - \dot{z}\dot{f}}{2zf^2}. \tag{70}$$

Equality of the first two members of (70), together with that of the last two, yields the pair of equations

$$\frac{y\ddot{x} - x\ddot{y}}{y\dot{x} - x\dot{y}} = \frac{\dot{f}}{f} = \frac{z\ddot{y} - y\ddot{z}}{z\dot{y} - y\dot{z}},$$

or, if we ignore the middle member,

$$\frac{\dfrac{d}{dt}(y\dot{x} - x\dot{y})}{y\dot{x} - x\dot{y}} = \frac{\dfrac{d}{dt}(z\dot{y} - y\dot{z})}{z\dot{y} - y\dot{z}}.$$

Integrating, we obtain

$$\log (y\dot{x} - x\dot{y}) = \log (z\dot{y} - y\dot{z}) + \log C_1,$$

or

$$y\dot{x} - x\dot{y} = C_1(z\dot{y} - y\dot{z});$$

whence

$$\frac{\dot{x} + C_1\dot{z}}{x + C_1 z} = \frac{\dot{y}}{y}.$$

A second integration thus yields

$$\log (x + C_1 z) = \log y + \log C_2$$

or

$$x - C_2 y + C_1 z = 0$$

—the equation of a plane through the center of the sphere (origin of coordinates) whose intersection with the sphere (69) is the great circle arrived at in 3-5(c).

Although integrating the differential equations (68) presents a simple task in the special case (69), the integration problem is in general quite difficult. The major advantage of the method of (c) above is that it leads quite directly to an important theoretical result from the standpoint of differential geometry. This result is given explicitly under exercise 10 at the end of this chapter.

EXERCISES

1. Suppose that, in the solution of a specific isoperimetric problem, computation of the Lagrange multiplier yields the result $\lambda = 0$. What is the significance of this result?

2. (a) Demonstrate the following reciprocity relationship for the simple isoperimetric problem: The particular function which renders I an extremum with respect to functions which give J a prescribed value also renders J an extremum with respect to functions which give I a prescribed value. (The relationship does not, however, apply to the special circumstance referred to in exercise 1.)

(b) Use part (a), together with the result of 4-3(a), to establish the result: Of all simple closed curves enclosing a given area, the least perimeter is possessed by the circle.

3. (a) Carry out in detail the procedure outlined in 4-2(a) in order to achieve the result stated there. Why is it necessary to introduce an $(N + 1)$-parameter family of comparison functions?

(b) Bring the problems of 4-2(b,c) to a point where the results of 3-10 may be directly shown to apply as stated.

(c) Carry out the details of the argument required to achieve the stated result of 4-2(d).

4. Why do we not apply equation (25) to the problem of 4-3(a)? HINT: Compare exercise 10, Chap. 3.

5. Carry out the integration of the equations (31) by direct means with the aid of the suggestion given in 4-3(a) directly below (31).

6. Work out the problem of 4-3(b) by using (24) of 4-2(e) rather than (14) of 4-1(b).

7. (a) A rope of given length L hangs in equilibrium between two fixed points (x_1,y_1) and (x_2,y_2) in such fashion that the distribution of its mass M is *uniform with respect to the horizontal*; that is, $(dM/dx) = \alpha$, a given constant, *in the equilibrium configuration*. Show, by means of methods developed in the foregoing chapter, that the shape of the hanging rope must be parabolic. HINT: A certain quick, thoughtless attack upon the problem yields a circular shape; this is of course wrong! A second, swindling approach makes use of equation (24) of 4-2(e) to obtain a parabolic shape, but this is likewise wrong. A thoughtful approach takes into account the precise nature of the comparison of the potential energy of the rope's equilibrium configuration with other configurations consistent with the constraints; this leads to the required answer

$$C_1(y - C_3) = \tfrac{1}{2}\alpha(x - C_2)^2. \tag{71}$$

(b) Although the result (71) is apparently devoid of any dependence upon a Lagrange multiplier—since C_1, C_2, C_3 are introduced directly as integration constants —show that

$$C_1 = \alpha C_3 + \lambda, \tag{72}$$

where λ is the Lagrange multiplier introduced to fulfill the requirement that the length of the rope be the same in all its comparison configurations. HINT: Prove and use the fact that $(ds/dx) = 1$ at $x = C_2, y = C_3$.

What relation between C_1 and C_2 replaces (72) if we require that the total mass, rather than the length, of the rope be kept constant and so introduce the multiplier λ'? ANSWER: $C_1 = \alpha(C_3 + \lambda')$.

8. (a) Work out the problem of 4-5(b) by rewriting the integrand of (59) as $f(y,\dot{y},\dot{z},t)$. ANSWER: Result (62) unchanged.

(b) Generalize the method of 4-5(b) so as to solve exercise 7(c), Chap. 3. ANSWER:

$$\frac{\partial f}{\partial y} - \frac{d}{dt}\left(\frac{\partial f}{\partial \dot{y}}\right) + \frac{d^2}{dt^2}\left(\frac{\partial f}{\partial \ddot{y}}\right) - \cdots + (-1)^n \frac{d^n}{dt^n}\left(\frac{\partial f}{\partial y^{(n)}}\right) = 0,$$

where $y^{(n)} = (d^n y/dt^n)$.

(c) Derive the condition which must be satisfied at an end point ($t = t_1$ or $t = t_2$) at which any one of the functions $x, y, \ldots, z$ introduced at the start of 4-5 is not required to have a prescribed value. ANSWER: If, for example, $x(t)$ is not prescribed at $t = t_1$ (or $t = t_2$), we have $(\partial F/\partial \dot{x}) = 0$ at $t = t_1$ (or $t = t_2$).

(*d*) Apply the result of part (*c*) to exercise 7(*b*), Chap. 3 as an extension of the method of 4-5(*b*).

9. (*a*) Derive the differential equation which must be satisfied by the function which extremizes the integral

$$I = \int_{x_1}^{x_2} f(x,y,y',y'')dx$$

with respect to twice-differentiable functions $y = y(x)$ for which

$$J = \int_{x_1}^{x_2} g(x,y,y',y'')dx$$

possesses a given prescribed value, and with y and y' both prescribed at $x = x_1$ and $x = x_2$.

(i) Use the method of 4-1 to show, first, that

$$\int_{x_1}^{x_2} \left(\eta \frac{\partial f^*}{\partial y} + \eta' \frac{\partial f^*}{\partial y'} + \eta'' \frac{\partial f^*}{\partial y''} \right) dx = 0, \tag{73}$$

where $f^* = f + \lambda g$, and η is arbitrary to within consistency with the end-point conditions.

(ii) Combine the method of 4-2(*d*) with that of 1-5(*b*) to achieve the required result.

(*b*) Show that leaving y unspecified at either end point leads to the condition

$$\frac{\partial f^*}{\partial y'} - \frac{d}{dx} \left(\frac{\partial f^*}{\partial y''} \right) = 0 \tag{74}$$

at that end point.

(*c*) Show that leaving y' unspecified at either end point leads to the condition

$$\frac{\partial f^*}{\partial y''} = 0 \tag{75}$$

at that end point.

[The results (73), (74), (75) are required below in Chap. 10.]

10. Use the final form of (67) to show that (68) implies

$$\frac{d^2x/ds^2}{\partial G/\partial x} - \frac{d^2y/ds^2}{\partial G/\partial y} = \frac{d^2z/ds^2}{\partial G/\partial z}.$$

In the language of differential geometry this result demonstrates that the *principal normal* to any point of a geodesic arc lies along the normal to the survace $G(x,y,z) = 0$ at that point.

11. (*a*) It is required to extremize

$$I = \int_{x_1}^{x_2} f(x,y,y')dx + F(w)$$

with respect to functions $y(x)$ and values of the quantity w for which

$$J = \int_{x_1}^{x_2} g(x,y,y')dx + G(w)$$

has a prescribed value, with y prescribed at x_1 and x_2; F and G are given differentiable functions of w. Show that the required extremum is achieved if

$$\frac{\partial f^*}{\partial y} - \frac{d}{dx} \left(\frac{\partial f^*}{\partial y'} \right) = 0 \quad \text{and} \quad F^{*\prime}(w) = 0, \tag{76}$$

where $f^* = f + \lambda g$ and $F^* = F + \lambda G$. HINT: Introduce, in addition to the two-parameter family (3) of 4-1(a), the variable $W = w + \epsilon_1 \gamma_1 + \epsilon_2 \gamma_2$, where γ_1 and γ_2 are arbitrary constants, and so form $I(\epsilon_1, \epsilon_2)$ and $J(\epsilon_1, \epsilon_2)$. Etc.

(b) Apply part (a) to the following problem: A perfectly flexible uniform rope of length L hangs in (unstable) equilibrium, with one end fixed at (x_1, y_1), so that it passes over a frictionless pin at (x_2, y_2). It is clear from the first of (76) and 4-4 that the portion of the rope extended between the two given points hangs in the form of the catenary (41). What is the position of the free end of the rope? ANSWER: $(x_2, -\lambda/\sigma g)$.

CHAPTER 5

GEOMETRICAL OPTICS: FERMAT'S PRINCIPLE

The brachistochrone problem (3-2,6) was first solved by Johann Bernoulli through application of the laws of geometrical (or ray) optics. His method of solution has its basis in the principle of Fermat, which states that *the time elapsed in the passage of light between two fixed points is an extremum with respect to possible paths connecting the points.* In this chapter we accept Fermat's principle as the fundamental characterization of geometrical optics and so develop the ideas underlying the Bernoulli solution.

In what follows we consider only those light paths which lie in a plane—$z = 0$, for the sake of definiteness.

5-1. Law of Refraction (Snell's Law)

(a) Fermat's principle clearly implies that the light path between two points in an (optically) homogeneous medium is a straight line connecting the points. For since the velocity of light is the same at all points of such a medium,[1] the extremum (minimum) of time is equivalent to the extremum (minimum) of path length. Thus, in studying the passage of light between points in two contiguous homogeneous media, we need to consider as possible paths only those which consist of a pair of connected straight-line segments, with the point of connection at the common boundary of the media.

We apply Fermat's principle to the passage of light from the point (x_1,y_1) in a homogeneous medium M_1 to the point (x_2,y_2) in a homogeneous medium M_2 which is separated from M_1 by the line[†] $y = y_0$ $(x_1 < x_2)$. The respective light velocities in the two media are u_1 and u_2 (see Fig. 5-1). If we designate the point of intersection of an arbitrary two-segment path with $y = y_0$ as (x,y_0), the time of light passage along the path would be

$$T = \frac{\sqrt{(x - x_1)^2 + (y_0 - y_1)^2}}{u_1} + \frac{\sqrt{(x_2 - x)^2 + (y_2 - y_0)^2}}{u_2}.$$

---

[1] In fact the constancy of velocity *defines* the "optically homogeneous medium."

[†] Actually, a *plane* separates the two media. Since we confine our attention to the plane $z = 0$, however, it is more convenient to speak of a line as separating them.

According to Fermat's principle, therefore, the actual light path is characterized by the value of x for which

$$\frac{dT}{dx} = \frac{x - x_1}{u_1 \sqrt{(x - x_1)^2 + (y_0 - y_1)^2}} - \frac{x_2 - x}{u_2 \sqrt{(x_2 - x)^2 + (y_2 - y_0)^2}} = 0,$$

or

$$\frac{\sin \phi_1}{u_1} = \frac{\sin \phi_2}{u_2}, \tag{1}$$

where ϕ_1 is the angle between the normal to the interface $y = y_0$ and the path in M_1, and ϕ_2 is the corresponding angle in M_2. The relation (1) is

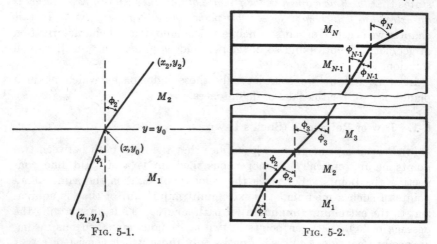

FIG. 5-1.   FIG. 5-2.

known as Snell's law of the refraction of light at the interface of two homogeneous media.   Experimentally, it is established beyond all doubt.

(b) We consider now a set of N contiguous parallel-faced homogeneous media $M_1, M_2, \ldots, M_N$ (in order of position), where the interfaces are lines of constant y (see Fig. 5-2); the light velocity in M_j is denoted by u_j $(j = 1,2, \ldots ,N)$. Since the interfaces are parallel, the angle ϕ_j which a given light ray makes with the normal to one boundary of M_j is equal to the angle it makes with the normal to the opposite boundary of M_j $(j = 1,2, \ldots ,N)$. Thus the Snell's-law relation (1) may be applied to the successive interfaces as follows:

$$\frac{\sin \phi_1}{u_1} = \frac{\sin \phi_2}{u_2} = \cdots = \frac{\sin \phi_N}{u_N},$$

or

$$\frac{\sin \phi_j}{u_i} = K \qquad (j = 1,2, \ldots ,N), \tag{2}$$

where K is a constant for any given light path. (The value of K is determined by the orientation of the path segment within any one of the M_j—within M_1, for example.)

(c) Next we consider a single optically inhomogeneous medium M in which the light velocity is a single-valued continuous function of the y coordinate; *i.e.*, we have

$$u = u(y).$$

(The medium is assumed optically isotropic: The velocity at any point is independent of the direction of the light path through the point.) To arrive at the law which describes the configuration of a light ray connecting any two points of such a medium, we first approximate M by a sequence of parallel-faced *homogeneous* media $M_1, M_2, \ldots, M_N$ having the character of the arrangement described in (b) above. The light velocity u_j within M_j is chosen to be equal to $u(y)$ evaluated at some point between the ($y = $ constant) lines which bound M_j.

The light path through the sequence of subdividing media is a polygonal line, the orientation of each of whose segments is described by the extended form of Snell's law (2). The smaller the width of the individual subdivisions and the larger their number N, the closer is the approximating arrangement to the actual medium M; and, therefore, the closer an approximation is a given polygonal light path through $M_1, M_2, \ldots, M_N$ to an actual light path through M. As we improve the degree of approximation indefinitely by letting N increase without limit and having the width of each subdivision approach zero, the relation (2) applies at every stage of the process. In the limit in which the approximation is perfect, (2) describes the direction, at any point, of the tangent line to an actual light path in M. We therefore rewrite (2) as

$$\frac{\sin \phi}{u} = K, \tag{3}$$

where ϕ and u are continuous functions of y.

If $y = y(x)$ is the equation of a light path in M, we have (see Fig. 5-3) that $y'(x) = \cot \phi$, so that

$$\sin \phi = \frac{1}{\sqrt{1 + y'^2}}. \tag{4}$$

Thus, for a medium whose optical properties are described by the given velocity function $u(y)$, (3) reads, with the aid of (4),

$$\frac{1}{u \sqrt{1 + y'^2}} = K \tag{5}$$

—a first-order differential equation whose solution is directly found to be

$$x = \pm K \int \frac{u \, dy}{\sqrt{1 - K^2 u^2}}. \tag{6}$$

The constant K and the constant of integration are fixed by specifying two points through which the light path is required to pass, provided such a path actually exists.

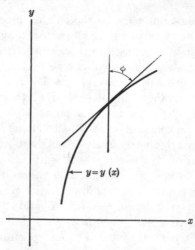

FIG. 5-3.

5-2. Fermat's Principle and the Calculus of Variations

In this section we traverse a second path of reasoning to achieve the results (5) and (6) for a light path within a medium in which the light velocity varies continuously as a function of one cartesian coordinate. Here we express Fermat's principle directly as applied to an inhomogeneous medium: If the velocity of light is given by the continuous function $u = u(y)$, the actual light path connecting the points (x_1, y_1) and (x_2, y_2) is one which extremizes the time integral

$$I = \int_{(x_1, y_1)}^{(x_2, y_2)} \frac{ds}{u} = \int_{x_1}^{x_2} \frac{\sqrt{1 + y'^2}}{u} \, dx. \tag{7}$$

(This statement of the principle is correct even if $u = u(x, y)$.)

According to 3-3 we thus have that $y = y(x)$—the equation of the actual light path—must satisfy the Euler-Lagrange equation (25) of that section, with

$$f = \frac{\sqrt{1 + y'^2}}{u(y)}, \tag{8}$$

the integrand of (7). Since f is explicitly independent of x, however, we may employ directly the first integral (28) of 3-4(b)—namely,

$$y' \frac{\partial f}{\partial y'} - f = C_1. \tag{9}$$

With (8), equation (9) reads

$$\frac{y'^2}{u \sqrt{1 + y'^2}} - \frac{\sqrt{1 + y'^2}}{u} = C_1,$$

or

$$\frac{1}{u \sqrt{1 + y'^2}} = -C_1. \tag{10}$$

Clearly, (10) is identical with the result (5), derived on the basis of Snell's law, with the constant C_1 identified with $-K$. Thus (10) also leads directly to (6), with the same identification of constants.

We have in (6), therefore, the solution of a problem in the calculus of variations—the problem of finding the function for which the integral (7) is an extremum—obtained by methods other than those which are peculiar to the calculus of variations. The methods employed to achieve this solution are those of geometrical optics, together with a limiting process following upon the approximation of an optically inhomogeneous medium by a sequence of homogeneous media.

If, for example, we choose the velocity function to be

$$u(y) = \sqrt{2g(y - y_0)},$$

where y_0 and $g(> 0)$ are given constants, the time integral (7) becomes identical with the integral (16) of 3-2(c), whose extremization results in the solution of the brachistochrono problem. Through the procedure of 5-1, therefore, the solution of the brachistochrone problem is effected by means of geometrical optics. In essence, this is the method employed by Johann Bernoulli to solve his brachistochrone problem at the end of the seventeenth century.

EXERCISES

1. (a) Write down the integral which must be extremized, according to Fermat's principle, if the light paths are not restricted to plane curves, and with $u = u(x,y,z)$. Let x be the independent variable.

(b) Write down the pair of Euler-Lagrange equations (again with x as independent variable) which describe light paths in three dimensions if $u = u(x,y,z)$.

2. Describe the plane paths of light in the (two-dimensional) media in which the light velocities are given respectively by (i) $u = ay$; (ii) (a/y); (iii) $ay^{\frac{1}{2}}$; (iv) $ay^{-\frac{1}{2}}$; where $a > 0$, $y > 0$.

CHAPTER 6

DYNAMICS OF PARTICLES

The material of the ensuing chapter is based upon an assumed knowledge of only the most elementary concepts of particle dynamics.[1] Adequate comprehension of the subject matter should therefore require negligible background in physics. On the other hand this chapter cannot be considered as a suitable introduction to an intensive study of particle dynamics. It is meant, rather, to provide a glimpse of the role played by the calculus of variations in a small segment of dynamics and to serve as a springboard for several of the problems considered in chapters following. The discussion is confined to nonrelativistic, or "classical," phenomena.

6-1. Potential and Kinetic Energies. Generalized Coordinates

(a) We consider a system of p particles subject to given geometric constraints and otherwise influenced by forces which are functions only of the positions of the particles. (The geometric constraints, which do not vary in time, may consist, for example, of the confinement of certain of the particles to given curves or surfaces, or of the constancy of the distance separating certain pairs of the particles, etc.) Specifically, the force acting upon the jth particle (at x_j,y_j,z_j) of the system (apart from the forces of constraint) has the cartesian components $F_x^{(j)}$, $F_y^{(j)}$, $F_z^{(j)}$ $(j = 1,2, \ldots ,p)$, which are functions of the $3p$ position coordinates $x_1, y_1, z_1, \ldots , x_p, y_p, z_p$ of the particles of the system.

In all but the final section of this chapter we confine consideration to the special type of force system—"conservative" system—for which there exists a single function $V = V(x_1,y_1,z_1, \ldots ,x_p,y_p,z_p)$ from which we may derive the $3p$ force components as

$$F_x^{(j)} = - \frac{\partial V}{\partial x_j}, \qquad F_y^{(j)} = - \frac{\partial V}{\partial y_j}, \qquad F_z^{(j)} = - \frac{\partial V}{\partial z_j} \qquad (j = 1,2, \ldots ,p).$$

$$(1)$$

The function V is called the *potential energy* of the system; we do not concern ourselves here with questions of its existence or determination in specific physical situations: For purposes of this chapter the statement of

[1] In particular the terms *mass* and *force* are employed here without definition.

a conservative-dynamics problem involves three *given* elements: (i) the number and respective masses of the particles, (ii) the geometric constraints upon the particles, and (iii) the potential-energy function V.

(b) The *kinetic energy* of a particle is defined as the quantity

$$\tfrac{1}{2}m \, (ds/dt)^2 = \tfrac{1}{2}m(\dot{x}^2 + \dot{y}^2 + \dot{z}^2),$$

where m is the mass of the particle and[1] $(ds/dt)^2 = (\dot{x}^2 + \dot{y}^2 + \dot{z}^2)$ is the square of the velocity of the particle. For a system of p particles the kinetic energy is defined as the sum

$$T = \tfrac{1}{2} \sum_{j=1}^{p} m_j(\dot{x}_j^2 + \dot{y}_j^2 + \dot{z}_j^2), \tag{2}$$

where m_j is the mass of the jth particle. Since the mass is never negative, we have the inequality $T \geqq 0$, with equality holding only if the system is at rest.

(c) The effect of constraints upon a system of p particles is to reduce the number of independent coordinates describing the positions of the particles. If the constraints are completely specified by the k ($< 3p$) consistent and independent equations

$$\phi_i(x_1, y_1, z_1, \ldots, x_p, y_p, z_p) = 0 \qquad (i = 1, 2, \ldots, k), \tag{3}$$

the number of independent coordinate variables is $(3p - k)$; the equations (3) may be used, at least in principle, to eliminate the remaining k variables from the problem.

It is more convenient, however, to introduce a set of $(3p - k) = N$ independent variables $q_1, q_2, \ldots, q_N$ through which the positions of all p particles are described. Thus the equations of constraint (3) are in effect replaced by the equivalent system of $3p$ equations

$$x_j = x_j(q_1, \ldots, q_N), \quad y_j = y_j(q_1, \ldots, q_N), \quad z_j = z_j(q_1, \ldots, q_N), \tag{4}$$

for $j = 1, 2, \ldots, p$. The variables $q_1, q_2, \ldots, q_N$ are known as *generalized coordinates;* specification of their respective values establishes, through (4), the positions of the p particles—and always consistently with the geometric constraints imposed upon them.

The choice of a set of generalized coordinates for the description of the positions of a particular system of particles subject to given constraints is not unique; but the number of such coordinates which must be employed is perfectly definite: It is the smallest number of variables required to describe completely the position configuration of the system when the constraints are known. For example, a particle confined to a given sur-

---

[1] Throughout this chapter, and frequently in chapters following, we employ the superior dot to indicate differentiation with respect to the *time* variable t.

face has associated with it two generalized coordinates; a convenient choice would be a pair of surface-coordinate variables. A particle constrained to move along a given curve would require a single coordinate to describe its position. An unconstrained particle requires three coordinates for the description of its position; it is frequently convenient to employ the three cartesian coordinates in this case.

(*d*) To express the kinetic energy (2) in terms of the generalized coordinates $q_1, q_2, \ldots, q_N$ we differentiate each of the $3p$ equations (4) with respect to time:

$$\dot{x}_j = \sum_{i=1}^{N} \frac{\partial x_j}{\partial q_i} \dot{q}_i, \qquad \dot{y}_j = \sum_{i=1}^{N} \frac{\partial y_j}{\partial q_i} \dot{q}_i, \qquad \dot{z}_j = \sum_{i=1}^{N} \frac{\partial z_j}{\partial q_i} \dot{q}_i. \tag{5}$$

Substituting (5) into (2), we obtain the important result: *The kinetic energy is a homogeneous function, of degree two, of the "generalized velocity components"* $\dot{q}_1, \dot{q}_2, \ldots, \dot{q}_N$. More specifically, the kinetic energy is a *quadratic form* in the generalized velocity components; the coefficients in this form are functions of the generalized coordinates.

We assume, in the remainder of this chapter, that the potential-energy function V associated with any problem is expressed solely in terms of the $q_1, q_2, \ldots, q_N$; the corresponding kinetic energy is, until 6-3, assumed expressed solely in terms of the $q_1, q_2, \ldots, q_N, \dot{q}_1, \dot{q}_2, \ldots, \dot{q}_N$. When these functions are so expressed it is useful to define the lagrangian function—or, simply, the *lagrangian*—as

$$L(q_1, q_2, \ldots, q_N, \dot{q}_1, \dot{q}_2, \ldots, \dot{q}_N) = T - V. \tag{6}$$

6-2. Hamilton's Principle. Lagrange Equations of Motion

(*a*) Although Newton's laws of motion are the most fundamental mathematical description of mechanical phenomena in general, it best suits the purposes of our study to assume the validity of *Hamilton's principle* as the physical law which describes the motion of any system of the type considered in 6-1 above. The principle of Hamilton reads:

The actual motion of a system whose lagrangian is

$$(T - V) = L(q_1, q_2, \ldots, q_N, \dot{q}_1, \dot{q}_2, \ldots, \dot{q}_N)$$

is such as to render the (Hamilton's) *integral*

$$I = \int_{t_1}^{t_2} (T - V) dt = \int_{t_1}^{t_2} L \, dt, \tag{7}$$

where t_1 and t_2 are two arbitrary instants of time, an extremum with respect to continuously twice-differentiable functions $q_1(t), q_2(t), \ldots, q_N(t)$ *for* *which $q_i(t_1)$ and $q_i(t_2)$ are prescribed for all $i = 1, 2, \ldots, N$.* We accept Hamilton's principle as applicable to the motion of any conservative system.

Although it is in some places stated that Hamilton's principle may be used to replace Newton's laws of motion as the fundamental starting point for mechanical systems possessing a lagrangian, it should be realized that Newton's laws are implicitly employed in the preceding paragraphs in at least two ways: (i) The definition of mass resides in Newton's third law. (ii) In the tacit assumption that our system of coordinates is fixed relative to an inertial frame of reference, we make use of Newton's first law, by means of which an *inertial frame* is defined.

(*b*) With the result (57) of 3-8(*a*) we conclude from Hamilton's principle that the generalized coordinates describing the motion of a system of particles must satisfy the set of Euler-Lagrange equations

$$\frac{\partial L}{\partial q_i} - \frac{d}{dt}\left(\frac{\partial L}{\partial \dot{q}_i}\right) = 0 \qquad (i = 1, 2, \ldots, N). \tag{8}$$

The equations (8), Lagrange's equations of motion, constitute a set of N simultaneous second-order differential equations, whose solution yields the functions $q_1(t), q_2(t), \ldots, q_N(t)$. The $2N$ constants involved in the general solution of (8) are evaluated when the initial ($t = 0$, for example) values of all the q_i and $\dot{q}_i$ ($i = 1, 2, \ldots, N$) are given. Once the initial state of the system is thus prescribed, its future motion is described in detail by the functions obtained through the solution of (8).

Application of Lagrange's equations to specific mechanical systems is found in the exercises at the end of this chapter.

(*c*) Since the lagrangian L does not explicitly involve the time variable t, the equations of motion (8) lead, according to (59) of 3-8(*b*), to the first integral

$$\sum_{i=1}^{N} \dot{q}_i \frac{\partial L}{\partial \dot{q}_i} - L = E, \tag{9}$$

where E is a constant. To interpret (9) we note first that $(\partial V/\partial \dot{q}_i) = 0$, so that, according to (6) of 6-1(*d*), $(\partial L/\partial \dot{q}_i) = (\partial T/\partial \dot{q}_i)$ for all $i = 1, 2, \ldots, N$. Since, according to 6-1(*d*), T is a homogeneous function, of degree two, of $\dot{q}_1, \dot{q}_2, \ldots, \dot{q}_N$, we therefore have from Euler's theorem (2-5) that $\Sigma \dot{q}_i(\partial L/\partial \dot{q}_i) = \Sigma \dot{q}_i(\partial T/\partial \dot{q}_i) = 2T$. Thus, with (6), equation (9) becomes $2T - (T - V) = E$, or

$$T + V = E. \tag{10}$$

That is, the motion of a conservative system is characterized by the constancy of the sum of the potential and kinetic energies (whence the appellation conservative). The constant E, the *total energy* of the system, is determined when the initial values of all the q_i and $\dot{q}_i$ are assigned.

6-3. Generalized Momenta. Hamilton Equations of Motion

(*a*) Dealing with a system whose position configuration is completely described by the generalized coordinates $q_1, q_2, \ldots, q_N$ and whose lagrangian is L, we define the set $p_1, p_2, \ldots, p_N$ of *generalized momenta* as

$$p_i = \frac{\partial T}{\partial \dot{q}_i} \qquad (i = 1, 2, \ldots, N). \tag{11}$$

Since, according to 6-1(*d*), the kinetic energy T is a quadratic form in the generalized velocity components $\dot{q}_i$, it follows from the definition (11) that each p_i is a linear homogeneous function of $\dot{q}_1, \dot{q}_2, \ldots, \dot{q}_N$. Conversely, solution of the N equations (11) must yield each of the $\dot{q}_i$ as a linear homogeneous function[1] of $p_1, p_2, \ldots, p_N$.

(*b*) Using the equations (11) to eliminate $\dot{q}_1, \dot{q}_2, \ldots, \dot{q}_N$ from the lagrangian L—which is thus expressed solely as a function of $q_1, q_2, \ldots, q_N, p_1, p_2, \ldots, p_N$—we define the *hamiltonian* H of the system through the identity

$$H(q_1, q_2, \ldots, q_N, p_1, p_2, \ldots, p_N) = \sum_{i=1}^{N} p_i \dot{q}_i - L, \tag{12}$$

where the appearance of each $\dot{q}_i$ in the right-hand member represents the solution of (11) for this quantity in terms of the generalized momenta. Since (12) is an identity in the p_i and q_i (via (11)), we may form the partial derivative with respect to p_j—whereby all the p_i, with $i \neq j$, together with all the q_i, are held constant—and so obtain

$$\frac{\partial H}{\partial p_j} = \dot{q}_j + \sum_{i=1}^{N} p_i \frac{\partial \dot{q}_i}{\partial p_j} - \sum_{i=1}^{N} \frac{\partial L}{\partial \dot{q}_i} \frac{\partial \dot{q}_i}{\partial p_j}$$

$$= \dot{q}_j + \sum_{i=1}^{N} \frac{\partial \dot{q}_i}{\partial p_j} \left(p_i - \frac{\partial T}{\partial \dot{q}_i} \right) = \dot{q}_j \qquad (j = 1, 2, \ldots, N), \tag{13}$$

---

[1] This follows directly from Cramer's rule for the solution of linear equations by means of determinants, provided the determinant of the coefficients of $\dot{q}_1, \dot{q}_2, \ldots, \dot{q}_N$ in (11) does not vanish. We accept here the nonvanishing of this determinant, which is a consequence of the *positive definite* character of T as a quadratic form in the $\dot{q}_i$.

since $(\partial L/\partial \dot{q}_i) = (\partial T/\partial \dot{q}_i)$ for all i, and because of the definition (11) of p_i. The N equations (13) clearly represent the explicit solution of the set (11) for each $\dot{q}_j$ in terms of the p_i.

Substituting (11) into (12) and making use of the result

$$2T = \sum_{i=1}^{N} \dot{q}_i \frac{\partial T}{\partial \dot{q}_i}$$

arrived at in 6-2(c), we have

$$H = \sum_{i=1}^{N} \frac{\partial T}{\partial \dot{q}_i} \dot{q} - L = 2T - (T - V) = T + V, \tag{14}$$

with the aid of the definition (6) of L. That is, the hamiltonian of a system is the sum of the potential energy and the kinetic energy, when the latter quantity is expressed in terms of the q_i and the p_i, rather than in terms of the $\dot{q}_i$. Thus the most convenient method of forming the hamiltonian of a given system is the following: (i) We write down the potential energy in terms of the q_i, the kinetic energy in terms of the q_i and $\dot{q}_i$; (ii) form and solve the N equations (11)—since the explicit solution (13) is not available *prior* to formation of the hamiltonian!—for the $\dot{q}_i$; (iii) substitute for the $\dot{q}_i$ in T and so obtain $H = T + V$ in terms of $q_1, q_2, \ldots, q_N, p_1, p_2, \ldots, p_N$.

(c) In terms of the hamiltonian the integral (7) of 6-2(a) whose extremization leads, according to Hamilton's principle, to the equations of motion of a mechanical system is given by

$$I = \int_{t_1}^{t_2} \left(\sum_{i=1}^{N} p_i \dot{q}_i - H \right) dt, \tag{15}$$

on substitution for L through (12). The extremization must be effected with respect to the $2N$ continuously differentiable functions $q_1, q_2, \ldots, q_N, p_1, p_2, \ldots, p_N$, among which there obtains the set of N relations

$$\dot{q}_i - \frac{\partial H}{\partial p_i} = 0 \qquad (i = 1,2, \ldots ,N), \tag{16}$$

according to (13).

To derive the set of differential equations called for by the extremization, we employ the method of 4-5(a). With (15) and (16) respectively representing specific cases of (42) and (43) of the earlier section we form the function—(52) of 4-5(a)—

$$F = \sum_{i=1}^{N} p_i \dot{q}_i - H + \sum_{i=1}^{N} \mu_i(t) \left(\dot{q}_i - \frac{\partial H}{\partial p_i} \right), \tag{17}$$

where $\mu_1, \mu_2, \ldots, \mu_N$ are undetermined functions. Substituting (17) into (57) of 4-5(a), with appropriate changes of notation, we obtain, since $(\partial F/\partial \dot{p}_j) = 0$ (identically) for all j,

$$\dot{q}_j - \frac{\partial H}{\partial p_j} - \sum_{i=1}^{N} \mu_i(t) \frac{\partial^2 H}{\partial p_j \, \partial p_i} = 0 \qquad (j = 1,2,\ldots,N) \tag{18}$$

and

$$-\frac{\partial H}{\partial q_j} - \sum_{i=1}^{N} \mu_i(t) \frac{\partial^2 H}{\partial q_j \, \partial p_i} - \frac{d}{dt}(p_j + \mu_j) = 0 \qquad (j = 1,2,\ldots,N). \tag{19}$$

Because of (16) the N equations (18) read

$$\sum_{i=1}^{N} \mu_i(t) \frac{\partial^2 H}{\partial p_j \, \partial p_i} = 0 \qquad (j = 1,2,\ldots,N), \tag{20}$$

for which an obvious solution[1] is $\mu_i = 0$ for all $i = 1, 2, \ldots, N$. Substituting this result into (19), we obtain $\dot{p}_j = -(\partial H/\partial q_j)$ for all j. This set of equations, taken in conjunction with the relations (16), supplies the system of $2N$ equations—the Hamilton equations of motion—

$$\dot{p}_i = -\frac{\partial H}{\partial q_i}, \qquad \dot{q}_i = \frac{\partial H}{\partial p_i} \qquad (i = 1,2,\ldots,N). \tag{21}$$

The system (21) constitutes $2N$ first-order ordinary differential equations. Their general solution is accomplished in the attainment of $2N$ finite equations which relate $q_1, q_2, \ldots, q_N, p_1, p_2, \ldots, p_N$ and the time variable t, and which involve $2N$ arbitrary constants of integration. These constants become determinate when initial ($t = 0$, for example) values are assigned to $q_1, q_2, \ldots, q_N$ and to $p_1, p_2, \ldots, p_N$—or, equivalently, to $q_1, q_2, \ldots, q_N, \dot{q}_1, \dot{q}_2, \ldots, \dot{q}_N$. Thus, through (21), knowledge of the motion of a mechanical system is completely determined if the hamiltonian function H is known, along with the initial state of the system.

---

[1] That this solution is *unique* follows from the nonvanishing of the determinant of the coefficients of the μ_i in (20). We accept here the nonvanishing of this determinant —a direct consequence of the positive definite character of T as a quadratic form in $p_1, p_2, \ldots, p_N$ (see footnote 1, p. 76 and also end-chapter exercise 8).

Application of the Hamilton equations (21) to specific problems is left for the end-chapter exercises.

6-4. Canonical Transformations

(a) We consider a mechanical system whose hamiltonian function $H(q_1,q_2, \ldots ,q_N,p_1,p_2, \ldots ,p_N)$ is known and propose a change of variables which has the following character:

(i) By means of $2N$ finite equations we define a system of $2N$ *new* variables $Q_1, Q_2, \ldots , Q_N, P_1, P_2, \ldots , P_N$ as functions of the original set of variables $q_1, q_2, \ldots , q_N, p_1, p_2, \ldots , p_N$. The possibility that the time variable t may appear explicitly in the equations of transformation is admitted.

(ii) There is no functional relationship among the variables $q_1, q_2, \ldots , q_N, Q_1, Q_2, \ldots , Q_N$ which is completely independent of all the p_i and P_i.

(iii) The equations of motion, written in terms of the new variables, must possess the same form as the Hamilton equations (21) in the sense that there exists a function $K = K(Q_1,Q_2, \ldots ,Q_N,P_1,P_2, \ldots ,P_N,t)$ such that the transformed equations of motion read

$$\dot{P}_i = - \frac{\partial K}{\partial Q_i}, \qquad \dot{Q}_i = \frac{\partial K}{\partial P_i} \qquad (i = 1,2, \ldots ,N). \qquad (22)$$

That a transformation which satisfies (i) and (iii) is always possible should be obvious if we take into account the Hamilton's-principle derivation of (21) in 6-3(c) and the result of 3-4(c) which allows the addition to an integrand of an "exact derivative" without alteration of the resulting Euler-Lagrange equation.[1] Thus if we effect a transformation through the *identities*

$$\sum_{i=1}^{N} p_i \dot{q}_i - H = \sum_{i=1}^{N} P_i \dot{Q}_i - K + \frac{dS}{dt}, \qquad \dot{Q}_i = \frac{\partial K}{\partial P_i}$$

$$(i = 1,2, \ldots ,N), \qquad (23)$$

where S is *any* continuously differentiable function of $q_1, q_2, \ldots , q_N, Q_1, Q_2, \ldots , Q_N, t$, we should expect Hamilton's principle to lead directly to (22).   (The explicit equations of transformation are derived from (23) in (b) below.)

To prove this assertion we use the first of (23)—with $q_1, q_2, \ldots , q_N$ eliminated from S in terms of $Q_1, Q_2, \ldots , Q_N, P_1, P_2, \ldots , P_N, t$— to substitute for the integrand of (15).   We proceed to extremize I with

---

[1] The result of 3-4(c) must be extended to apply to the present case.   The extension is implicit in the derivation below of (22).   We merely use 3-4(c) here as a *guide*.

respect to the new variables P_i, Q_i, which are linked by the second of (23) (N equations). Following the procedure of 6-3(c), we replace (17), according to (23), by

$$F = \sum_{i=1}^{N} P_i \dot{Q}_i - K + \frac{dS}{dt} + \sum_{i=1}^{N} \mu_i(t) \left(\dot{Q}_i - \frac{\partial K}{\partial P_i} \right). \tag{24}$$

Substituting (24) into (57) of 4-5(a), with appropriate changes of notation, we obtain

$$\dot{Q}_j - \frac{\partial K}{\partial P_j} + \frac{\partial}{\partial P_j}\left(\frac{dS}{dt}\right) - \sum_{i=1}^{N} \mu_i \frac{\partial^2 K}{\partial P_j\, \partial P_i} = \frac{d}{dt}\left[\frac{\partial}{\partial \dot{P}_j}\left(\frac{dS}{dt}\right) \right]$$

$$(j = 1,2,\ldots,N) \quad (25)$$

and

$$-\frac{\partial K}{\partial Q_j} + \frac{\partial}{\partial Q_j}\left(\frac{dS}{dt}\right) - \sum_{i=1}^{N} \mu_i \frac{\partial^2 K}{\partial Q_j\, \partial P_i} = \frac{d}{dt}\left[P_j + \frac{\partial}{\partial \dot{Q}_j}\left(\frac{dS}{dt}\right) + \mu_j \right]$$

$$(j = 1,2,\ldots,N). \quad (26)$$

Since S is supposed expressed in terms of $Q_1, Q_2, \ldots, Q_N, P_1, P_2, \ldots, P_N, t$, in (24), (25), and (26), we have

$$\frac{dS}{dt} = \frac{\partial S}{\partial t} + \sum_{i=1}^{N}\left(\frac{\partial S}{\partial Q_i}\dot{Q}_i + \frac{\partial S}{\partial P_i}\dot{P}_i \right), \tag{27}$$

so that

$$\frac{d}{dt}\left[\frac{\partial}{\partial \dot{P}_j}\left(\frac{dS}{dt}\right) \right] = \frac{d}{dt}\left(\frac{\partial S}{\partial P_j}\right) = \frac{\partial^2 S}{\partial t\, \partial P_j} + \sum_{i=1}^{N}\left(\frac{\partial^2 S}{\partial Q_i\, \partial P_j}\dot{Q}_i + \frac{\partial^2 S}{\partial P_i\, \partial P_j}\dot{P}_i \right)$$

$$= \frac{\partial}{\partial P_j}\left[\frac{\partial S}{\partial t} + \sum_{i=1}^{N}\left(\frac{\partial S}{\partial Q_i}\dot{Q}_i + \frac{\partial S}{\partial P_i}\dot{P}_i \right) \right] = \frac{\partial}{\partial P_j}\left(\frac{dS}{dt}\right). \quad (28)$$

Because of (28) and the second of (23) the system (25) reads

$$\sum_{i=1}^{N} \mu_i \frac{\partial^2 K}{\partial P_j\, \partial P_i} = 0 \qquad (j = 1,2,\ldots,N), \tag{29}$$

for which the solution is† $\mu_i = 0$, for all $i = 1, 2, \ldots, N$. Further, it

† See end-chapter exercise 8.

follows from (27), in the manner of achieving (28), that[1]

$$\frac{\partial}{\partial Q_j}\left(\frac{dS}{dt}\right) = \frac{d}{dt}\left[\frac{\partial}{\partial Q_j}\left(\frac{dS}{dt}\right)\right], \tag{30}$$

so that, with the vanishing of all the μ_i, (26) reduces to

$$-(\partial K/\partial Q_j) = \dot{P}_j \qquad (j = 1,2, \ldots ,N).$$

We therefore have, with the second of (23), the required set (22) of the *transformed* Hamilton equations of motion.

(b) The result of (a) above demonstrates that any change of variables of the type (i) for which the identities (23) hold necessarily satisfies the requirement (iii). We proceed to derive from the first of (23), with application of the restriction (ii), the actual equations of transformation:

Multiplying the first of (23) by dt and expanding the differential of $S = S(q_1,q_2, \ldots ,q_N,Q_1,Q_2, \ldots ,Q_N,t)$, we get

$$\sum_{i=1}^{N} p_i\,dq_i - H\,dt = \sum_{i=1}^{N} P_i\,dQ_i - K\,dt + \frac{\partial S}{\partial t}dt + \sum_{i=1}^{N}\left(\frac{\partial S}{\partial q_i}dq_i + \frac{\partial S}{\partial Q_i}dQ_i\right),$$

or

$$\sum_{i=1}^{N}\left(p_i - \frac{\partial S}{\partial q_i}\right)dq_i - \sum_{i=1}^{N}\left(P_i + \frac{\partial S}{\partial Q_i}\right)dQ_i + \left(K - H - \frac{\partial S}{\partial t}\right)dt = 0. \tag{31}$$

Since the first of (23)—and therefore also (31)—is an identity in the variables involved, and because of the requirement (ii) of (a) above (whereby the $dq_1, dq_2, \ldots , dq_N, dQ_1, dQ_2, \ldots , dQ_N$ may be assigned arbitrary values), we conclude from (31) that

(i) $K = H + \dfrac{\partial S}{\partial t},$ (ii) $p_i = \dfrac{\partial S}{\partial q_i},$ (iii) $P_i = - \dfrac{\partial S}{\partial Q_i}$ $(i = 1,2, \ldots ,N).$ (32)

Thus if S is any continously differentiable function of the q_i, the Q_i, and t, (32) generates a transformation—a so-called *canonical* transformation—of the character called for at the opening of (a) above. The $2N$ equations (32,ii,iii) express the actual relations among the new (Q_i,P_i) and the old (q_i,p_i) variables, with t playing the role of parameter in the transformation; (32,i) provides the function K, which plays the role of hamiltonian in the transformed equations of motion (22) and which, for sake of brevity, we call the *kamiltonian*.

---

[1] See end-chapter exercise 7.

We may therefore choose at random any suitable function S and so generate a canonical transformation (see end-chapter exercise 6). It is of course hoped, in the execution of any such transformation, that it leads to a set of equations of motion (22) whose integration is less difficult than that of the original set (21). In the section following we consider a method for choosing the function S so that the integration of (22) may be accomplished with maximum simplicity.

6-5. The Hamilton-Jacobi Differential Equation

(a) The most easily integrated set of transformed equations of motion (22) is arrived at by a canonical transformation which leads to a kamiltonian K which is identically zero. For, in such an event, (22) reads $\dot{P}_i = \dot{Q}_i = 0$ for all i, so that the solutions are simply

$$P_i = \beta_i,\, Q_i = a_i \qquad (i = 1, 2, \ldots, N),$$

where the β_i and a_i are two sets of arbitrary constants. With these solutions obtained we may then solve the transformation equations (32,ii,iii) and so obtain the p_i and q_i, for all i, as functions of t and the $2N$ arbitrary constants $\beta_1, \beta_2, \ldots, \beta_N, a_1, a_2, \ldots, a_N$.

For a canonical transformation to lead to a kamiltonian K identically zero, it follows from (32,i) that the function S which generates the transformation must be such that $H + (\partial S/\partial t) = 0$. Or if

$$H = H(q_1, q_2, \ldots, q_N, p_1, p_2, \ldots, p_N)$$

is the hamiltonian of the system under study, we therefore have from (32,ii) that S must satisfy the partial differential equation

$$H\left(q_1, q_2, \ldots, q_N, \frac{\partial S}{\partial q_1}, \frac{\partial S}{\partial q_2}, \ldots, \frac{\partial S}{\partial q_N}\right) + \frac{\partial S}{\partial t} = 0 \qquad (33)$$

—the so-called *Hamilton-Jacobi equation*.

The equation (33) has an infinity of solutions, of which our interest lies solely in the *complete* solutions—those which involve N independent arbitrary constants $\alpha_1, \alpha_2, \ldots, \alpha_N$, aside from the one additive constant of integration. (It is clear that $(S + C)$ is a solution of (33) if S is a solution, where C is any constant; this constant C is *not* included among the N constants of a complete solution.)

We suppose that $S = S(q_1, q_2, \ldots, q_N, \alpha_1, \alpha_2, \ldots, \alpha_N, t)$ is a complete solution of the Hamilton-Jacobi equation (33). Since the α_i $(i = 1, 2, \ldots, N)$ are constants only in so far as they are independent of $q_1, q_2, \ldots, q_N, t$, we may effect the identification $\alpha_i = Q_i$ $(i = 1,2, \ldots ,N)$ and so obtain

$$S = S(q_1,q_2, \ldots ,q_N,Q_1,Q_2, \ldots ,Q_N,t) \qquad (34)$$

as the function which generates, through (32), a canonical transformation. But since (34) satisfies (33), we have, as planned, that $K = 0$, by (32,i), so that (22) yields the results $P_i = \beta_i$, $Q_i = a_i$, for each i, as required. To obtain the original variables $q_1, q_2, \ldots, q_N, p_1, p_2, \ldots, p_N$ as functions of t and the $2N$ arbitrary constants, we employ (as stated in the opening paragraph of this section) the transformation equations (32,ii,iii), with the arbitrary constant values substituted for the P_i and Q_i.

Since, however, the a_i, as well as the α_i, constitute a set of N independent arbitrary constants, we may make the identification $a_i = \alpha_i$ $(i = 1,2, \ldots ,N)$, bypass the substitution $\alpha_i = Q_i$ in the complete solution $S = S(q_1,q_2, \ldots ,q_N,\alpha_1,\alpha_2, \ldots ,\alpha_N,t)$ of (33), and directly rewrite the set of transformation equations (32,ii,iii) as

$$p_i = \frac{\partial S}{\partial q_i}, \qquad \beta_i = -\frac{\partial S}{\partial \alpha_i} \qquad (i = 1,2, \ldots ,N), \qquad (35)$$

where $\alpha_1, \alpha_2, \ldots, \alpha_N$ are the independent constants of the complete solution of (33), and $\beta_1, \beta_2, \ldots, \beta_N$ are a second set of arbitrary constants substituted for $P_1, P_2, \ldots, P_N$ in (32,iii). *The solution of the $2N$ finite equations (35) for the q_i and p_i ($i = 1,2, \ldots ,N$) constitutes the general solution of the original Hamilton equations of motion (21).* Thus the solution of the $2N$ ordinary differential equations (21) is reduced to the achievement of a complete solution of the single *partial* differential equation (33).

(b) By writing

$$S = S^* - Et, \qquad (36)$$

where S^* is independent of t and E is an arbitrary constant, we see that S is a solution of (33) if S^* satisfies the time-independent *reduced* Hamilton-Jacobi equation

$$H\left(q_1,q_2, \ldots ,q_N,\frac{\partial S^*}{\partial q_1},\frac{\partial S^*}{\partial q_2}, \cdots ,\frac{\partial S^*}{\partial q_N}\right) = E. \qquad (37)$$

Since, according to (14) of 6-3(b), $H = T + V$, and $(T + V)$ is a constant—the total energy—during the motion of a given conservative system, according to 6-2(c), the arbitrary constant E in (36) and (37) must be identified with the *total energy* of the system whose hamiltonian is H. (Since the total energy of a system is determined only when the $2N$ constants of integration of the equations of motion are assigned definite values, E maintains its character as an *arbitrary* constant.)

To obtain a complete solution of the Hamilton-Jacobi equation (33) one usually first determines a complete solution of the reduced equation (37), after which (36) is employed as the solution of (33). A complete solution of (37) involves $(N - 1)$ arbitrary constants α_2, α_3, . . . , α_N, as well as E (aside from the trivial additive constant). Thus, with E properly regarded as an arbitrary constant—equal to α_1, say—the solution of (33) given by (36) is the required complete solution.

(c) In the case of a single particle of mass m moving under the influence of a conservative force, but completely free of geometric constraints, we may use the cartesian coordinates as the generalized coordinates— namely, $q_1 = x$, $q_2 = y$, $q_3 = z$. Thus, according to (2) of 6-1(b), the kinetic energy is given by $T = \frac{1}{2}m(\dot{x}^2 + \dot{y}^2 + \dot{z}^2) = \frac{1}{2}m(\dot{q}_1^2 + \dot{q}_2^2 + \dot{q}_3^2)$. From the definition (11) of 6-3(a) we have the generalized momenta $p_1 = m\dot{q}_1$, $p_2 = m\dot{q}_2$, $p_3 = m\dot{q}_3$, so that $T = (1/2m)(p_1^2 + p_2^2 + p_3^2)$. From (14) of 6-3(b), therefore, the hamiltonian of the single-particle system is $H = (1/2m)(p_1^2 + p_2^2 + p_3^2) + V(x,y,z)$, where V is the potential energy of the particle. Accordingly, the reduced Hamilton-Jacobi equation (37) reads, in this important special case,

$$\frac{1}{2m}\left[\left(\frac{\partial S^*}{\partial x}\right)^2 + \left(\frac{\partial S^*}{\partial y}\right)^2 + \left(\frac{\partial S^*}{\partial z}\right)^2\right] + V(x,y,z) = E, \qquad (38)$$

since $q_1 = x$, $q_2 = y$, $q_3 = z$.

(d) To illustrate the use of the Hamilton-Jacobi method of determining the motion of a system we consider the special case of the unconstrained single particle in which V depends only on z—namely, $V = V(z)$. In accordance with a general mode of procedure we seek a solution of (38) of the form

$$S^* = X(x) + Y(y) + Z(z), \qquad (39)$$

whence (38), with $V = V(z)$, becomes

$$\left(\frac{dX}{dx}\right)^2 + \left(\frac{dY}{dy}\right)^2 + \left(\frac{dZ}{dz}\right)^2 = 2m[E - V(z)].$$

An obvious complete solution is achieved by letting each of the first two terms equal arbitrary constants, so that

$$X = \alpha_1 \sqrt{2m}\,x, \quad Y = \alpha_2 \sqrt{2m}\,y, \quad Z = \sqrt{2m}\int [\alpha_3 - V(z)]^{\frac{1}{2}}\,dz, \quad (40)$$

where α_3 is written for $(E - \alpha_1^2 - \alpha_2^2)$. With (39) and (40), together with (36), we therefore have

$$S = \sqrt{2m} \{\alpha_1 x + \alpha_2 y + \int [\alpha_3 - V(z)]^{\frac{1}{2}} dz\} - (\alpha_3 + \alpha_1^2 + \alpha_2^2)t,$$

whence the solutions (35) of the equations of motion read (since $q_1 = x$, $q_2 = y$, $q_3 = z$) $p_1 = \sqrt{2m}\,\alpha_1$, $p_2 = \sqrt{2m}\,\alpha_2$, $p_3 = \sqrt{2m(\alpha_3 - V)}$, and

$$\beta_1 = \sqrt{2m}\,x - 2\alpha_1 t, \qquad \beta_2 = \sqrt{2m}\,y - 2\alpha_2 t,$$

$$\beta_3 = -t + \int \frac{\sqrt{\frac{1}{2}m}\,dz}{[\alpha_3 - V(z)]^{\frac{1}{2}}}. \qquad (41)$$

The more important trio of equations (41) may be solved for x, y, and z—upon performance of the integration when $V(z)$ is given explicitly—in terms of t and the six arbitrary constants α_i, β_i.

Additional discussion of the Hamilton-Jacobi method, including further treatment of the foregoing problem in exercise 10, is reserved for the end-chapter exercises.

6-6. Principle of Least Action

(a) We denote by the symbol C_1 the configuration—*i.e.*, the aggregate of the positions of the individual particles—exhibited by a given system of particles at an instant $t = t_1$; C_2 denotes the configuration at a later instant $t = t_2$. The aggregate of all the paths traversed by the individual particles when the system pursues its course from the configuration C_1 to the configuration C_2 we call the *configuration path*, or *orbit*, of the system from C_1 to C_2. The actual, or *dynamical*, orbit of a system between two given configurations clearly depends upon the geometric constraints imposed upon the system and the forces which influence the motion. It is useful, also, to speak of *possible* orbits between two given configurations; these are configuration paths which are merely *geometrically*, while not necessarily *dynamically*, feasible within the limitations of the constraints. For example, we consider a single particle which is constrained to lie in a fixed plane; it moves, under the influence of a given force, along a certain arc connecting the points P_1 and P_2 in this plane. Its *actual* orbit between P_1 and P_2 is that arc; but *any* (smooth) curve which lies in the fixed plane and which connects P_1 and P_2 is considered a *possible* orbit between these points.

We consider a given conservative system whose kinetic energy is

$$T(q_1,q_2, \ldots ,q_N,\dot{q}_1,\dot{q}_2, \ldots ,\dot{q}_N),$$

whose potential energy is $V(q_1,q_2, \ldots ,q_N)$, and which pursues its dynamical orbit O_d from configuration C_1 to configuration C_2 with the constant value† E of the total energy $(T + V)$. We next conceive of

† See 6-2(c).

the system as pursuing an arbitrary *possible* orbit O_p from C_1 to C_2 in the following manner:

(i) The system starts from C_1 at the same instant $t = t_1$ at which the *actual* motion starts from C_1.

(ii) The motion along O_p is characterized by the same constant value E of the total energy $(T + V)$ as that which characterizes the actual motion along O_d. (In general, the instant of arrival at C_2 is not the same $t = t_2$ at which the pursuit of O_d is completed at C_2.)

We evaluate the so-called *action* integral

$$I^* = 2 \int_{t_1}^{t_2} T \, dt \qquad (42)$$

for the actual motion from C_1 to C_2 along O_d, and for the motions—*as described by* (i) *and* (ii)—along all possible orbits O_p which connect C_1 and C_2, where t_2 represents the time of arrival at C_2 (different, in general, for each choice of O_p) for each individual motion. In (*b*) below we demonstrate the validity of the *principle of least action:*

The actual motion from C_1 to C_2 is characterized by an extremum of the action (42) with respect to possible motions from C_1 to C_2 for which the total energy is constant and equal to the actual total energy E.

(*b*) To prove the validity of the least-action principle, we show that it leads to a set of equations identical with Lagrange's equations of motion (8) of 6-2(*b*).

Since the upper limit t_2 is not prescribed for the extremization of (42), the proof is greatly simplified through introduction of a parameter $u = u(t)$, which plays the role of independent variable in (42). This parameter must be chosen differently for each possible orbit, but in such fashion that $u_1 = u(t_1)$ and $u_2 = u(t_2)$ have the same pair of values for every possible orbit. Thus the extremization of (42) is reduced, since $dt = (du/\dot{u})$, to the extremization of

$$I^* = 2 \int_{u_1}^{u_2} T \, \frac{du}{\dot{u}}, \qquad (43)$$

in which both u_1 and u_2 have *fixed* values.

To complete the elimination of the variable t we write, for each $i = 1,$ $2, \ldots, N$, $\dot{q}_i = (dq_i/du)\dot{u} = q_i'\dot{u} = q_i'w$, where the prime indicates differentiation with respect to u, and $w = w(u) = \dot{u}$ is introduced for sake of convenience. Thus we have

$$T = T(q_1, \ldots, q_N, q_1'w, \ldots, q_N'w) = w^2 T(q_1, \ldots, q_N, q_1', \ldots, q_N'), \quad (44)$$

since T is a homogeneous function, of degree two, in $\dot{q}_1, \dot{q}_2, \ldots, \dot{q}_N$, according to 6-1(*d*). Writing $T^* = T(q_1, \ldots, q_N, q_1', \ldots, q_N')$, we use

(44) to express (43) as

$$I^* = 2 \int_{u_1}^{u_2} wT^* \, du. \tag{45}$$

Finally, we may rewrite the constancy-of-energy condition $T + V = E$ as

$$w^2 T^* + V = E. \tag{46}$$

With the transformations of the preceding paragraph we may restate the principle of least action briefly as follows: The actual orbit is characterized by an extremum of (45) with respect to the functions $q_1(u)$, $q_2(u)$, . . . , $q_N(u)$, $w(u)$ which satisfy the auxiliary condition (46), and for which q_1, q_2, . . . , q_N are prescribed at $u = u_1$ and $u = u_2$.

We proceed to effect the extremization indicated using the method of 4-5(a). With (45) and (46) respectively representing specific cases of (42) and (43) of 4-5(a) we form the function [(52) of 4-5(a)]

$$F = 2wT^* + \mu(u)(w^2 T^* + V - E), \tag{47}$$

where μ is an undetermined function. Substituting (47) into (57) of 4-5(a), with appropriate change of notation, we obtain, since $(\partial F/\partial w') = 0$ (identically),

$$2T^* + 2\mu w T^* = 0, \tag{48}$$

and, since $(\partial V/\partial q_i') = 0$ (identically) for all i,

$$2w \frac{\partial T^*}{\partial q_i} + \mu \left(w^2 \frac{\partial T^*}{\partial q_i} + \frac{\partial V}{\partial q_i} \right) = \frac{d}{du} \left[(2w + \mu w^2) \frac{\partial T^*}{\partial q_i'} \right]$$
$$(i = 1, 2, \ldots, N). \tag{49}$$

From (44) it follows, since $w = \dot{u}$, that

$$\frac{\partial T^*}{\partial q_i'} = \frac{1}{w^2} \frac{\partial(w^2 T^*)}{\partial q_i'} = \frac{1}{w^2} \frac{\partial T}{\partial q_i'} = \frac{1}{w^2} \frac{\partial T}{\partial \dot{q}_i} \frac{\partial \dot{q}_i}{\partial q_i'} = \frac{1}{\dot{u}} \frac{\partial T}{\partial \dot{q}_i}$$

and $w^2(\partial T^*/\partial q_i) = (\partial T/\partial q_i)$. With the aid of these results, and in conjunction with $\mu = -(1/w) = -(1/\dot{u})$ from (48), equation (49) reads, on multiplication by $w = \dot{u}$,

$$\frac{\partial(T - V)}{\partial q_i} = \dot{u} \frac{d}{du} \left(\frac{\partial T}{\partial \dot{q}_i} \right) = \frac{d}{dt} \left[\frac{\partial}{\partial \dot{q}_i} (T - V) \right] \quad (i = 1, 2, \ldots, N), \tag{50}$$

since $\dot{u}(d/du) = (d/dt)$ and $(\partial V/\partial \dot{q}_i) = 0$ (identically) for all i. Comparison of (50) with (8) of 6-2(b) reveals, since $(T - V) = L$, that the principle of least action does indeed lead to Lagrange's equations for the motion of a system of particles. The validity of the principle is hereby proved.

(c) In the case of a single particle of mass m without geometric constraint the principle of least action leads directly to the differential equations of the particle's orbit, with the time variable t eliminated. In this case we have

$$T = \frac{1}{2} m(\dot{x}^2 + \dot{y}^2 + \dot{z}^2) = \frac{1}{2} m \left(\frac{ds}{dt}\right)^2, \qquad V = V(x,y,z), \qquad (51)$$

so that the constancy of energy (10) of 6-2(c) reads

$$\frac{1}{2} m \left(\frac{ds}{dt}\right)^2 = E - V. \qquad (52)$$

With the first of (51), and on substitution from (52), the action (42) is given by

$$I^* = m \int_{t_1}^{t_2} \left(\frac{ds}{dt}\right)^2 dt = m \int_{s_1}^{s_2} \frac{ds}{dt} ds = \sqrt{2m} \int_{s_1}^{s_2} \sqrt{E - V} \, ds. \qquad (53)$$

If, for example, x is used as the independent variable in the equations describing the particle's orbit, we write $ds = \sqrt{1 + y'^2 + z'^2} \, dx$, where the primes indicate differentiation with respect to x, so that (53) reads

$$I^* = \sqrt{2m} \int_{x_1}^{x_2} \sqrt{E - V} \sqrt{1 + y'^2 + z'^2} \, dx. \qquad (54)$$

In the substitution from (52) into (53) the constancy-of-energy requirement of the least-action principle is taken care of, so that the extremization of (54) with respect to functions $y(x)$ and $z(x)$, prescribed at $x = x_1$ and $x = x_2$, is effected by the particular $y(x)$ and $z(x)$ which describe the actual orbit of the particle between a given pair of fixed points. According to (57) of 3-8(a), therefore—with appropriate changes of notation—the differential equations of the actual orbit are the pair of Euler-Lagrange equations

$$\frac{\partial f}{\partial y} - \frac{d}{dx}\left(\frac{\partial f}{\partial y'}\right) = 0, \qquad \frac{\partial f}{\partial z} - \frac{d}{dx}\left(\frac{\partial f}{\partial z'}\right) = 0, \qquad (55)$$

where f is the integrand of (54).

Application of the preceding result to specific examples is left for the end-chapter exercises.

6-7. The Extended Hamilton's Principle

For application to dynamical systems which involve certain types of forces not derivable from a potential-energy function, we have recourse to a form of Hamilton's principle somewhat more general than the statement given in 6-2(a).

We consider a system of p particles subject to given geometric constraints and to a set of forces which are only in part (if at all) derivable from a potential energy V. That is, the three cartesian components of the force influencing the motion of the jth particle are given by

$$-\frac{\partial V}{\partial x_j} + F_x^{(j)}, \qquad -\frac{\partial V}{\partial y_j} + F_y^{(j)}, \qquad -\frac{\partial V}{\partial z_j} + F_z^{(j)} \qquad (j = 1,2, \ldots ,p),$$

(56)

where $V = V(x_1,y_1,z_1, \ldots ,x_p,y_p,z_p)$, and the components of the nonconservative part of the force acting upon each particle are functions of the coordinates $x_1, y_1, z_1, \ldots , x_p, y_p, z_p$ of the system and the time variable t. If the generalized coordinates which describe the configuration of the system are $q_1, q_2, \ldots , q_N$, we define the set of *generalized force components*

$$G_i = \sum_{j=1}^{p} \left(F_x^{(j)} \frac{\partial x_j}{\partial q_i} + F_y^{(j)} \frac{\partial y_j}{\partial q_i} + F_z^{(j)} \frac{\partial z_j}{\partial q_i} \right) \qquad (i = 1,2, \ldots ,N). \quad (57)$$

We accept as applicable to the dynamical motion of a system of particles under the influence of the forces described by (56) the *extended Hamilton's principle*:

The actual motion of the given system is such as to render the integral

$$I = \int_{t_1}^{t_2} \left(T - V + \sum_{k=1}^{N} \int G_k \, dq_k \right) dt$$

(58)

an extremum with respect to continuously twice-differentiable functions $q_1(t), q_2(t), \ldots , q_N(t)$ *for which* $q_i(t_1)$ *and* $q_i(t_2)$ *are prescribed for all* $i = 1, 2, \ldots , N$. Here $T = T(q_1, \ldots ,q_N,\dot{q}_1, \ldots ,\dot{q}_N)$ is the kinetic energy of the system, and the G_k, expressed in terms of $q_1, q_2, \ldots , q_N$, t, are given by (57). The indefinite integrals in the integrand of (58) are to be regarded in such fashion that

$$\frac{\partial}{\partial q_i} \left(\sum_{k=1}^{N} \int G_k \, dq_k \right) = G_i \qquad (i = 1,2, \ldots ,N). \quad (59)$$

In the important special case in which the generalized force components are explicitly independent of the generalized coordinates, the indefinite integral $\int G_k \, dq_k$ may be replaced by $G_k q_k$ for each k; that is, the integral (58) may be rewritten as

$$I = \int_{t_1}^{t_2} \left(T - V + \sum_{k=1}^{N} G_k q_k \right) dt.$$

(60)

Because of (59), however, the equations of motion derived from the extremization of (60) are no different from those derived from the extremization of (58) [see end-chapter exercise 13(a)].

It is clear from (57) that the generalized force components are ordinary cartesian force components if we employ the cartesian coordinates as generalized coordinates. If, for example, we have $q_i = x_k$ for a particular pair of values of i and k, it follows from (57) that $G_i = F_x^{(k)}$.

EXERCISES

1. Consider a system of p particles moving under the influence of a set of forces as described in the opening paragraph of 6-1(a). In the course of its motion between two given configurations [see 6-6(a)] the system has done upon it by the given forces an amount of *work* defined by

$$W = \sum_{j=1}^{p} \int (F_x^{(j)} \, dx_j + F_y^{(j)} \, dy_j + F_z^{(j)} \, dz_j), \qquad (61)$$

where the jth line integral is computed over the path pursued by the jth particle.

(a) Show that the integral (61) is equal to the loss of potential energy of the system if (1) holds.

(b) Show that the work (61) is given, in terms of the generalized force components defined by (57) of 6-7, by

$$W = \sum_{i=1}^{N} \int_{q_i^{(1)}}^{q_i^{(2)}} G_i \, dq_i,$$

where $q_i^{(1)}$ and $q_i^{(2)} (i = 1, 2, \ldots, N)$ respectively describe the initial and final configurations of the system.

2. Show that a *necessary* condition for the equilibrium of a conservative system is

$$\frac{\partial V}{\partial q_i} = 0 \qquad (i = 1, 2, \cdots, N). \qquad (62)$$

HINT: Using the fact that T is a quadratic form in the $\dot{q}_i$, set all the $\dot{q}_i$ and $\ddot{q}_i$ equal to zero *after* carrying out the differentiations indicated in (8).

3. Introduce a convenient set of generalized coordinates and derive the (Lagrange) equations of motion for each of the following systems; a single particle of mass m is involved in each:

(a) A particle is constrained to lie on a given circle of radius R in a fixed vertical plane; $V = mgz$, where $g = $ positive constant, and $z = $ vertical coordinate measured upward from *any* convenient horizontal line in the plane (simple pendulum). HINT: Introduce the angular displacement (θ) from the vertical of the line from the center of circle to the particle; $V = mgR(1 - \cos \theta)$, $T = \frac{1}{2}mR^2\dot{\theta}^2$. ANSWER: $R\ddot{\theta} + g \sin \theta = 0$.

(b) A particle is constrained to lie on a given straight line; $V = \frac{1}{2}kx^2$, where $k = $ positive constant, and $x = $ displacement from a fixed point on the line (harmonic oscillator). ANSWER: $m\ddot{x} + kx = 0$.

(c) A particle is constrained to move on the surface of a given sphere of radius R; $V = mgz$, where $g = $ positive constant, and $z = $ vertical coordinate measured upward from any convenient horizontal plane (spherical pendulum). HINT: Introduce $x = R \sin \theta \cos \phi$, $y = R \sin \theta \sin \phi$, $z = -R \cos \theta$. $T = \frac{1}{2}mR^2(\dot{\theta}^2 + \dot{\phi}^2 \sin^2 \theta)$.

(d) A particle is unconstrained; $V = mgz$, where the symbols have the same meaning as in part (c) (projectile). ANSWER: $\ddot{x} = 0$, $\ddot{y} = 0$, $\ddot{z} = -g$.

4. For each of the systems listed in exercise 3:

(a) Determine the generalized momenta. ANSWER: For exercise 3(b): $p = m\dot{x}$.

(b) Write down the hamiltonian function. ANSWER: For exercise 3(b):

$$H = \left(\frac{p^2}{2m}\right) + \frac{1}{2}kx^2;$$

for exercise 3(d): $H = [(p_x^2 + p_y^2 + p_z^2)/2m] + mgz$.

(c) Construct the Hamilton equations of motion. ANSWER: For exercise 3(b): $(p/m) = \dot{x}$, $\dot{p} = -kx$; for exercise 3(d): $\dot{p}_x = 0$, $\dot{p}_y = 0$, $\dot{p}_z = -mg$, $\dot{x} = (p_x/m)$, $\dot{y} = (p_y/m)$, $\dot{z} = (p_z/m)$.

5. Use exercise 2 to determine the equilibrium positions, if any, for the systems of exercise 3. ANSWER: For exercise 2(a) $\theta = 0, \pi$; (b) $x = 0$; (c) $\theta = 0, \pi$; (d) none.

6. (a) Apply the canonical transformation generated by $S = -\frac{1}{2}\sqrt{km}\, x^2Q$ to the system of exercise 3(b); derive the transformed equations of motion. ANSWER: $p = -\sqrt{km}\, xQ$, $P = \frac{1}{2}\sqrt{km}\, x^2$, $K - \sqrt{k/m}\, P(Q^2 + 1)$, $Q = \sqrt{k/m}\,(Q^2 + 1)$, $\dot{P} = -2\sqrt{k/m}\, PQ$.

(b) Integrate the transformed equations of motion obtained in part (a) and use the transformation relations to obtain p and x as functions of t (and two arbitrary constants).

(c) Integrate the Hamilton equations of motion obtained in exercise 4 for the system of exercise 3(b). Show, by convenient designation of the constants of integration, that the results are identical with those obtained in part (b) of this exercise. ANSWER: $x = x_0 \cos \omega t + (p_0/\sqrt{km}) \sin \omega t$, $p = p_0 \cos \omega t - x_0 \sqrt{km} \sin \omega t$, where $\omega = \sqrt{k/m}$, and x_0, p_0, are arbitrary constants. (It may take a bit of juggling to get the result of part (b) into this form.)

7. Carry out the details of deriving (30) from (27).

8. Discuss the validity of the use of the solution $\mu_i = 0$ $(i - 1,2, \ldots, N)$ of (29) in the event it is not unique. HINT: Consider the specific purpose for the introduction of the μ_i into the extremization problem at hand. Answer the crucial question: Is this purpose fulfilled if all the μ_i are set equal to zero? ANSWER: Yes, automatically.

9. (a) Write down and solve the reduced Hamilton-Jacobi differential equation for the system of exercise 3(b). Thus write down a complete solution S of the time-dependent Hamilton-Jacobi equation. ANSWER:

$$S = \frac{1}{2}\sqrt{mk}\left[\alpha^2 \sin^{-1}\left(\frac{x}{\alpha}\right) + x\sqrt{\alpha^2 - x^2}\right] - \frac{1}{2}k\alpha^2 t,$$

where $\alpha = \sqrt{2E/k}$.

(b) Use the result of part (a) to derive, by means of (35), the solution of the Hamilton equations of motion of the system. Compare with the result of exercise 6(c).

10. (a) If S is given by (36), where S^* is a complete solution of (37), and if we choose $\alpha_1 = E$, what is the significance of the set of equations [from (35)] $\beta_i = -(\partial S/\partial \alpha_i)$ for $i = 2, 3, \ldots, N$? HINT: Go on to parts (b), (c) below.

(b) Carry through the work of 6-5(d) without introducing $\alpha_3 = (E - \alpha_1^2 - \alpha_2^2)$; instead, let $\alpha_3 = E$. In particular write down the equations $\beta_1 = -(\partial S/\partial \alpha_1)$,

$\beta_2 = -(\partial S/\partial \alpha_2)$. Interpret these equations to show that the orbit of the particle lies in a plane parallel to the z axis.

(c) Show that the result of part (b) is equivalent to the pair of equations obtained by eliminating t from the three equations of (41).

(d) In 6-5(d) consider the particular case $V = mgz$, with the initial conditions $y(0) = \dot{y}(0) = 0$, $x(0) = x_0$, $\dot{x}(0) = v_0 \cos \phi$, $z(0) = z_0$, $\dot{z}(0) = v_0 \sin \phi$, $(0 < \phi < \pi)$. Show that the orbit is the parabola

$$z = z_0 + (x - x_0) \tan \phi - \frac{g}{2v_0^2} (x - x_0)^2 \sec^2 \phi$$

in the plane $y = 0$.

11. (a) Apply the least-action principle to the unconstrained particle under the influence of $V = V(z)$. Show that the orbit equations are given by

$$y = \frac{c_2}{c_1} x + c_3, \qquad x = c_1 \int \frac{dz}{\sqrt{E - c_1^2 - c_2^2 - V}}.$$

HINT: Use the fact that the integrand of (54) is in this case explicitly independent of x. Apply (59) of 3-8(b), with appropriate change of notation.

(b) Show that the result of part (a) is identical with that of 6-5(c) when t is eliminated from (41) (see exercise 10(b,c) above).

12. (a) A particle of mass m constrained to lie in a given plane has a potential energy which is a function only of its distance r from a fixed point in the plane. Use the principle of least action to derive the equation

$$\phi = c_1 \int \frac{dr}{\sqrt{r^4(E - V) - c_1^2 r^2}}$$

of the particle's orbit, where (r, ϕ) are plane polar coordinates.

(b) Apply part (a) to the special case $V = -(k^2/r)$. Identify the orbit in each of the cases $E > 0$, $E = 0$, $E < 0$.

(c) Solve the problem of part (a), and subsequently that of part (b), by the Hamilton-Jacobi method. SOLUTION: $T = \frac{1}{2}m(\dot{r}^2 + r^2\dot{\phi}^2)$, $p_r = m\dot{r}$, $p_\phi = mr^2\dot{\phi}$,

$$H = \left(\frac{1}{2m}\right)\left[p_r^2 + \left(\frac{p_\phi}{r}\right)^2 \right] + V(r),$$

$$S^* = \alpha_1\phi + \int \sqrt{2m(E - V) - \left(\frac{\alpha_1}{r}\right)^2}\, dr;$$

use the second of (35) with $i = 1$.

13. (a) With the aid of (59) show that the extremization of both (58) and (60) leads to the equations of motion

$$\frac{\partial T}{\partial q_i} - \frac{d}{dt}\left(\frac{\partial T}{\partial \dot{q}_i}\right) = \frac{\partial V}{\partial q_i} - G_i \qquad (i = 1, 2, \cdots, N).$$

(b) Apply the result of part (a) to the problem of the harmonic oscillator of exercise 3(b) in which a force whose x component is $F(t)$ is applied to the particle. ANSWER: $m\ddot{x} + kx = F(t)$.

CHAPTER 7

TWO INDEPENDENT VARIABLES: THE VIBRATING STRING

7-1. Extremization of a Double Integral

(a) We consider the double integral[1]

$$I = \iint_D f(x,y,w,w_x,w_y)dx\, dy \tag{1}$$

carried out over a given domain D of the xy plane. The given function f is twice differentiable with respect to the indicated arguments. We proceed to derive the partial differential equation which must be satisfied by the function which renders (1) an extremum[2] with respect to continuously differentiable functions $w(x,y)$ which assume prescribed values at all points of the boundary curve C of the domain D.

To effect the extremization of (1) we employ the method of 3-3(b) whereby we introduce a one-parameter family of comparison functions

$$W(x,y) = w(x,y) + \epsilon\eta(x,y), \tag{2}$$

where $w(x,y)$ is assumed to be the actual extremizing function, and ϵ is the parameter of the family. Thus no matter what the choice of $\eta(x,y)$, arbitrary to within continuous differentiability and

$$\eta(x,y) = 0 \quad \text{on } C, \tag{3}$$

we have that the integral formed by replacing w by W in (1) is an extremum for $\epsilon = 0$. That is,

$$I'(0) = 0, \tag{4}$$

where

$$I(\epsilon) = \iint_D f(x,y,W,W_x,W_y)dx\, dy. \tag{5}$$

---

[1] In this and ensuing chapters we employ, whenever the usual notation becomes too cumbersome, subscripts to indicate partial differentiation. Thus we write w_x for $(\partial w/\partial x)$, w_{xy} for $(\partial^2 w/\partial y\, \partial x)$, etc.

[2] We use the term "extremum" here in the sense of 3-3(c), with obvious extension to the case of functions of two variables.

Using (2) to compute $(\partial W/\partial\epsilon) = \eta$, $(\partial W_x/\partial\epsilon) = \eta_x$, $(\partial W_y/\partial\epsilon) = \eta_y$, we differentiate (5) with respect to ϵ to form

$$I'(\epsilon) = \iint_D \left(\frac{\partial f}{\partial W} \eta + \frac{\partial f}{\partial W_x} \eta_x + \frac{\partial f}{\partial W_y} \eta_y \right) dx\, dy.$$

Since, according to (2), setting $\epsilon = 0$ is equivalent to replacing W by w, we therefore have

$$I'(0) = \iint_D \left(\frac{\partial f}{\partial w} \eta + \frac{\partial f}{\partial w_x} \eta_x + \frac{\partial f}{\partial w_y} \eta_y \right) dx\, dy = 0, \tag{6}$$

because of (4).

Applying Green's theorem (22) of 2-13 to the final two terms of the middle member of (6), we obtain

$$0 = \iint_D \eta \left[\frac{\partial f}{\partial w} - \frac{\partial}{\partial x}\left(\frac{\partial f}{\partial w_x}\right) - \frac{\partial}{\partial y}\left(\frac{\partial f}{\partial w_y}\right) \right] dx\, dy$$

$$+ \int_C \eta \left(\frac{\partial f}{\partial w_x} \frac{dy}{ds} - \frac{\partial f}{\partial w_y} \frac{dx}{ds} \right) ds \tag{7}$$

$$= \iint_D \eta \left[\frac{\partial f}{\partial w} - \frac{\partial}{\partial x}\left(\frac{\partial f}{\partial w_x}\right) - \frac{\partial}{\partial y}\left(\frac{\partial f}{\partial w_y}\right) \right] dx\, dy, \tag{8}$$

because of (3). From the basic lemma of 3-1(c) we therefore conclude that the extremizing function $w = w(x,y)$ must satisfy

$$\frac{\partial f}{\partial w} - \frac{\partial}{\partial x}\left(\frac{\partial f}{\partial w_x}\right) - \frac{\partial}{\partial y}\left(\frac{\partial f}{\partial w_y}\right) = 0 \tag{9}$$

everywhere in D.

(b) We may directly extend the result of (a) above to the case in which the functions eligible for the extremization of (1) are required to satisfy no special condition on the boundary C. The only alteration of the procedure of (a) is to remove the restriction (3) and so adopt the result (7).

Since the right-hand member of (7) must vanish for *all* choices of arbitrary differentiable $\eta(x,y)$, it must *in particular* vanish for those η which satisfy (3). For such functions η equation (7) reduces to (8), and we immediately conclude the applicability of (9). With (9) equation (7) becomes

$$\int_C \eta \left(\frac{\partial f}{\partial w_x} \frac{dy}{ds} - \frac{\partial f}{\partial w_y} \frac{dx}{ds} \right) ds = 0,$$

for η arbitrary along C.   Applying a form of the basic lemma of 3-1, we therefore have

$$\frac{\partial f}{\partial w_x}\frac{dy}{ds} - \frac{\partial f}{\partial w_y}\frac{dx}{ds} = 0 \text{ along } C \tag{10}$$

as the condition which must be fulfilled in case the functions eligible for the extremization are not prescribed on C.

In case the eligible functions are prescribed on a *portion* of C but are arbitrary on the remainder of C, it is clear from the preceding paragraph that (10) must hold along the remainder of C.   That is, every point of C is characterized by either the prescription of w or the fulfillment of (10) by the actual extremizing function w.

(c) By adapting the procedure of 4-1 we achieve the following result for the simple isoperimetric problem involving two independent variables: The function which extremizes (1) of (a) above with respect to functions w for which the integral

$$J = \iint_D g(x,y,w,w_x,w_y)dx\,dy$$

has a given prescribed value must satisfy the Euler-Lagrange equation

$$\frac{\partial f^*}{\partial w} - \frac{\partial}{\partial x}\left(\frac{\partial f^*}{\partial w_x}\right) - \frac{\partial}{\partial y}\left(\frac{\partial f^*}{\partial w_y}\right) = 0,$$

where $f^* = f + \lambda g$.   Along portions of C on which w is not prescribed, the condition (10), with f replaced by f^*, is fulfilled by the extremizing function w.

7-2. The Vibrating String

In this section we apply Hamilton's principle [6-2(a)] to a system involving a *continuous distribution* of mass—as distinguished from a discrete set of mass particles, to which our attention is confined in Chap. 6. The means for effecting this application is the simple device—employed with great success through the domain of "continuum mechanics"—of replacing sums over discrete particles by integrals over the continuous mass distributions.

(a) We consider a perfectly flexible elastic string stretched under constant tension τ along the x axis with its end points fixed at $x = 0$ and $x = L$.   This undistorted state is called the *equilibrium* configuration. Following the proper type of stimulus, the string is permitted to vibrate freely in a plane containing the x axis in such fashion that each particle of the string moves in a straight line perpendicular to the x axis; the amplitude of vibration is supposed so small that the slope (with respect to the x axis) of the string at any point is small compared with unity at all instants of time t. We further assume that there is no frictional (or other) damping, so that we deal with a conservative system.

The transverse displacement at time t of the particle whose equilibrium position is characterized by its distance x from the end of the string at $x = 0$ is denoted by the function $w = w(x,t)$; thus $w(x,t)$, with $0 \leq x \leq L$, describes the shape of the string during the course of the vibration. The slope of the string is given by $(\partial w/\partial x) = w_x(x,t)$ as a function of position x and time t. At time t the velocity of the particle at a particular value of x is denoted by $(\partial w/\partial t) = \dot{w}(x,t)$. The fact that the ends are fixed (with zero displacement) at $x = 0$ and $x = L$ supplies the end-point conditions $w(0,t) = w(L,t) = 0$, for all t.

Since the string is perfectly flexible, the amount of work which must be done upon it in order to effect a given distorted configuration *must be employed merely to increase the length* of the string relative to its equilibrium length L. Therefore, in order to compute the potential energy V of the string at an arbitrary instant of time we must compute merely the amount of work which is required to stretch it from the length L to its total length in the configuration exhibited at the given instant. Thus since the stretching force is equal to the tension† τ, the potential energy is given by

$$V = \tau \Big(\int_0^L \sqrt{1 + w_x^2}\, dx - L \Big), \tag{11}$$

where the integral is clearly the length of the string in its distorted configuration. With the assumption that $|w_x|$ is small compared with unity we may expand $\sqrt{1 + w_x^2} = (1 + \tfrac{1}{2}w_x^2 + \cdots)$ and neglect the higher powers of w_x^2 to obtain from (11)

$$V = \tau \Big[\int_0^L (1 + \tfrac{1}{2}w_x^2) dx - L \Big] = \tfrac{1}{2}\tau \int_0^L w_x^2\, dx. \tag{12}$$

We assume a distribution of mass along the string of density (mass per unit length) $\sigma(x)$, where $\sigma = \sigma(x)$ is a positive continuous function. Thus the mass contained in an element of length dx at x is $\sigma(x)dx$, with the associated kinetic energy $\tfrac{1}{2}\sigma(x)dx[\dot{w}(x,t)]^2$—or, simply, $\tfrac{1}{2}\sigma\dot{w}^2\, dx$. The total kinetic energy of the string, accordingly, is

$$T = \tfrac{1}{2} \int_0^L \sigma \dot{w}^2\, dx. \tag{13}$$

With (12) and (13) we apply Hamilton's principle [6-2(a)] to the vibration of the string. That is, the function which describes the actual

---

† It is tacitly assumed that the elongation is so slight that the tension remains constant throughout the stretching.

motion of the string is one which renders

$$I = \int_1^{t_2} (T - V)dt = \tfrac{1}{2} \int_1^{t_2} \int_0^L (\sigma \dot{w}^2 - \tau w_x^2)dx\, dt \qquad (14)$$

an extremum with respect to functions $w(x,t)$ which describe the actual configuration at $t = t_1$ and $t = t_2$ and which vanish, for all t, at $x = 0$ and $x = L$. (The instants t_1 and $t_2 > t_1$ are completely arbitrary.)

The extremization of (14) is accomplished through 7-1(a) above (with the replacement of y by t, w_y by $\dot{w}$) if we denote by D the "rectangle" $0 \leqq x \leqq L$, $t_1 \leqq t \leqq t_2$ in the "xt plane." According to the preceding paragraph the functions eligible for the extremization are prescribed everywhere on the boundary C of D. Thus we may apply (9) of 7-1(a) to the integrand $f = \tfrac{1}{2}(\sigma \dot{w}^2 - \tau w_x^2)$ of (14) to obtain

$$\frac{\partial^2 w}{\partial x^2} = \frac{\sigma(x)}{\tau} \frac{\partial^2 w}{\partial t^2} \qquad (15)$$

as the partial differential equation which describes the motion of our vibrating string.

(b) We consider also the case in which each end point of the vibrating string described in (a) above, instead of being maintained in fixed position, is allowed to move freely along a straight line perpendicular to the x axis and lying in the plane of vibration.[1] Mathematically the only change incident upon freeing the end points in this fashion is to remove the restriction that the functions w eligible for the extremization of (14) vanish at $x = 0$ and $x = L$. Since, because of Hamilton's principle, the eligible functions are still prescribed at $t = t_1$ and $t = t_2$, we may therefore apply the free-boundary condition (10) of 7-1(b), with appropriate change of notation, only along the "sides" $x = 0$, $x = L$ of the "rectangle" D in the "xt plane" described in the final paragraph of (a) above. Along these sides we have $(dt/ds) = \pm 1$ and $(dx/ds) = 0$. Thus (10) is reduced to $(\partial f / \partial w_x) = 0$; with the integrand $f = \tfrac{1}{2}(\sigma \dot{w}^2 - \tau w_x^2)$ of (14) it reads, simply,

$$\frac{\partial w}{\partial x} = 0 \qquad (16)$$

at $x = 0$, $x = L$. In case one end of the string is held fixed and the other is free, condition (16) holds at the free end only, of course, while $w = 0$ holds at the other end.

(c) Mathematically, the problem of the vibrating string is completely equivalent to the problem of the plane longitudinal vibrations of an

---

[1] Such an arrangement is, approximately at least, physically feasible. The reader is urged to devise schemes by which the "free-end string" may be set up.

elastic medium. Specifically, the foregoing results apply, for example, to the vibrations of the gas filling a long cylindrical tube. At a closed end of the tube the vibration amplitude w is required to vanish (the fixed-end problem of (a) above); at an open end, since there is no constraint upon the amplitude w, the condition (16) is applicable. The quantity σ appearing in (15) is the mass of the gas per unit length of the tube and is in general a constant; τ is a constant related to the compressibility of the gas. In the remainder of this chapter we speak only of the vibrating string; it should be understood, however, that our results are generally applicable to the gas-vibration problem as well.

7-3. Eigenvalue-Eigenfunction Problem for the Vibrating String

(a) The initial attack upon the vibrating-string equation (15) involves seeking a solution of the form

$$w = \phi(x)q(t), \tag{17}$$

where ϕ is independent of t and q is independent of x. From (17) it follows that $w_{xx} = \phi''(x)q(t) = \phi''q$ and $\ddot{w} = \phi(x)q''(t) = \phi\ddot{q}$, so that substitution into (15) gives, on division by $(\sigma\phi q/\tau)$,

$$\frac{\tau\phi''}{\sigma\phi} = \frac{\ddot{q}}{q}. \tag{18}$$

Since the left-hand member depends upon x alone and the right-hand member upon t alone, it follows that the only circumstance in which (18) can hold for all values of the *independent* variables x and t is that both members be equal to a constant[1] independent of x and t; we denote the constant, at this point undetermined, by $(-\lambda)$. Thus (18) implies the two ordinary differential equations

$$\text{(i)} \quad \tau\frac{d^2\phi}{dx^2} + \lambda\sigma(x)\phi = 0, \qquad \text{(ii)} \quad \frac{d^2q}{dt^2} + \lambda q = 0. \tag{19}$$

Determination of the values which may be assigned to λ depends upon the particular set of end-point conditions we happen to deal with. If both ends of the string are fixed, the conditions $w(0,t) = w(L,t) = 0$ lead, through (17), to the conditions $\phi(0) = \phi(L) = 0$ upon ϕ. On the other hand if one or both end points of the string are free, the vanishing of ϕ must be replaced, according to (16) and (17), by $(d\phi/dx) = 0$ at one or

---

[1] This line of argument, the basis of the so-called *method of separation of variables*, is employed repeatedly in chapters following.

both of $x = 0$, $x = L$.   In what follows, we suppose that we have to deal with one from among the possible sets of such end-point conditions.[1]

Thus we are faced with the problem of not only solving (19,i) but of fitting the general solution $\phi = \phi(x,\lambda)$ to the required end-point conditions.   It can be shown[2] that there exists only a discrete set of values of λ for which the end-point conditions are satisfied by $\phi(x,\lambda)$.   (It is quite obvious that $\phi = 0$, *identically in* $0 \leqq x \leqq L$, satisfies both (19,i) and any of the various sets of end-point conditions, for arbitrary λ.   This *trivial* solution must be ignored as irrelevant to our problem.)   This privileged set of values of λ we may list in the *increasing* order $\lambda_1, \lambda_2, \ldots ,$ $\lambda_n, \ldots$; it has infinitely many members, of which there is a smallest λ_1, but for which λ_n is unbounded as $n \rightarrow \infty$.   Any such value of λ— for which there exists a solution of (19,i) which conforms with the end-point conditions—is called an *eigenvalue* of λ; the corresponding solution is called an *eigenfunction* of (19,i) in conjunction with the particular end-point conditions.   Corresponding to any one eigenvalue there is one and *only one* linearly independent eigenfunction; the vibrating-string eigenvalues are therefore said to be *nondegenerate*.   Clearly, the totality of eigenvalues of λ associated with a given problem depends upon the values of L and τ, the function $\sigma(x)$, and the particular set of end-point conditions involved in the problem.

Since both the differential equation and the end-point conditions which the eigenfunctions are required to satisfy are linear and homogeneous, it follows that the product of an eigenfunction by any nonzero constant is also an eigenfunction corresponding to the same eigenvalue.   For this reason we may impose the convenient restriction

$$\int_0^L \sigma\phi^2 \, dx = 1 \tag{20}$$

for every eigenfunction we deal with.   (Because $\sigma > 0$, the left-hand member of (20) must be positive for any real function ϕ not identically zero.   In case the integral were not equal to unity but equal to c^2, say, the corresponding integral, with ϕ replaced by (ϕ/c), would be unity.)   Any function ϕ for which (20) holds is said to be normalized with respect to the weight function σ in the interval $0 \leqq x \leqq L$—or, briefly, *normalized*.

(b) There are no negative eigenvalues of λ.   To prove this fact we multiply (19,i) by ϕ and integrate the resulting equation from $x = 0$ to

---

[1] That is, we deal with a string with both ends fixed, both ends free, or one end fixed and the other free, with only *one* of these cases considered in any given discussion.

[2] The proof is beyond our present scope.   See, however, Ince, Chap. 10.   See also exercise 2 at the end of this chapter.

$x = L$. We thus obtain, with the aid of (20),

$$\lambda = -\tau \int_0^L \phi \frac{d^2\phi}{dx^2} \, dx = -\tau\phi \frac{d\phi}{dx} \Big]_0^L + \tau \int_0^L \left(\frac{d\phi}{dx}\right)^2 dx, \qquad (21)$$

on integration by parts. If ϕ is an eigenfunction, either $\phi = 0$ or $(d\phi/dx) = 0$ at $x = 0$ and $x = L$, so that the right-hand member of (21) reduces to its final term. Since $\tau > 0$, we therefore conclude that no eigenvalue of λ can be negative.

With the exceptional case $\lambda = 0$ left for end-chapter exercise 6(b) we employ the positivity of the eigenvalues of λ to solve the time-dependent equation (19,ii). If λ_n is an eigenvalue of λ, the general solution of (19,ii) is

$$q = q_n = A_n \cos \sqrt{\lambda_n}\, t + B_n \sin \sqrt{\lambda_n}\, t \qquad (n = 1,2,3, \ldots), \qquad (22)$$

where A_n and B_n are arbitrary constants. If $\phi_n = \phi_n(x)$ is the corresponding eigenfunction,

$$w = w_n = \phi_n(x)(A_n \cos \sqrt{\lambda_n}\, t + B_n \sin \sqrt{\lambda_n}\, t) \qquad (n = 1,2,3, \ldots) \qquad (23)$$

is therefore, for each n, a solution of the equation (15) describing the motion of the vibrating string under a given set of end-point conditions.

Clearly, the solution (23) is periodic in the time variable t; the period is given by $(2\pi/\sqrt{\lambda_n})$, the frequency by $(\sqrt{\lambda_n}/2\pi)$. Thus we conclude that the elastic string has the ability to vibrate with any of a discreet set of frequencies which are determined by the eigenvalues of λ. In other words evaluation of the set of eigenvalues associated with a given vibrating-string problem provides the set of *natural vibration frequencies* of the string. In particular the lowest eigenvalue λ_1 provides the so-called *fundamental frequency* $(\sqrt{\lambda_1}/\pi)$ of the string. In 7-5 below it is demonstrated that the general motion of a vibrating string is a linear superposition of the various single-frequency modes of vibration represented by (23).

7-4. Eigenfunction Expansion of Arbitrary Functions. Minimum Characterization of the Eigenvalue-Eigenfunction Problem

(a) We consider the sequence of normalized eigenfunctions ϕ_1, ϕ_2, . . . , ϕ_n, . . . and corresponding eigenvalues λ_1, λ_2, . . . , λ_n, . . . , arranged in increasing order, of a given vibrating-string problem; that is, for each $n = 1, 2, 3, \ldots$, the function ϕ_n satisfies

$$\tau\phi_n'' + \lambda_n \sigma \phi_n = 0 \qquad (0 \le x \le L) \qquad (24)$$

and either

$$\phi_n = 0 \qquad or \qquad \phi_.' = 0 \qquad (25)$$

at $x = 0$ and $x = L$. Rewriting (24) with $n = j$, we multiply this equation by $\phi_k(j \neq k)$. Reversing the indices j and k in the result so achieved, we obtain a second such result. Subtracting one from the other of these two results and integrating from $x = 0$ to $x = L$, we get

$$(\lambda_k - \lambda_j) \int_0^L \sigma\phi_j\phi_k \, dx = \tau \int_0^L (\phi_k\phi_j'' - \phi_j\phi_k'')dx$$

$$= \tau(\phi_k\phi_j' - \phi_j\phi_k') \Big]_0^L, \qquad (26)$$

on integration by parts of each term of the second member. Since ϕ_j and ϕ_k are eigenfunctions of the same problem, they satisfy the *same* conditions—either of (25)—at each of $x = 0$, $x = L$. It therefore follows that the final member of (26) vanishes. Also, since $j \neq k$—and therefore† $\lambda_j \neq \lambda_k$—we have

$$\int_0^L \sigma\phi_j\phi_k \, dx = 0 \qquad \text{for } j \neq k. \qquad (27)$$

Any two functions ϕ_j, ϕ_k which satisfy (27) are said to be orthogonal with respect to the weight function σ in the interval $0 \leqq x \leqq L$—or, briefly, *orthogonal*. A set of functions $\phi_1, \phi_2, \ldots, \phi_n, \ldots$ of which every two distinct members satisfy (27) is said to constitute a set of *orthogonal functions* (with respect to the weight function σ in the interval $0 \leqq x \leqq L$). In case all the functions of an orthogonal set satisfy the normalization condition (20) of 7-3(a)—with the same weight function and same interval—they are said to constitute an *orthonormal set*. Since the vibrating-string eigenfunctions are required to be normalized, the result (27) discloses the fact that they constitute an orthonormal set. This fact is best expressed through introduction of the Kronecker delta δ_{jk}, a symbol which denotes 0 when $j \neq k$ and 1 when $j = k$. Thus we have for the eigenfunctions of a given vibrating-string problem

$$\int_0^L \sigma\phi_j\phi_k \, dx = \delta_{jk}. \qquad (28)$$

(b) We state without proof the following theorem concerning the expansion of an arbitrary function in terms of the known set of eigenfunctions:

If the arbitrary function $g(x)$ is continuous and piecewise differentiable[1] in $0 \leqq x \leqq L$, the series

$$\sum_{n=1}^{\infty} c_n\phi_n(x), \qquad \text{with } c_n = \int_0^L \sigma\phi_n g \, dx,$$

converges uniformly to $g(x)$ in every subinterval of $0 \leqq x \leqq L$ in which $g(x)$ is continuous. We may therefore write

† See end-chapter exercise 3.
[1] See 2-1.

$$g(x) = \sum_{n=1}^{\infty} c_n \phi_n(x) \qquad \left(c_n = \int_0^L \sigma \phi_n g \, dx \right). \tag{29}$$

Moreover, in any subinterval in which $g'(x)$ is continuous, we may differentiate (29) term by term to obtain

$$g'(x) = \sum_{n=1}^{\infty} c_n \phi_n'(x), \tag{30}$$

and the convergence is uniform. (Possible exceptions at $x = 0$ and $x = L$ are mentioned below.)

Given that there exists a series expansion of the type shown in (29), we get the parenthetic part of (29) directly from the orthonormality relationship (28). The details are left for exercise 4 at the end of this chapter.

The matter of end-point conditions requires special discussion. If the eigenfunctions employed in the expansion of a given function $g(x)$ according to (29) satisfy the condition $\phi_n = 0$ at $x = 0$ (or $x = L$, or both), the series (29) clearly converges to zero at $x = 0$ (or $x = L$, or both). Thus, although the function $g(x)$ may be continuous in the neighborhood of $x = 0$ (or $x = L$, or both), the sum of the series (29) is discontinuous at $x = 0$ (or $x = L$, or both) in case $g(x)$ does not vanish at $x = 0$ (or $x = L$, or both). We encounter no difficulty from this fact in our study, however.

If, on the other hand, the eigenfunctions ϕ_n which appear in (29) satisfy the condition $\phi_n' = 0$ at $x = 0$ (or $x = L$, or both), the difficulty of the preceding paragraph does not arise at $x = 0$ (or $x = L$, or both); that is, if $g(x)$ is continuous at $x = 0$ (or $x = L$, or both), the series (29) is also continuous at $x = 0$ (or $x = L$, or both). The derivative series (30), however, is discontinuous at $x = 0$ (or $x = L$, or both), in case $g'(x)$ is continuous and different from zero at $x = 0$ (or $x = L$, or both).

(c) With the aid of the expansion theorem of (b) above we demonstrate the following minimum characterization of the eigenvalue-eigenfunction problem for a given vibrating string:

The kth eigenvalue[†] λ_k *is the minimum of the integral*

$$I = \tau \int_0^L \phi'^2 \, dx \tag{31}$$

with respect to those functions ϕ which satisfy the normalization condition

$$\int_0^L \sigma \phi^2 \, dx = 1 \tag{32}$$

[†] The totality of the eigenvalues is supposed arranged in the ascending order $\lambda_1 < \lambda_2 < \cdots < \lambda_k < \cdots$.

and the $(k - 1)$ *orthogonality relations*

$$\int_0^L \sigma \phi_j \phi \, dx = 0 \qquad (j = 1, 2, \ldots, k - 1), \tag{33}$$

where ϕ_j *is the eigenfunction that satisfies*

$$\tau \phi_j'' + \lambda_j \sigma \phi_j = 0 \qquad (j = 1, 2, 3, \ldots) \tag{34}$$

and the set of end-point conditions $(\phi_j = 0$ *or* $\phi_j' = 0$ *at* $x = 0$, $x = L)$ *associated with the problem.* Further, every function ϕ eligible for the minimization must be continuous everywhere in $0 \leqq x \leqq L$ and have a first derivative ϕ' which is piecewise continuous in $0 \leqq x \leqq L$.

In the problem of the string whose end at $x = 0$ (or $x = L$, or both) is held fixed, the *additional* restriction $\phi = 0$ at $x = 0$ (or $x = L$, or both) must be imposed upon the eligible functions. (No special restriction is imposed upon the eligible functions ϕ at $x = 0$ (or $x = L$, or both) if the end of the string at $x = 0$ (or $x = L$, or both) is free.)

The minimum λ_k *of I under the stated restrictions is achieved when* $\phi = \phi_k$.

(We note, in particular, that the functions ϕ eligible for the *first* minimization of I, whereby the minimum is the lowest eigenvalue λ_1, are required to satisfy *no* orthogonality condition (30).)[1]

To prove the stated characterization we expand the arbitrary function ϕ eligible for the kth minimization of (31) in accordance with (29) and (30) of (b) above:

$$\phi = \sum_{n=1}^{\infty} c_n \phi_n, \qquad \phi' = \sum_{n=1}^{\infty} c_n \phi_n' \qquad \left(c_n = \int_0^L \sigma \phi_n \phi \, dx \right). \tag{35}$$

(The eigenfunctions employed in the expansion are associated with the particular vibrating-string problem under discussion. Thus, according to the above statement of possible end-point restrictions upon the eligible functions ϕ, every ϕ is *required* to vanish at $x = 0$ (or $x = L$, or both) if and only if every ϕ_n vanishes at $x = 0$ (or $x = L$, or both). We therefore avoid the end-point-discontinuity difficulty mentioned in the penultimate paragraph of (b) above. On the other hand, at a free end point, the second series of (35) may be discontinuous (according to the final paragraph of (b) above), since each ϕ_n' must vanish there, while ϕ' is arbitrary; this fact involves no difficulty in the proof which follows, however.) From the parenthetic portion of (35) it follows that the orthogonality

---

[1] A reading of 9-9(c) at this point (with a few obvious minor changes of wording) should be extremely helpful in achieving a fuller understanding of the foregoing characterization.

conditions (33) are satisfied only if

$$c_1 = c_2 = \cdots = c_{k-1} = 0. \tag{36}$$

Substituting the first of (35) for one factor of (32), we obtain[1]

$$\int_0^L \sigma\phi^2 \, dx = \sum_{n=1}^{\infty} c_n \int_0^L \sigma\phi_n\phi \, dx = \sum_{n=1}^{\infty} c_n^2 = 1, \tag{37}$$

with the aid of the parenthetic part of (35).

Substituting the second of (35) for one factor of (31), we obtain

$$I = \tau \sum_{n=1}^{\infty} c_n \int_0^L \phi_n'\phi' \, dx = \tau \sum_{n=1}^{\infty} c_n \left\{ \left[\phi_n'\phi\right]_0^L - \int_0^L \phi_n''\phi \, dx \right\}, \tag{38}$$

on integration by parts.[2]   Since either $\phi = 0$ or $\phi_n' = 0$ at both of† $x = 0$, $x = L$, the integrated portion of every term of (38) must vanish.   With the aid of (34)—with j replaced by n—(38) therefore becomes

$$I = - \sum_{n=1}^{\infty} c_n \int_0^L (\tau\phi_n'')\phi \, dx = \sum_{n=1}^{\infty} c_n\lambda_n \int_0^L \sigma\phi_n\phi \, dx = \sum_{n=1}^{\infty} \lambda_n c_n^2, \tag{39}$$

according to the parenthetic part of (35).

Taking into account (36) and (37), we may rewrite (39) as

$$I = \sum_{n=k}^{\infty} \lambda_n c_n^2 = \lambda_k \sum_{n=k}^{\infty} c_n^2 + \sum_{n=k}^{\infty} (\lambda_n - \lambda_k)c_n^2 = \lambda_k + \sum_{n=k}^{\infty} (\lambda_n - \lambda_k)c_n^2.$$

Since $\lambda_n > \lambda_k$ if $n > k$, it therefore follows that $I \geqq \lambda_k$; the equality sign holds if $c_k = 1$ and $c_{k+1} = c_{k+2} = c_{k+3} = \cdots = 0$.   But according to the first of (35), this choice of the set of coefficients c_n (clearly consistent with (37)), taken in conjunction with (36), implies $\phi = \phi_k$.   The stated

---

[1] The interchange of summation and integration, carried out in the sequel without explicit statement of justification, is justified by the *uniform* convergence of the series expansions, as stated in (*b*) above.   Since ϕ' is merely *piecewise* continuous, the second series of (35) may be discontinuous at a finite number of points.   The consequent nonuniformity of convergence is confined, however, to a finite number of arbitrarily narrow subintervals whose contribution to any term-by-term integration over $0 \leqq x \leqq L$ can be made arbitrarily small—and therefore zero.

[2] It is this integration by parts which requires the *continuity* of the eligible functions ϕ (see 2-4).

† See the parenthetic remark of the preceding paragraph.   We may, of course, have both $\phi = 0$ and $\phi_n' = 0$ at one or both end points.

minimum characterization of the vibrating-string eigenvalue-eigenfunction problem is hereby proved.   Application is found in 7-6 below.

7-5.   General Solution of the Vibrating-string Equation

(a) By means of the expansion theorem enunciated in 7-4(b) we may obtain a solution of the vibrating-string equation (15) of 7-2(a), together with an appropriate set of end-point conditions, which is sufficiently general to cover at least those cases which are of physical interest.   The method we employ bypasses the equation (15) itself, but instead returns to the integral (14)—namely,

$$I = \tfrac{1}{2} \int_{t_1}^{t_2} \int_0^L (\sigma \dot{w}^2 - \tau w_x^2) dx \, dt \tag{40}$$

—whose extremization according to Hamilton's principle leads directly to (15).   Further, the method presupposes the prior solution of the eigenvalue-eigenfunction problem associated with the given vibrating-string problem; $i.e.$, we have at our disposal the sequence of orthonormal eigenfunctions $\phi_1, \phi_2, \ldots, \phi_m, \ldots$ and corresponding eigenvalues $\lambda_1, \lambda_2, \ldots, \lambda_m, \ldots$ for which

$$\tau \phi_m'' + \lambda_m \sigma \phi_m = 0 \qquad (0 \leq x \leq L) \tag{41}$$

and either $\phi_m = 0$ or $\phi_m' = 0$ at $x = 0$ and $x = L$, for each m.   (At a free end point of the string $\phi_m' = 0$; at a fixed end $\phi_m = 0$.)

We suppose that $w = w(x,t)$ is arbitrary for all t in $0 \leq x \leq L$, to within the following limitations: (i) continuity of w and $\dot{w}$ with respect to both x and t, (ii) piecewise continuity of w_x and $\dot{w}_x$ with respect to x, (iii) $w(x,t)$ describes the actual vibrating-string configuration at two arbitrary instants $t = t_1$ and $t = t_2$ (requirement of Hamilton's principle). Finally, if the string under consideration is fixed at one end point (or both), $w(x,t)$ vanishes at that end point (or both), for all t.   Regarded as a function of x, $w(x,t)$ clearly satisfies the requirements of the theorem of 7-4(b) for expansion in terms of the eigenfunctions associated with the given vibrating-string problem.   Since $w(x,t)$ depends upon t, however, the coefficients in the expansion must depend upon t.

We thus write, according to (29) of 7-4(b),

$$w = \sum_{m=1}^{\infty} c_m(t) \phi_m(x) \qquad \left(c_m = \int_0^L \sigma \phi_m w \, dx \right). \tag{42}$$

We may also expand $\dot{w}(x,t)$ as

$$\dot{w} = \sum_{m=1}^{\infty} d_m(t) \phi_m(x) \qquad \left(d_m = \int_0^L \sigma \phi_m \dot{w} \, dx \right), \tag{43}$$

according to (29). Since $\dot{w}$ is continuous with respect to x, it follows from the parenthetic parts of (42) and (43), and from the rule [2-3(a)] for differentiation of an integral, that $d_m = \dot{c}_m$. Thus since we may differentiate the series of (42) term by term with respect to x, we have the two expansions

$$\dot{w} = \sum_{m=1}^{\infty} \dot{c}_m(t)\,\phi_m(x), \qquad w_x = \sum_{m=1}^{\infty} c_m(t)\,\phi'_m(x). \tag{44}$$

Substituting the appropriate series of (44) for one factor of each term of the integrand of (40), we obtain

$$I = \tfrac{1}{2} \sum_{m=1}^{\infty} \int_{t_1}^{t_2} \left(\dot{c}_m \int_0^L \sigma\phi_m\dot{w}\,dx - \tau c_m \int_0^L w_x\phi'_m\,dx \right) dt. \tag{45}$$

Integrating by parts the second integral over x, and using the fact that either w (at a fixed end) or ϕ'_m (at a free end) must vanish at $x = 0$ and $x = L$, we obtain (45) in the form

$$I = \tfrac{1}{2} \sum_{m=1}^{\infty} \int_{t_1}^{t_2} \left(\dot{c}_m \int_0^L \sigma\phi_m\dot{w}\,dx + c_m \int_0^L w(\tau\phi''_m)dx \right) dt$$

$$= \tfrac{1}{2} \sum_{m=1}^{\infty} \int_{t_1}^{t_2} \left(\dot{c}_m \int_0^L \sigma\phi_m\dot{w}\,dx - \lambda_m c_m \int_0^L \sigma\phi_m w\,dx \right) dt$$

$$= \tfrac{1}{2} \sum_{m=1}^{\infty} \int_{t_1}^{t_2} (\dot{c}_m^2 - \lambda_m c_m^2)dt, \tag{46}$$

with the aid of (41), the parenthetic portions of (42) and (43), and the fact that $d_m = \dot{c}_m$.

Thus the extremization of the integral (40) with respect to the arbitrary function $w(x,t)$ is reduced to the extremization of the simple integral (46) with respect to the infinite set of functions $c_m(t)$. To avoid any possible difficulty involved in the circumstance of the appearance of infinitely many dependent variables $c_m(t)$ in the integrand of (46), we suppose that all but one—$c_n(t)$, say—are correctly determined as extremizing functions. With this we may apply the Euler-Lagrange equation (25) of 3-3(b)—replacing y by c_n and x by t, and with $f = \tfrac{1}{2} \Sigma (\dot{c}_m^2 - \lambda_m c_m^2)$ —to obtain $\ddot{c}_n + \lambda_n c_n = 0$ as the differential equation which must be satisfied by the extremizing function† $c_n(t)$. Since the choice of n is arbitrary, however, this equation must hold for all $n = 1, 2, 3, \ldots$.

---

† Since the arbitrary eligible function $w(x,t)$ is required to describe the *actual* string configuration at $t = t_1$ and $t = t_2$, we must suppose that $c_n(t)$ is prescribed at $t = t_1$ and $t = t_2$. There are thus no special "end-point" conditions on $c_n(t)$ at $t = t_1$ and $t = t_2$.

The general solution of the above equation for $c_n(t)$ is—with the exceptional case in which $\lambda_1 = 0$ left for end-chapter exercise $6(c)$—

$$c_n(t) = A_n \cos \sqrt{\lambda_n}\, t + B_n \sin \sqrt{\lambda_n}\, t \qquad (n = 1,2,3,\ldots), \qquad (47)$$

where the A_n and B_n constitute two sets of arbitrary constants. With (47) we conclude from (42) that the general solution of the vibrating-string problem is

$$w(x,t) = \sum_{n=1}^{\infty} \phi_n(x)(A_n \cos \sqrt{\lambda_n}\, t + B_n \sin \sqrt{\lambda_n}\, t). \qquad (48)$$

(b) The result (48) justifies the frequently employed analysis of any given state of vibration of an elastic string as a linear superposition of vibrations, each of which is characterized by a single frequency. Comparison with (23) of 7-3(b) reveals that each term of (48) represents one of the single-frequency modes of vibration which the string is capable of executing.

The infinite sets of arbitrary constants A_n, B_n may be determined if initial ($t = 0$) conditions are prescribed for w and $\dot{w}$; discussion of this point is reserved for end-chapter exercise 7.

7-6. Approximation of the Vibrating-string Eigenvalues and Eigenfunctions (Ritz Method)

Since precise analytical methods are not available in all cases, one must in general resort to methods for *approximating* the eigenvalues and eigenfunctions associated with a given vibrating-string problem.[1] One such method, generally known as the Ritz method, is a direct consequence of the minimum characterization developed above in 7-4(c).[2]

(a) According to 7-4(c) substitution into

$$I = \tau \int_0^L \phi'^2 \, dx \qquad (49)$$

of any continuous, piecewise differentiable function $\phi(x)$—for which $\phi(0) = \phi(L) = 0$ and

$$\int_0^L \sigma\phi^2 \, dx = 1 \qquad (50)$$

—bestows upon I a value *no less than the lowest eigenvalue* λ_1 associated with the *fixed*-end vibrating-string problem to which L, τ, σ pertain. Accordingly, I provides an *upper bound* for λ_1.

---

[1] A few problems in which it is possible to write down explicit expressions for the eigenvalues and eigenfunctions are handled in the end-chapter exercises.

[2] See 10-10 below, second paragraph.

Further, we may substitute into (49) an eligible function $\phi(x)$ which depends upon one or more parameters $k_1, k_2, \ldots, k_N$. Thus I is computed as a function of $k_1, k_2, \ldots, k_N$, with respect to which parameters the integral is subsequently minimized. The minimum so achieved is accordingly the *lowest* upper bound to λ_1 obtainable through the class of functions defined by ϕ and the sets of values assumed by the N parameters. The larger the number N, the wider is the class of functions so defined, and so, in general, the lower is the computed upper bound for λ_1. (In the possible fortunate circumstance in which ϕ, for some particular set of values of $k_1, k_2, \ldots, k_N$, coincides with the actual eigenfunction ϕ_1, the upper bound computed by minimizing I with respect to these parameters is exactly λ_1.) The essence of the Ritz method lies in the acceptance of the minimum of I with respect to the N parameters as an *approximation* to λ_1. (In (c) below we consider approximation of the higher eigenvalues $(\lambda_2, \lambda_3, \ldots)$.)

The closeness of the approximation of course depends upon the selection of the parameter-laden function ϕ. Although the criteria for the accuracy in any given application of the Ritz method to a vibrating-string problem are by no means clear-cut, it is generally (except, of course, when the approximation happens to be perfect, as described parenthetically in the preceding paragraph) possible to improve a given approximate computation at the expense of increased labor. It is possible, in fact, to refine the method into a convergent procedure [see (c) below] although the degree of accuracy is uncertain at every stage and the difficulty of computation increases inordinately with each improvement of the approximation. Justification for the expenditure of such labor can lie, of course, only in the degree of urgency resident in any particular computation.

(It should be emphasized at this point that although one may perhaps be inclined to dismiss as insignificant the problem of the vibrating string and the application thereto of the Ritz method, the concepts, ideas, and techniques involved here are of enormous significance in their extension and adaptation to problems of possibly greater importance. A complete understanding of the work of the present chapter is an almost indispensable prerequisite to comprehension of much of the subject matter which is found in the final four chapters of this volume.)

(b) Although we may employ the Ritz method as described in (a) above to achieve, in a particular case, a useful approximation to λ_1, it should be recognized that the parameter-laden function ϕ which leads to this approximation is not necessarily a correspondingly useful approximation to the precise eigenfunction $\phi_1(x)$, even with the parameters $k_1, k_2, \ldots, k_N$ set at their minimizing values. In spite of this fact it is convenient to label as "corresponding approximate eigenfunction" the function ϕ that renders the integral (49) equal to a value which we accept as an approximation to a particular eigenvalue. (This usage is maintained without further comment in chapters following.)

It is very often far more important to know the few lowest eigenvalues associated with a given problem than it is to possess knowledge of any of the eigenfunctions. As in the case of the vibrating string, the few lowest eigenvalues involved in any vibration problem determine the few lowest, and most important, natural vibration frequencies; in the Schrödinger problem of Chap. 11 the lowest eigenvalues are the lowest, and most important, energy levels of a quantum-mechanical system. In fact one is often satisfied to know merely the lowest eigenvalue associated with a given problem. For this reason we touch only lightly, in this and following chapters, upon the possible approximation of eigenfunctions. Instead, we devote our main effort toward the development of methods which are readily amenable to the numerical approximation of the first few eigenvalues.

(c) The Ritz method for approximating the first few—say s—eigenvalues of a given vibrating-string problem may be formulated by means of a rephrasing of the statement of the minimum characterization given at the opening of 7-4(c):

The approximation Λ_k to the kth eigenvalue λ_k is the minimum of the integral

$$I = \tau \int_0^L \psi'^2 \, dx \tag{51}$$

with respect to those functions ψ (belonging to the special class introduced directly below) which satisfy the normalization condition

$$\int_0^L \sigma \psi^2 \, dx = 1 \tag{52}$$

and the $(k-1)$ orthogonality relations

$$\int_0^L \sigma \psi_m \psi \, dx = 0 \qquad (m = 1, 2, \ldots, k-1), \tag{53}$$

where ψ_m is the mth approximate eigenfunction—*i.e.*, a function which renders I equal to Λ_m—for $m = 1, 2, 3, \ldots$.

It is clear, through comparison of the preceding formulation with 7-4(c), that inclusion in the class represented by ψ of *all* the functions ϕ which are eligible for the minimization in the *precise* minimum characterization would provide the result $\Lambda_k = \lambda_k$ for all k. In any single application of the Ritz method, however, we create a special *subclass* of eligible functions with respect to which the minimization of (51) is carried out: We suppose that $\Phi_1(x)$, $\Phi_2(x)$, $\ldots$, $\Phi_s(x)$ are s conveniently given continuous functions continuously differentiable in $0 \leqq x \leqq L$. If we deal with a string whose end at $x = 0$ (or $x = L$, or both) is fixed, each of the Φ_j $(j = 1,2, \ldots ,s)$ must be chosen so as to vanish at $x = 0$ (or $x = L$, or both). The special subclass consists of all functions ψ which exhibit the form

$$\psi = c_1\Phi_1 + c_2\Phi_2 + \cdots + c_s\Phi_s = \sum_{j=1}^{s} c_j\Phi_j, \qquad (54)$$

where $c_1, c_2, \ldots , c_s$ are arbitrary constants consistent with the normalization condition (52).

We immediately arrive at the inequality $\lambda_1 \leqq \Lambda_1$, since λ_1 is the minimum of (51) with respect to a class of normalized functions which is much wider than the class represented by (54). It is not at all obvious, however, that $\lambda_2 \leqq \Lambda_2$, since it is not generally possible to determine whether or not the members of (54) eligible for the *second* minimization of (51) form a subclass of the class of functions with respect to which the (second) minimum of (51) is the precise eigenvalue λ_2. For, according to (33) of 7-4(c), every member of the latter class is orthogonal to the *precise* eigenfunction ϕ_1; on the other hand, each member of (54) eligible for the second minimization of (51) is orthogonal merely to the *approximate* eigenfunction ψ_1, according to (53). By the same token we cannot tell at a glance whether or not $\lambda_3 \leqq \Lambda_3$, $\lambda_4 \leqq \Lambda_4$, etc. Nevertheless, the inequality $\lambda_k \leqq \Lambda_k$ does hold for $k = 1, 2, \ldots , s$, but the proof is not at this point within our reach. In Chap. 9, however, a method is developed by means of which $\lambda_k \leqq \Lambda_k$ is readily established; a proof is called for in exercise 23 at the end of that chapter. We therefore borrow the result: Each Λ_k is an *approximation from above* to the corresponding eigenvalue λ_k; or, every Λ_k is an *upper bound* for the corresponding λ_k.

The larger we choose the integer s in (54), the wider is the class of functions represented by ψ, and the more accurate are the approximations we achieve. If $\Phi_1, \Phi_2, \ldots , \Phi_s$ are the first s of an infinite sequence of functions for which there exists an expansion theorem such as the one stated in 7-4(b),[1] the approximations become perfect in the limit $s \to \infty$, for then the class represented by ψ includes *all* the functions ϕ eligible for the precise minimizations. Unfortunately, however, the difficulties of computation generally multiply tremendously with increase of s.

Substituting (54) into (52), we obtain

$$\int_0^L \sigma\psi^2\, dx = \sum_{i=1}^{s} \sum_{j=1}^{s} c_ic_j \int_0^L \sigma\Phi_i\Phi_j\, dx = \sum_{i=1}^{s} \sum_{j=1}^{s} c_ic_j\sigma_{ij} = 1, \qquad (55)$$

---

[1] In which case the sequence is said to be "closed."

where we define

$$\sigma_{ij} = \sigma_{ji} = \int_0^L \sigma \Phi_i \Phi_j \, dx. \tag{56}$$

In accordance with (54) we write the approximate eigenfunctions as

$$\psi_m = \sum_{j=1}^{s} c_j^{(m)} \Phi_j \qquad (m = 1, 2, \ldots, s), \tag{57}$$

so that the problem of finding each ψ_m is equivalent to that of determining the set of values $c_1^{(m)}$, $c_2^{(m)}$, $\ldots$, $c_s^{(m)}$ for the coefficients c_1, c_2, $\ldots$, c_s, respectively, in (54). Substituting (54) and (57) into (53), we obtain, with the aid of the definition (56),

$$\int_0^L \sigma \psi_m \psi \, dx = \sum_{i=1}^{s} \sum_{j=1}^{s} c_i c_j^{(m)} \sigma_{ij} = 0 \qquad (m = 1, 2, \ldots, k-1). \tag{58}$$

Finally, if we define

$$\Gamma_{ij} = \Gamma_{ji} = \tau \int^L \Phi_i' \Phi_j' \, dx, \tag{59}$$

substitution of (54) into (51) gives

$$I = \sum_{i=1}^{s} \sum_{j=1}^{s} c_i c_j \Gamma_{ij} \tag{60}$$

for the quantity whose minima we seek. In particular since $I = \Lambda_k$ when $\psi = \psi_k$, it follows from (57), with $m = k$, that

$$\Lambda_k = \sum_{i=1}^{s} \sum_{j=1}^{s} c_i^{(k)} c_j^{(k)} \Gamma_{ij}. \tag{61}$$

To minimize (60) under the restrictions (55) and (58) we use the method of Lagrange multipliers (2-6), whereby we form the quantity

$$I^* = \sum_{i=1}^{s} \sum_{j=1}^{s} c_i c_j \Gamma_{ij} - \Lambda^{(k)} \sum_{i=1}^{s} \sum_{j=1}^{s} c_i c_j \sigma_{ij} - \sum_{m=1}^{k-1} \lambda_k^{(m)} \sum_{i=1}^{s} \sum_{j=1}^{s} c_i c_j^{(m)} \sigma_{ij},$$

where $\Lambda^{(k)}$, $\lambda_k^{(1)}$, $\lambda_k^{(2)}$, $\ldots$, $\lambda_k^{(k-1)}$ are undetermined multipliers. (In our quest for the kth minimum Λ_k of I, we suppose that $c_j^{(1)}$, $c_j^{(2)}$, $\ldots$, $c_j^{(k-1)}$ are known for all $j = 1, 2, \ldots, s$.) According to 2-6 the kth minimum of I is characterized by the s conditions[1]

---

[1] We make use of the relations $\Gamma_{ij} = \Gamma_{ji}$ and $\sigma_{ij} = \sigma_{ji}$, here and below, without explicit mention thereof.

$$\frac{\partial I^*}{\partial c_i} = 2 \sum_{j=1}^{s} c_j \Gamma_{ij} - 2\Lambda^{(k)} \sum_{j=1}^{s} c_j \sigma_{ij} - \sum_{m=1}^{k-1} \lambda_k^{(m)} \sum_{j=1}^{s} c_j^{(m)} \sigma_{ij} = 0,$$

or

$$\sum_{j=1}^{s} (\Gamma_{ij} - \Lambda^{(k)} \sigma_{ij}) c_j = \tfrac{1}{2} \sum_{m=1}^{k-1} \lambda_k^{(m)} \sum_{j=1}^{s} c_j^{(m)} \sigma_{ij} \qquad (i = 1,2, \ldots ,s). \qquad (62)$$

This set of s equations is satisfied by $c_1 = c_1^{(k)}$, $c_2 = c_2^{(k)}$, $\ldots$, $c_s = c_s^{(k)}$.

In (d) below we prove that the multipliers $\lambda_k^{(1)}$, $\lambda_k^{(2)}$, $\ldots$, $\lambda_k^{(k-1)}$ all vanish. Accepting this fact here, we conclude from (62) that

$$\sum_{j=1}^{s} (\Gamma_{ij} - \Lambda^{(k)} \sigma_{ij}) c_j^{(k)} = 0 \qquad (i = 1,2, \ldots ,s). \qquad (63)$$

Since (63) constitutes a system of s linear homogeneous equations in the s quantities $c_1^{(k)}$, $c_2^{(k)}$, $\ldots$, $c_s^{(k)}$, which, according to (55), do not all vanish, the determinant of the coefficients must vanish.[1] That is, $\Lambda^{(k)}$ must be a root of the algebraic equation in Λ of degree s

$$\begin{vmatrix} \Gamma_{11} - \Lambda\sigma_{11} & \Gamma_{12} - \Lambda\sigma_{12} & \ldots & \Gamma_{1s} - \Lambda\sigma_{1s} \\ \Gamma_{21} - \Lambda\sigma_{21} & \Gamma_{22} - \Lambda\sigma_{22} & \ldots & \Gamma_{2s} - \Lambda\sigma_{2s} \\ \cdots & \cdots & \cdots & \cdots \\ \Gamma_{s1} - \Lambda\sigma_{s1} & \Gamma_{s2} - \Lambda\sigma_{s2} & \ldots & \Gamma_{ss} - \Lambda\sigma_{ss} \end{vmatrix} = 0. \qquad (64)$$

Under the assumption that $\Lambda = \Lambda^{(k)}$ satisfies (64) and therefore that the system (63) is satisfied, we multiply the ith equation of (63) by $c_i^{(k)}$, for all $i = 1, 2, \ldots , s$, and add the resulting equations—sum over i, that is—to obtain

$$\sum_{i=1}^{s} \sum_{j=1}^{s} (\Gamma_{ij} - \Lambda^{(k)} \sigma_{ij}) c_j^{(k)} c_i^{(k)} = 0. \qquad (65)$$

Solving (65) for $\Lambda^{(k)}$, we find, with the aid of the normalization (55)—with c replaced by $c^{(k)}$ for all subscripts—

$$\Lambda^{(k)} = \sum_{i=1}^{s} \sum_{j=1}^{s} \Gamma_{ij} c_j^{(k)} c_i^{(k)} = \Lambda_k, \qquad (66)$$

according to (61).

Since the determinantal equation (64) has no explicit dependence upon the specific choice of the index k, we must conclude from (66) that its

[1] See 2-8(b).

s roots, arranged in ascending order, are the approximate eigenvalues Λ_1, $\Lambda_2, \ldots, \Lambda_s$—the quantities whose values we seek. Thus application of the Ritz method to the approximation of the first few eigenvalues associated with a given vibrating-string problem involves merely (i) choosing a suitable set of functions $\Phi_1, \Phi_2, \ldots, \Phi_s$, (ii) computing the sets of integrals defined by (56) and (59), and (iii) solving the algebraic equation (64). Examples of the application are found in the exercises at the end of this chapter.

(d) The Lagrange multipliers $\lambda_k^{(m)}$ are introduced in (c) above to ensure fulfillment of the relations (58) by the set of coefficients $c_1^{(k)}, c_2^{(k)}, \ldots, c_s^{(k)}$ which satisfy the system (62) of linear equations, with $c_j = c_j^{(k)}$ for all $j = 1, 2, \ldots, s$. For the actual solutions of (62) equation (58) reads, in conjunction with (55),

$$\sum_{i=1}^{s} \sum_{j=1}^{s} c_i^{(k)} c_j^{(m)} \sigma_{ij} = \delta_{km} \qquad (m = 1,2, \ldots, k \leqq s), \qquad (67)$$

where δ_{km} is the Kronecker delta introduced in the final paragraph of 7-4(a) above. We proceed to show that the multipliers $\lambda_k^{(1)}, \lambda_k^{(2)}, \ldots, \lambda_k^{(k-1)}$ vanish for all $k = 2, 3, \ldots, s$. To do this we must deduce the result, for $n = 1, 2, \ldots, s$,

$$\sum_{j=1}^{s} (\Gamma_{ij} - \Lambda^{(n)} \sigma_{ij}) c_j^{(n)} = 0 \qquad (68)$$

from (62) and (67); we employ the method of *complete induction*.

Surely (68) holds for $n = 1$, as we find by setting $k = 1$ in (62) (written with $c_j = c_j^{(k)}$), since then the sum over m in the right-hand member is empty. We now suppose that (68) holds for $n = 1, 2, \ldots, k - 1$: Multiplying the ith equation of (62) (with $c_j = c_j^{(k)}$) by $c_i^{(n)}$ ($n < k$), for all $i = 1, 2, \ldots, s$, and adding the resulting s equations, we obtain

$$\sum_{i=1}^{s} \sum_{j=1}^{s} \Gamma_{ij} c_j^{(k)} c_i^{(n)} - \Lambda^{(k)} \sum_{i=1}^{s} \sum_{j=1}^{s} \sigma_{ij} c_j^{(k)} c_i^{(n)} = \tfrac{1}{2} \sum_{m=1}^{k-1} \lambda_k^{(m)} \sum_{i=1}^{s} \sum_{j=1}^{s} \sigma_{ij} c_j^{(m)} c_i^{(n)}. \qquad (69)$$

Because of (67) the coefficient of $\Lambda^{(k)}$ in (69) vanishes, since $n < k$. For the same reasons, the only term in the sum over m which does not vanish is the one for which $m = n$, and the coefficient of $\lambda_k^{(n)}$ is $\tfrac{1}{2}$. Thus (69) becomes, on slight rearrangement of the left-hand member and use of the fact that $\Gamma_{ij} = \Gamma_{ji}$,

$$\tfrac{1}{2}\lambda_k^{(n)} = \sum_{j=1}^{s} c_j^{(k)} \sum_{i=1}^{s} \Gamma_{ji} c_i^{(n)} = \Lambda^{(n)} \sum_{j=1}^{s} c_j^{(k)} \sum_{i=1}^{s} \sigma_{ji} c_i^{(n)}$$

$$= \Lambda^{(n)} \sum_{i=1}^{s} \sum_{j=1}^{s} c_i^{(n)} c_j^{(k)} \sigma_{ji},$$

(70)

with the aid of (68) (with indices i, j reversed), which is assumed valid for $n < k$. Since $n < k$, (67) dictates that the final member of (70) vanishes, so that we have $\lambda_k^{(n)} = 0$ for $n = 1, 2, \ldots, k - 1$; that is, every term of the right-hand member of (62)—with $c_j = c_j^{(k)}$—must vanish, and (68) holds also for $n = k$.

The required proof is complete: Assumption that (68) holds for $n = 1, 2, \ldots, k - 1$ implies that (68) holds also for $n = k$. But (68) is known to hold, as pointed out above, for $n = 1$; it therefore holds also for $n = 2$. Since it holds for $n = 1, 2$, it must hold also for $n = 3$, and— by continuation of the same argument (the usual argument of "complete induction")—for *all* $n = 1, 2, \ldots, s$. The direct passage from (62) to (63) in (c) above is hereby justified.

7-7. Remarks on the Distinction between Imposed and Free End-point Conditions

In the statement in 7-4(c) of the minimum characterization of the eigenvalue-eigenfunction problem for the vibrating string, we note a distinction between the two types of end-point conditions which does not appear in the characterization of the problem through the differential equation (19,i) of 7-3(a). When one deals with the differential equation, on the one hand, the cases in which the (fixed) end-point condition $\phi = 0$ applies are handled in much the same manner as those in which the (free) end-point condition $\phi' = 0$ applies. In the minimum characterization, on the other hand, we observe the following exceedingly important difference between the two cases:

At a fixed end point we must require that *every function ϕ eligible for the minimization* must vanish at that end point. At a free end point, however, there is *no* special restriction which must be placed upon the functions ϕ eligible for the minimization; the vanishing of ϕ' at a free end point arises as a "natural" end-point condition which turns out to be necessarily satisfied by the minimizing functions. That is, an eigenfunction in a free end-point problem effects the minimization not merely with respect to eligible functions which satisfy the end-point condition $\phi' = 0$ but with respect to eligible functions which satisfy *arbitrary* end-point conditions. An eigenfunction in a fixed end-point problem, however, effects the minimization *only* with respect to the class of eligible functions which satisfy the end-point condition $\phi = 0$.

Although the foregoing remarks are of course redundant with respect to 7-4(c), they appear for the sake of reemphasis and because the same type of distinction plays a role of extreme importance in chapters following—especially in Chap. 9. We must keep clearly in mind the essential difference between *imposed* boundary conditions and so-called *natural* boundary conditions.

EXERCISES

1. Assuming the existence of a surface $z = z(x,y)$ for which the area enclosed by a fixed space curve is a minimum, show that

$$\frac{\partial}{\partial x}\left(\frac{z_x}{\sqrt{1 + z_x^2 + z_y^2}}\right) + \frac{\partial}{\partial y}\left(\frac{z_y}{\sqrt{1 + z_x^2 + z_y^2}}\right) = 0$$

on the minimal surface.

2. (a) Let $y = \phi(x)$ be a curve in the xy plane. Show that $(\phi''/\phi) > 0$ implies that the curve is everywhere *convex* toward the x axis, and that $(\phi''/\phi) < 0$ implies it is everywhere *concave* toward the x axis.

(b) Use part (a) to show that *no* function ϕ which satisfies (19,i)—in which $\tau > 0$, $\sigma(x) > 0$—can (i) vanish for two distinct values of x if $\lambda \leqq 0$, (ii) have a vanishing derivative for two distinct values of x if $\lambda < 0$, (iii) vanish at one value of x while its derivative vanishes at a second value if $\lambda \leqq 0$. (Thus we have a geometric-intuitive proof of the fact that there are no negative vibrating-string eigenvalues.)

(c) Let $\phi = \phi(x,\lambda)$ be *the* solution of (19,i) for which $\phi(0,\lambda) = 0$, $\phi'(0,\lambda) = 1$, for arbitrary λ. It can be shown that $\phi(x,\lambda)$ and $\phi'(x,\lambda)$ are continuous functions of λ. HINT: Let $\psi(x) = \phi(x,\lambda + \Delta\lambda) - \phi(x,\lambda)$, so that $\psi(0) = \psi'(0) = 0$. Show that $\tau\psi'' + \lambda\sigma\psi = -\Delta\lambda\sigma\phi(x,\lambda + \Delta\lambda)$. Assuming ϕ to be known, use the method of *variation of parameters* to solve the equation for ψ, with the given conditions at $x = 0$, and so show that ψ and ψ' are proportional to $\Delta\lambda$.

With the knowledge of this continuity develop a geometric-intuitive proof that there exists a smallest positive value λ_1 of λ for which $\phi(L,\lambda) = 0$. HINT: By part (b) above, $\phi(L,0) > 0$. As λ increases continuously from zero, the curvature of $y = \phi(x,\lambda)$, which is *concave* toward the x axis, increases. For sufficiently large λ, the curve must cross the x axis at $x = L$.

(d) Extend the method of part (c) to show that there exists an infinite unbounded sequence of positive values $\lambda_1, \lambda_2, \lambda_3, \ldots$ of λ such that $\phi(L,\lambda) = 0$. The demonstration should be such as to make evident the fact that $\phi(x,\lambda_n)$ vanishes $(n - 1)$ times in the *open* interval $0 < x < L$, for all $n = 1, 2, 3, \ldots$ (Thus we have a geometric-intuitive proof of the existence of an unbounded positive sequence of eigenvalues associated with a vibrating string having both ends fixed. The restriction $\phi'(x,0) = 1$ is unessential; each function $\phi(x,\lambda_n)$ may be multiplied by any nonzero constant without altering the essence of the argument.)

(e) Adapt the method of parts (c) and (d) to demonstrate the existence of the eigenvalues associated with the string having (i) both ends free and (ii) one end—say $x = 0$—free and the other fixed. HINT: In each case let $\phi(x,\lambda)$ be the solution of (19,i) for which $\phi(0,\lambda) = 1$, $\phi'(0,\lambda) = 0$, etc.

In both cases the nth eigenfunction vanishes $(n - 1)$ times in the *open* interval $0 < x < L$.

3. Let $\phi(x)$ and $\phi^*(x)$ be two distinct solutions of (19,i) for the *same* value of λ. Prove that the wronskian $[2\text{-}8(e)]\, w = (\phi^*\phi' - \phi\phi^{*\prime})$ is a constant. HINT: Use (19,i) to prove $(dw/dx) = 0$.

Thus show that, if both ϕ and ϕ^*, or both ϕ' and $\phi^{*\prime}$, vanish at a single value of x, the wronskian is identically zero. (Since ϕ and ϕ^* are therefore linearly dependent, we have here a proof that there corresponds only one linearly independent eigenfunction to each vibrating-string eigenvalue.)

4. Given the series expansion (29), use (28) to derive the parenthetic part of (29). HINT: Multiply both sides of (29) by $\sigma\phi_m$, for $m = 1, 2, 3, \ldots$, and integrate term by term from $x = 0$ to $x = L$. In the resulting right-hand member, only the term for which $n = m$ is different from zero.

5. Use 4-1 and 4-2(b) to characterize the vibrating-string eigenvalue-eigenfunction problem as an isoperimetric problem: The eigenfunctions extremize the integral $I = \tau \displaystyle\int_0^L \phi'^2\, dx$ with respect to functions ϕ for which the integral $J = \displaystyle\int_0^L \sigma\phi^2\, dx$ has a prescribed value; at fixed end points the eligible functions must vanish; at free end points the eligible functions are arbitrary.

6. (a) Prove in three ways that $\lambda = 0$ is an eigenvalue in (19,i) if and only if we have $\phi'(0) = \phi'(L) = 0$: (i) Use (21). (ii) Use exercise 2 above. (iii) Solve (19,i) explicitly. Show that the corresponding eigenfunction is $\phi = $ constant.

(b) How must (22), and accordingly (23), be modified in the event $\lambda_1 = 0$? HINT: Solve (19,ii) with $\lambda = 0$.

(c) Rewrite (48) so that it applies to the problem of the string with both ends free.

7. Show that the sets of coefficients A_m, B_m in (48) are evaluated as

$$A_m = \int_0^L \sigma\phi_m w(x,0)dx, \qquad B_m = \frac{1}{\sqrt{\lambda_m}} \int_0^L \sigma\phi_m \dot{w}(x,0)dx,$$

where the initial shape $w(x,0)$ and initial velocity distribution $\dot{w}(x,0)$ are prescribed arbitrarily. HINT: For A_m, set $t = 0$ in (48); then use (29). For B_m, use (48) to from $\dot{w}(x,0)$; then use (29).

8. A vibrating string is subjected to a nonconservative transverse force per unit length given by the expression $F(x,t)$. (That is, an element of length dx at x experiences the externally applied force $F(x,t)dx$ perpendicular to the x axis in the plane of vibration.)

(a) Use the extended Hamilton's principle of 6-7 to show that the equation of motion of the string so influenced is derived by extremizing the integral

$$I = \int_{t_1}^{t_2} \int_0^L [\tfrac{1}{2}(\sigma\dot{w}^2 - \tau w_x^2) + Fw]dx\, dt.$$

Thus derive the equation of motion

$$\sigma\frac{\partial^2 w}{\partial t^2} = \tau\frac{\partial^2 w}{\partial x^2} + F(x,t),$$

as well as the condition $(\partial w/\partial x) = 0$ at a possible free end point. (We impose $w = 0$ at a fixed end point from the outset.)

(b) Extend the method of 7-5(a) to show that the general solution of the problem at hand is $w = \sum_{m=1}^{\infty} c_m(t)\phi_m(x)$, where

$$\frac{d^2c_m}{dt^2} + \lambda_m c_m = \int_0^L F\phi_m\, dx \qquad (m = 1,2,3, \cdots)$$

and the λ_m and ϕ_m constitute the sets of eigenvalues and corresponding *normalized* eigenfunctions associated with the same string in the absence of the external force distribution.

9. Throughout this exercise we deal with the uniform vibrating string of density $\sigma = \sigma_0$, a constant.

(a) Show that the general solution of (19,i) is, for $\sigma = \sigma_0$,

$$\phi = C \cos\left(\sqrt{\frac{\lambda\sigma_0}{\tau}}\, x\right) + D \sin\left(\sqrt{\frac{\lambda\sigma_0}{\tau}}\, x\right), \qquad (71)$$

where C, D are arbitrary constants. For the uniform string with fixed ends show that $C = 0$ and $\sin(\sqrt{\lambda\sigma_0/\tau}\, L) = 0$. Thus show that the eigenvalues are given by $\lambda_n = (n^2\pi^2\tau/L^2\sigma_0)$, with the corresponding eigenfunctions $\phi_n = D_n \sin(n\pi x/L)$, for $n = 1, 2, 3, \ldots$.

Show that the normalization (20) is satisfied if $D_n = \sqrt{2/\sigma_0 L}$ for all n.

What are the natural vibration frequencies of the uniform string with both ends fixed? ANSWER: $(\sqrt{\lambda_n}/2\pi) = (n/2L)\sqrt{\tau/\sigma_0}$.

(b) List the eigenvalues and corresponding normalized eigenfunctions for the uniform string fixed at $x = 0$ and free at $x = L$. HINT: Use (71). Show that $C = 0$ and $\cos(\sqrt{\lambda\sigma_0/\tau}\, L) = 0$.

(c) List the eigenvalues and corresponding normalized eigenfunctions for the uniform string free at both $x = 0$, $x = L$. ANSWER: $\lambda_{n+1} = (n^2\pi^2\tau/L^2\sigma_0)$ for $n = 0$, $1, 2, \ldots$; $\phi_1 = (1/\sqrt{\sigma_0 L})$, $\phi_{n+1} = \sqrt{2/\sigma_0 L} \cos(n\pi x/L)$ for $n = 1, 2, 3, \ldots$.

(d) Write down the general solution (48) explicitly for the string of part (a).

10. (a) With $\phi = A(x/L)^k[1 - (x/L)]$ $(k > \frac{1}{2})$ use the method of 7-6(a) to show that the first eigenvalue of the fixed-end uniform vibrating-string problem in which $\sigma = \sigma_0$, a constant, satisfies the inequality $\lambda_1 \leq [\tau k(k + 1)(2k + 3)/\sigma_0 L^2(2k - 1)]$. (We note that the requirement $\phi(0) = \phi(L) = 0$ is fulfilled.) HINT: First show that the condition (50), with $\sigma = \sigma_0$, demands that $A^2 = [(2k + 1)(k + 1)(2k + 3)/\sigma_0 L]$; then substitute into (49).

(b) Show that the "best" value for k in part (a) is the solution of

$$(2k + 3)(2k - 1)^2 = 6$$

for which $k > \frac{1}{2}$—namely, $k = 1.04$, approximately. That is, verify that the upper bound given for λ_1 is a minimum when $k = 1.04$. Thus show that $\lambda_1 \leq 9.98\,(\tau/\sigma_0 L^2)$. (Compare with the precise value $\lambda_1 = \pi^2(\tau/\sigma_0 L^2)$ derived in exercise 9(a).)

(c) With $\phi = A(x/L)^k$ $(k > \frac{1}{2})$ adapt the method of 7-6(a) to show that the first eigenvalue of the uniform-string problem with the end at $x = 0$ fixed and the end at $x = L$ free satisfies the inequality $\lambda_1 \leq [\tau k^2(2k + 1)/\sigma_0 L^2(2k - 1)]$. (We note that the single end-point requirement, $\phi(0) = 0$, is fulfilled.) HINT: First use (50), with $\sigma = \sigma_0$, to derive $A^2 = [(2k + 1)/\sigma_0 L]$; then use (49).

Show that the "best" choice is approximately $k = 0.8$, so that $\lambda_1 \leq 2.8(\tau/\sigma_0 L^2)$. (Compare with the precise value $\lambda_1 = (\pi^2/4)(\tau/\sigma_0 L^2)$ derived in exercise 9(b).)

11. (a) With $\Phi_j = (x/L)^j[1 - (x/L)]$ $(j = 1,2)$ use the method of 7-6(c), with $s = 2$, to approximate λ_1 and λ_2 for the fixed-end uniform $(\sigma = \sigma_0)$ string. ANSWER: $\Lambda_1 = (10\tau/\sigma_0 L^2)$, $\Lambda_2 = (42\tau/\sigma_0 L^2)$. Compare with λ_1 and λ_2 derived in exercise 9(a).

(b) With $\Phi_j = (x/L)^j$ $(j = 1,2)$ use the method of 7-6(c), with $s = 2$, to approximate λ_1 and λ_2 for the uniform $(\sigma = \sigma_0)$ string that has its end at $x = 0$ fixed and its end at $x = L$ free. ANSWER: $\Lambda_1 = (2.49\tau/\sigma_0 L^2)$, $\Lambda_2 = (32.2\tau/\sigma_0 L^2)$. Compare with λ_1 and λ_2 derived in exercise 9(b).

(c) With $\Phi_j = (x/L)^{j-1}$ $(j = 1,2,3)$ use the method of 7-6(c), with $s = 3$, to approximate λ_1, λ_2, λ_3 for the uniform $(\sigma = \sigma_0)$ string with both ends free. ANSWER: $\Lambda_1 = \lambda_1 = 0$, $\Lambda_2 = (12\tau/\sigma_0 L^2)$, $\Lambda_3 = (60\tau/\sigma_0 L^2)$. Compare with λ_2 and λ_3 derived in exercise 9(c).

CHAPTER 8

THE STURM-LIOUVILLE
EIGENVALUE-EIGENFUNCTION PROBLEM

In this chapter we consider a slight generalization of the eigenvalue-eigenfunction problem met in the theory of the vibrating string (Chap. 7). To achieve this generalization we appeal to no physical problem for a starting point but, instead, deal with a problem formulated in purely analytic terms.

8-1. Isoperimetric Problem Leading to a Sturm-Liouville System

(a) We consider the problem of extremizing the quantity

$$I = \int_{x_1}^{x_2} (\tau \phi'^2 - \mu \phi^2) dx + a_1 [\phi(x_1)]^2 + a_2 [\phi(x_2)]^2 \tag{1}$$

with respect to continuously differentiable functions $\phi(x)$ which satisfy the normalization condition

$$\int_{x_1}^{x_2} \sigma \phi^2 \, dx = 1. \tag{2}$$

The given functions $\tau(x)$ and $\sigma(x)$ are continuous *positive* functions, with $\tau(x)$ continuously differentiable in $x_1 \leq x \leq x_2$; $\mu(x)$ is given as continuous in the interval. The given constants a_1 and a_2 are nonnegative.

In one aspect of the problem no conditions are imposed upon the eligible functions ϕ at the given end points x_1 and x_2; we call this the "free-end-point problem." In a second aspect we require that the eligible functions ϕ vanish at one (the "free-fixed problem") or both (the "fixed-end-point problem") end points. (If $x = x_j$ is a fixed end point, the term involving a_j in (1) does not appear for $j = 1$, 2, or both.)

(b) To facilitate bringing the problem of (a) above within the scope of the isoperimetric problem considered in 4-1 and 4-2(b) we introduce the continuously differentiable function $a = a(x)$ which is arbitrary to within the limitations $a(x_1) = -a_1$ and $a(x_2) = a_2$. We may thus rewrite (1) as

$$I = \int_{x_1}^{x_2} \left[\tau \phi'^2 - \mu \phi^2 + \frac{d}{dx} (a \phi^2) \right] dx. \tag{3}$$

Using the method of 4-1, we form, from the integrands of (3) and (2), the function

$$f^* = \tau\phi'^2 - \mu\phi^2 + \frac{d}{dx}(a\phi^2) - \lambda\sigma\phi^2, \tag{4}$$

where $-\lambda$ is an undetermined multiplier. Substitution of (4) into (14) of 4-1(b), with y replaced by ϕ, provides[1]

$$\frac{d}{dx}(\tau\phi') + (\mu + \lambda\sigma)\phi = 0 \tag{5}$$

as the differential equation which must be satisfied by any extremizing function for the problem of (a).

In the free-end-point problem, substitution of (4) into the general free end-point condition (18) of 4-2(b) yields the result $\tau\phi' + a\phi = 0$ at $x = x_1$, $x = x_2$, or—since by definition $a(x_1) = -a_1$, $a(x_2) = a_2$—

(i) $\tau_1\phi'(x_1) - a_1\phi(x_1) = 0,$
(ii) $\tau_2\phi'(x_2) + a_2\phi(x_2) = 0,$ (6)

where we write $\tau(x_1) = \tau_1$, $\tau(x_2) = \tau_2$. In a fixed-end-point problem we replace (6), of course, by the conditions

(i) $\phi(x_1) = 0,$ (ii) $\phi(x_2) = 0.$ (7)

In a free-fixed problem one condition from each of (6) and (7)—(i) from one, (ii) from the other—applies.

Equation (5), with the functions τ, σ, μ given, is called the *Sturm-Liouville differential equation*. This equation, together with an appropriate set of end-point conditions from among (6) and (7), constitutes a *Sturm-Liouville system*. Such a system is linear and homogeneous: If any function ϕ satisfies the system, so also does the function $k\phi$, where k is an arbitrary constant. Since $\sigma(x) > 0$ for $x_1 \leq x \leq x_2$, any Sturm-Liouville ϕ (not identically zero) may therefore be supposed, when necessary, to satisfy the normalization condition (2). The solution of a Sturm-Liouville system is an eigenvalue-eigenfunction problem, of which the main problem encountered in Chap. 7 is a special case. (In Chap. 7 we suppose $a_1 = a_2 = 0$, $\mu(x) = 0$, $\tau = $ constant.) The eigenvalues are those values of λ for each of which (5) has a solution that meets the specific end-point conditions of the problem; the corresponding solutions are the eigenfunctions.

[1] It follows from the result of 3-4(c) that the term $(d/dx)(a\phi^2)$ in (4) has no influence upon the Euler-Lagrange equation generated upon f^*. (We may, of course, verify this fact by direct substitution into (14) of 4-1(b).) Thus the specific character of $a(x)$ plays no role here.

It can be shown[1] that the eigenvalues constitute a discrete unbounded set $\lambda_1 < \lambda_2 < \cdots < \lambda_n < \cdots$, with one and only one[2] linearly independent eigenfunction corresponding to each. The expansion theorem of 7-4(b) applies to the Sturm-Liouville eigenfunctions with no alteration other than replacing $0 \leqq x \leqq L$ by $x_1 \leqq x \leqq x_2$. From this theorem we may develop a minimum characterization of the Sturm-Liouville eigenvalue-eigenfunction problem analogous to the characterization given in 7-4(c) for the vibrating-string problem (see exercise 5 at the end of this chapter). Proofs of the orthogonality of the Sturm-Liouville eigenfunctions are reserved for end-chapter exercises 2 and 3.

8-2. Transformation of a Sturm-Liouville System

(a) For various purposes it is often necessary to effect a change in the form exhibited by a differential equation. According to what sort of change is required, it may be accomplished by change of independent variable, change of dependent variable, or both, or merely by multiplying the equation by a suitable factor. As an example of the latter case we demonstrate the possibility of writing any linear homogeneous equation of second order in the so-called *self-adjoint form*—namely, in the form

$$\frac{d}{dx}(\tau\phi') + g(x)\phi = 0 \tag{8}$$

—through multiplication by a suitable factor.

We consider the given equation

$$p(x)\phi'' + q(x)\phi' + h(x)\phi = 0, \tag{9}$$

which we multiply by the function $s = s(x)$, at this point undetermined. Comparison with the expanded form of (8) reveals that (9) becomes self-adjoint if and only if $(d/dx)(sp) = sq$, whence $(s'/s) = (q/p) - (p'/p)$. Integrating, we obtain directly

$$s = \frac{1}{p}e^{\int \frac{q}{p}dx} \tag{10}$$

and so conclude that (9) becomes the self-adjoint equation

$$\frac{d}{dx}(sp\phi') + sh\phi = 0,$$

where s is given by (10). (We note that the Sturm-Liouville equation (5) possesses the self-adjoint form. See end-chapter exercise 1.)

---

[1] See Ince, Chap. 10, and also exercise 6 at the end of this chapter.

[2] See end-chapter exercise 4 for a proof of the nondegeneracy.

(b) The transformation of a given differential equation through a change of independent variable, a change of dependent variable, or both, may be effected, of course, through direct substitution. Quite often the relationships expressed in the transformation equations are not completely known but remain to be determined in a fashion designed to bring the differential equation into a particular desired form. In the case of the Sturm-Liouville equation (5), or any other equation of the form (8), a great simplification in the determination of the required transformation relationships is achieved through carrying out all changes of variable in the integral whose extremization leads to the differential equation. The simplification lies in the fact that, while the differential equation involves the second derivative, the integrand function involves only the first derivative of the dependent variable. Moreover, any transformation of end-point conditions arising from a change of variable may be effected simultaneously with the change of form of the differential equation if the integral, rather than the differential equation itself, serves as target for the substitutions.

According to (4) of 8-1(b) the integral whose extremization leads to the Sturm-Liouville system represented by (5) and (6) is

$$I^* = \int_{x_1}^{x_2} \left[\tau \phi'^2 - (\mu + \lambda \sigma) \phi^2 + \frac{d}{dx} (a\phi^2) \right] dx, \qquad (11)$$

where $a(x_1) = -a_1$, $a(x_2) = a_2$. We note in passing that the end-point conditions (6) are expressed in terms of the function $a(x)$—arbitrary to within differentiability and the assigned end-point values—as

$$\tau(x)\phi' + a(x)\phi = 0 \qquad \text{at } x = x_1, x = x_2. \qquad (12)$$

The most general type of transformation we seek to effect here involves the change of independent variable $x = x(z)$, with z the new independent variable, and the simultaneous change of dependent variable

$$\phi = u(z)w(z), \qquad (13)$$

with w the new dependent variable; the role of $u(z)$ is discussed in the paragraph following. The change $x = x(z)$ is restricted to functions $x(z)$ whose derivative with respect to z—denoted by $\dot{x}(z)$—is strictly positive, so that z is a single-valued function of x. Further, we make the convenient, although unessential, requirement that the change of independent variable be such that $x_1 = x(0)$ and $x_2 = x(\pi)$. Thus, as z increases continuously from 0 to π, x increases continuously from x_1 to x_2, and vice versa.

In the change of dependent variable (13) from ϕ to w the function $u(z)$ plays the role of a tool, as does the function $x(z)$, in our quest for any given new form of the Sturm-Liouville system.   We make the restriction $u(z) > 0$ for $0 \leqq z \leqq \pi$.   One consequence of this requirement, according to (13), is that the vanishing of $w(z)$ for a particular value of z in $0 \leqq z \leqq \pi$ is equivalent to the vanishing of $\phi(x)$ for the corresponding value of x in $x_1 \leqq x \leqq x_2$; the total number of zeros (vanishing points) possessed by $w(z)$ in $0 \leqq z \leqq \pi$ equals the total number of zeros possessed by $\phi(x)$ in $x_1 \leqq x \leqq x_2$.

Using the superior dot to indicate differentiation with respect to z, we obtain, according to (13), $\phi' = (\dot\phi/\dot x) = [(u\dot w + \dot u w)/\dot x]$.   Substituting this result, with (13), $x = x(z)$, and $dx = \dot x\, dz$, into (11), we obtain

$$I^* = \int_0^\pi \left\{ \frac{\tau u^2}{\dot x}\, \dot w^2 + \frac{2\tau u\dot u}{\dot x}\, w\dot w + \left[\frac{\tau \dot u^2}{\dot x} - \dot x u^2(\mu + \lambda\sigma) \right] w^2 \right.$$
$$\left. + \frac{d}{dz}\, (au^2 w^2) \right\}\, dz,   (14)$$

after a slight regrouping of terms.   In order to evolve (14) into the same *form* as (11) we employ the identity

$$\frac{2\tau u\dot u}{\dot x}\, w\dot w + \frac{\tau \dot u^2}{\dot x}\, w^2 = \frac{d}{dz}\left(\frac{\tau u\dot u}{\dot x}\, w^2 \right) - w^2 u \frac{d}{dz}\left(\frac{\tau \dot u}{\dot x} \right);$$

with this, (14) becomes

$$I^* = \int_0^\pi \left\{ \frac{\tau u^2}{\dot x}\, \dot w^2 - \left[u \frac{d}{dz}\left(\frac{\tau \dot u}{\dot x} \right) + \dot x u^2(\mu + \lambda\sigma) \right] w^2 \right.$$
$$\left. + \frac{d}{dz}\left[\left(au^2 + \frac{\tau u\dot u}{\dot x} \right) w^2 \right] \right\}\, dz.   (15)$$

With the definitions[1]

$$\text{(i)} \qquad T(z) = \frac{\tau u^2}{\dot x}, \qquad \text{(ii)}\ \ M(z) = u \frac{d}{dz}\left(\frac{\tau \dot u}{\dot x} \right) + \dot x u^2 \mu,$$

$$\text{(iii)} \qquad S(z) = \dot x u^2 \sigma, \qquad \text{(iv)}\ \ A(z) = au^2 + \frac{\tau u\dot u}{\dot x}, \tag{16}$$

equation (15) becomes

$$I^* = \int_0^\pi \left[T\dot w^2 - (M + \lambda S)w^2 + \frac{d}{dz}\, (Aw^2) \right] dz, \tag{17}$$

which, in *form*, is identical with (11).   Accordingly, the Sturm-Liouville system required for the extremization of (17) may be written down imme-

---

[1] It is assumed that μ, σ, τ are all expressed in terms of z through $x = x(z)$.

diately by inserting the appropriate changes of nomenclature into (5) of 8-1(b) and (12):

$$\frac{d}{dz}(T\dot{w}) + (M + \lambda S)w = 0, \tag{18}$$

$$T\dot{w} + Aw = 0 \qquad \text{at } z = 0, z = \pi. \tag{19}$$

(In a fixed-end-point problem the functions $a(x), A(z)$ play no role whatever. Because of (13) and $u(z) > 0$ the conditions $\phi(x_1) = \phi(x_2) = 0$ automatically become the conditions $w(0) = w(\pi) = 0$. If merely one of the conditions $\phi(x_1) = 0$ *or* $\phi(x_2) = 0$ is imposed (the free-fixed problem), we are concerned with the specific values $a(x_2) = a_2$ and $A(\pi)$ if the former is required, with $a(x_1) = -a_1$ and $A(0)$ if the latter is required.)

Explicit justification for the above method of transforming (5) into (18), and (12) into (19)—namely, proof that the transformed integrand function leads to the same differential equation and end-point conditions we should achieve on *direct* substitution of $x = x(z)$, $\phi = uw$ into (5) and (12)—is reserved for (*e*) below. There follow directly specific examples of the above transformations.

(*c*) It is of some importance[1] to obtain the transformed Sturm-Liouville equation (18) in the special form in which $T = 1$ and $S = K^2$, where K is some conveniently chosen positive constant. According to (16,i,iii) the desired transformation may be effected through the simultaneous satisfaction of $(\tau u^2/\dot{x}) = 1$ and $\dot{x}u^2\sigma = K^2$, from which follow

$$u = \frac{\sqrt{K}}{(\sigma\tau)^{\frac{1}{4}}}, \qquad z = \frac{1}{K}\int_{x_1}^{x}\sqrt{\frac{\sigma}{\tau}}\,dx, \qquad K = \frac{1}{\pi}\int_{x_1}^{x_2}\sqrt{\frac{\sigma}{\tau}}\,dx, \tag{20}$$

since $\dot{x} = (dx/dz)$, $x(0) = x_1$, and $x(\pi) = x_2$.

With the transformation (20) the Sturm-Liouville equation (5) reads, through (16) and (18),

$$\frac{d^2w}{dz^2} + \left[\frac{K^2\mu}{\sigma} - \frac{\frac{d^2}{dz^2}(\sigma\tau)^{\frac{1}{4}}}{(\sigma\tau)^{\frac{1}{4}}} + K^2\lambda\right]w = 0. \tag{21}$$

The relation between the old and the new dependent variable is, according to (13) and the first of (20), $\phi = [\sqrt{K}/(\sigma\tau)^{\frac{1}{4}}]w$. (We are not concerned here with the transformed end-point conditions (19), although explicit representation is readily achieved through (20).)

A tacit special requirement for the transformation to (21) is clearly the differentiability of the product $\sigma\tau$.

---

[1] See Ince, pp. 270–273, for example.

(d) Discussion of the properties of a Sturm-Liouville system—in particular, proof of the existence of an infinite unbounded sequence of eigenvalues of λ—is greatly simplified[1] if we deal with a transformed system in which $T = 1$ and $A(0) = A(\pi) = 0$. The latter condition is ensured by requiring $A(z) = 0$ identically in $0 \leqq z \leqq \pi$. According to (16,i,iv) we may effect the desired transformation by setting

$$\frac{\tau u^2}{\dot{x}} = 1, \qquad au^2 + \frac{\tau u\dot{u}}{\dot{x}} = 0, \tag{22}$$

from which it follows directly that $(\dot{u}/u^3) + a = 0$, or

$$u = \left(2 \int a \, dz\right)^{-\frac{1}{2}} \quad \text{and} \quad \int_{x_1}^{x} \frac{dx}{\tau} = \frac{1}{2} \int_0^z \left(\int a \, dz\right)^{-1} dz, \tag{23}$$

since $\dot{x} = (dx/dz)$ and $x(0) = x_1$.

We note that the function $a(x)$ which appears in (22) and (23) is arbitrary, except in so far as it is continuously differentiable and fulfills the end-point conditions $a(x_1) = -a_1$, $a(x_2) = a_2$, where a_1 and a_2 are given nonnegative constants. Since $z = 0$ when $x = x_1$, and $z = \pi$ when $x = x_2$, the end-point restrictions may be expressed as $a = -a_1$ when $z = 0$, and $a = a_2$ when $z = \pi$. The simplest—but by no means unique—form of a as a function of z which satisfies these requirements is the linear function

$$a = \frac{1}{\pi} [a_1(z - \pi) + a_2 z]; \tag{24}$$

we use the representation (24) in all that follows. Thus we have the indefinite integral

$$\int a \, dz = \frac{1}{2\pi} [a_1(z - \pi)^2 + a_2 z^2] + C$$
$$= \frac{a_1 + a_2}{2\pi} \left(z - \frac{a_1 \pi}{a_1 + a_2}\right)^2 + \frac{a_1 a_2 \pi}{2(a_1 + a_2)} + C, \tag{25}$$

where the constant of integration C is chosen in such fashion that $x = x_2$ corresponds with $z = \pi$ in the second of (23). We have, namely,

$$\int_{x_1}^{x_2} \frac{dx}{\tau} = \int_0^{\pi} \frac{dz}{\dfrac{a_1 + a_2}{\pi} \left(z - \dfrac{a_1 \pi}{a_1 + a_2}\right)^2 + \dfrac{a_1 a_2 \pi}{a_1 + a_2} + 2C}, \tag{26}$$

according to (23), (25), and $x(\pi) = x_2$.

[1] A rigorous existence proof is found in Ince, Chap. 10. See also exercise 6 at the end of this chapter.

That there exists a constant C which satisfies the relation (26) is clear from the following considerations: Since $\tau(x) > 0$ and $x_1 < x_2$, the left-hand member of (26) is positive. By taking C sufficiently large, the right-hand member can be made arbitrarily close to zero. By taking C sufficiently close to, but greater than $[-a_1a_2\pi/2(a_1 + a_2)]$, the right-hand member of (26) can be made arbitrarily large.   Since the right-hand member is a positive continuous function of C for $C > [-a_1a_2\pi/2(a_1 + a_2)]$, it therefore follows that there exists a value of C for which (26) is satisfied. (We note that the required value of C renders (25) positive—a condition made necessary by the first relation of (23).)

In any specific problem the value of C may be computed from (26) through some procedure of numerical approximation and x may be determined as a function of z, or vice versa, from the second of (23).   For our present purpose it is sufficient merely to demonstrate the existence of a transformation of the type introduced in (b) above for which the transformed differential equation (18) reads

$$\frac{d^2w}{dz^2} + (M + \lambda S)w = 0, \tag{27}$$

with the transformed end-point conditions (19) reading

$$\dot{w}(0) = \dot{w}(\pi) = 0. \tag{28}$$

(e) We provide here the explicit justification—called for in the final paragraph of (b) above—of the transformation method developed in (b) and exemplified in (c) and (d).   Into the integral

$$I = \int_{x_1}^{x_2} f^*(x,\phi,\phi')dx \tag{29}$$

we introduce the changes of variable $x = x(z)$ and $\phi = uw$ having the characteristics of the transformations described in (b). Thus since $dx = \dot{x}\, dz$, $\phi' = [(u\dot{w} + \dot{u}w)/\dot{x}]$, $x_1 = x(0)$, and $x_2 = x(\pi)$, the integral (29) becomes

$$I = \int_0^\pi f^*\left(x(z),uw,\frac{u\dot{w} + \dot{u}w}{\dot{x}}\right)\dot{x}\, dz = \int_0^\pi g^*(z,w,\dot{w})dz, \tag{30}$$

where $g^* = f^*\dot{x}$.

To construct the Euler-Lagrange equation required for the extremization of (30) we employ the relations $\phi = uw$, $\phi' = [(u\dot{w} + \dot{u}w)/\dot{x}]$ to form the derivatives

$$\frac{\partial g^*}{\partial w} = \frac{\partial(f^*\dot{x})}{\partial\phi}\frac{\partial\phi}{\partial w} + \frac{\partial(f^*\dot{x})}{\partial\phi'}\frac{\partial\phi'}{\partial w} = \dot{x}\left(\frac{\partial f^*}{\partial\phi}u + \frac{\partial f^*}{\partial\phi'}\frac{\dot{u}}{\dot{x}}\right) \tag{31}$$

and

$$\frac{d}{dz}\left(\frac{\partial g^*}{\partial \dot{w}}\right) = \dot{x}\frac{d}{dx}\left[\frac{\partial(f^*\dot{x})}{\partial \phi'}\frac{\partial \phi'}{\partial \dot{w}}\right] = \dot{x}\frac{d}{dx}\left(\frac{\partial f^*}{\partial \phi'}u\right) \tag{32}$$

Since $(\dot{u}/\dot{x}) = (du/dx)$, we have directly from (31) and (32) that

$$\frac{\partial g^*}{\partial w} - \frac{d}{dz}\left(\frac{\partial g^*}{\partial \dot{w}}\right) = \dot{x}u\left[\frac{\partial f^*}{\partial \phi} - \frac{d}{dx}\left(\frac{\partial f^*}{\partial \phi'}\right)\right]. \tag{33}$$

Because, as required, $\dot{x}u \neq 0$ in $0 \leq z \leq \pi$, it follows that the vanishing of the left-hand member of (33)—satisfaction of the Euler-Lagrange equation associated with the transformed integral (30)—is a concomitant of the vanishing of the bracketed portion of the right-hand member—satisfaction of the Euler-Lagrange equation associated with the original integral (29).

Moreover, since $(\partial g^*/\partial \dot{w}) = u(\partial f^*/\partial \phi')$, the transformation of the end-point conditions is correctly carried out through the integral substitution. For in the free-end-point problem we obtain $(\partial f^*/\partial \phi') = 0$ at $x = x_1, x_2$ as necessary for the extremization of (29), $(\partial g^*/\partial \dot{w}) = 0$ at $z = 0, \pi$ as necessary for the extremization of (30). (In the fixed-end-point problem the question of transformation of the end-point conditions does not arise.)

8-3. Two Singular Cases: Laguerre Polynomials, Bessel Functions

(a) In 8-1(a) requirements are set forth concerning the functions $\tau(x)$, $\mu(x)$, $\sigma(x)$ which appear in a given Sturm-Liouville equation. In particular we have $\tau(x) > 0$, $\mu(x)$ continuous (and therefore bounded) and $\sigma(x) > 0$ in the closed interval of definition $x_1 \leq x \leq x_2$. Moreover, the end points x_1 and x_2 are assumed to be finite. In this section we treat two cases, of use in chapters following, in which not all of these conditions are met; such cases are characterized as *singular*. We deal here with extremum problems whose solutions lead respectively to the well-known Laguerre polynomials and Bessel functions. Since these are treated adequately in the literature,[1] many of the results concerning them are stated below without proof. It is our main purpose merely to demonstrate the possibility of defining these functions within the framework of the calculus of variations.

In the singular cases considered, $\tau(x)$ vanishes at one or both end points. Since we choose the function $a(x)$, introduced in 8-1(b), to be zero at such end points, the applicable end-point condition (6) reads $\tau\phi' = 0$. It would therefore appear, at first glance, that the end-point vanishing of τ automatically effects the satisfaction of this condition for all values of

[1] See, for example, Jackson.

the parameter λ. It happens, however, that the vanishing of τ for a particular value of x introduces a singularity into the Sturm-Liouville differential equation, so that we cannot be certain *a priori* that a solution $\phi(x)$, or its derivative $\phi'(x)$, is bounded at that value of x. It becomes expedient, therefore, to rewrite the end-point condition $\tau\phi' = 0$ at $x = x_1$ and/or $x = x_2$ as

$$\lim_{x \to x_1} (\tau\phi') = 0 \quad \text{and/or} \quad \lim_{x \to x_2} (\tau\phi') = 0. \tag{34}$$

The eigenvalues are those values of λ for which solutions of the differential equation exist satisfying one or both of (34) as required, as well as any other independent end-point condition which is applicable.

It happens that conditions of the type (34) are equivalent, in the cases we consider below, to (i) the requirement that the eigenfunctions be bounded in the interval of definition, and (ii) the requirement of the existence of the integral whose extremization leads to the differential equation. Requirement (i) is crucial in the problems of the circular membrane (exercise 13, Chap. 9) and the circular plate (exercise 29, Chap. 10), for example; requirement (ii) finds its greatest significance in the study of quantum mechanics (Chap. 11).

(*b*) We consider the singular Sturm-Liouville system defined by

$$\tau = x^{k+1}e^{-x}, \quad \mu = 0, \quad \sigma = x^k e^{-x}, \quad a_1 = a_2 = 0, \quad x_1 = 0, \quad x_2 = +\infty,$$

where $k > -1$. The integral whose extremization leads to the system is

$$I^* = \int_0^\infty x^k e^{-x}(x\phi'^2 - \lambda\phi^2)dx. \tag{35}$$

The corresponding Euler-Lagrange equation (5) reads, on division by $x^k e^{-x}$,

$$x\phi'' + (k + 1 - x)\phi' + \lambda\phi = 0; \tag{36}$$

the end-point condition (34) at $x = \infty$ is

$$\lim_{x \to \infty} (x^{k+1}e^{-x}\phi') = 0. \tag{37}$$

Equation (36) has a solution for which (37) holds if and only if $\lambda = n$ ($n = 0,1,2, \ldots$); this solution (unnormalized) is

$$\phi_n = L_n^{(k)}(x) = \frac{e^x x^{-k}}{n!} \frac{d^n}{dx^n} (e^{-x}x^{n+k}) \qquad (k > -1; n = 0,1,2, \ldots). \tag{38}$$

The function $L_n^{(k)}(x)$ is clearly a polynomial—the Laguerre polynomial[1] of degree n (with upper index k). It is obvious that (38) satisfies the

---

[1] Some authors omit the $n!$ in the denominator of (38) in the definition of $L_n^{(k)}$.

left-hand end-point condition of (34)—namely, $\lim\limits_{x\to 0} (x^{k+1}e^{-x}\phi') = 0$—
since $k > -1$.

(c) We consider the singular Sturm-Liouville system in which

$$\tau = x, \quad \mu = -\frac{n^2}{x}, \quad \sigma = x, \quad a_1 = 0, \quad a_2 \geqq 0, \quad x_1 = 0, \quad x_2 > 0,$$

where n is a given nonnegative constant. The integral whose extremization leads to this system is

$$I^* = \int_0^{x_2} \left[x\phi'^2 + \left(\frac{n^2}{x} - \lambda x\right) \phi^2 + \frac{d}{dx} (a\phi^2) \right] dx, \tag{39}$$

where $a(0) = 0$, $a(x_2) = a_2$. We effect the change of independent variable $z = \sqrt{\lambda}\,x$, whence (39) becomes

$$I^* = \int_0^{x_2\sqrt{\lambda}} \left[z\dot\phi^2 + \left(\frac{n^2}{z} - z\right) \phi^2 + \frac{d}{dz} (a\phi^2) \right] dz; \tag{40}$$

the superior dot indicates differentiation with respect to z. The extremization of (40) leads to the Bessel differential equation

$$z^2 \frac{d^2\phi}{dz^2} + z \frac{d\phi}{dz} + (z^2 - n^2)\phi = 0 \qquad (0 \leqq z \leqq x_2\sqrt{\lambda}). \tag{41}$$

For given n, the Bessel equation (41) has only one linearly independent solution which satisfies the left-hand end-point condition $\lim\limits_{z\to 0} (z\phi) = 0$; this solution, the nth-order Bessel function of the first kind, is given by the infinite series

$$J_n(z) = \sum_{k=0}^{\infty} (-1)^k \frac{(\frac{1}{2}z)^{n+2k}}{k!\,\Gamma(n+k+1)}, \tag{42}$$

convergent for all z. (If n is integral, $\Gamma(n+k+1) = (n+k)!$; in general,

$$\Gamma(n+k+1) = \int_0^\infty t^{n+k}e^{-t}\, dt.\Big)$$

The applicable (unnormalized) solution of (41) is, accordingly,

$$\phi = J_n(z) = J_n(\sqrt{\lambda}\,x).$$

The eigenvalues of λ are determined through application of (6,ii) of 8-1(b)—namely, $x_2\phi'(x_2) + a_2\phi(x_2) = 0$, or, since $\phi = J_n(\sqrt{\lambda}\,x)$,

$$x_2\sqrt{\lambda}\,J_n'(\sqrt{\lambda}\,x_2) + a_2 J_n(\sqrt{\lambda}\,x_2) = 0. \tag{43}$$

The equation (43) has an infinite unbounded sequence of positive solutions $\lambda = \lambda_1, \lambda_2, \ldots, \lambda_m, \ldots$, which constitute the eigenvalues of the problem. In the important special case $a_2 = 0$, $\lambda_m = (j'_{nm}/x_2)^2$ for $m = 1, 2, 3, \ldots$, where j'_{nm} is the mth positive zero of $J'_n(z)$.

If, however, we impose upon the functions ϕ eligible for the extremization of (39) the condition $\phi(x_2) = 0$, the relation (43) is clearly replaced by $J_n(\sqrt{\lambda}\, x_2) = 0$ as the equation from which the sequence of eigenvalues of λ is determined. In this problem the eigenvalues are given by $\lambda_m = (j_{nm}/x_2)^2$, where j_{nm} is the mth positive zero of $J_n(z)$.

EXERCISES

1. Show that the Euler-Lagrange equation derived from any integrand of the form

$$f = q(x)\phi'^2 + 2r(x)\phi\phi' + p(x)\phi^2 + \frac{d}{dx}g(x,\phi)$$

is self-adjoint.

2. When f^* is given by (4) of 8-1(b), the relation (12) of 4-1(b), with y replaced by ϕ, holds for arbitrary differentiable η_j if ϕ is a Sturm-Liouville eigenfunction and λ the corresponding eigenvalue. Let ϕ_i and ϕ_k be eigenfunctions of a given Sturm-Liouville system which correspond to the distinct eigenvalues λ_i and λ_k. With f^* given by (4) of 8-1(b), write down (12) of 4-1(b) twice—once with $\phi = \phi_i$, $\lambda = \lambda_i$, $\eta_j = \phi_k$; and then with $\phi = \phi_k$, $\lambda = \lambda_k$, $\eta_j = \phi_i$. Combine the two results to prove the orthogonality

$$\int_{x_1}^{x_2} \sigma\phi_i\phi_k\, dx = 0 \qquad (i \neq k). \tag{44}$$

(Give explicit justification for these substitutions, particularly in the fixed-end case in which η_j is required to vanish at one or both end points.)

3. (a) Let ϕ_i, ϕ_k, λ_i, λ_k have the same meanings as in exercise 2. Use the Sturm-Liouville differential equation to prove that

$$\left[\tau(\phi_i\phi_k' - \phi_k\phi_i')\right]_{x_1}^{x_2} = (\lambda_i - \lambda_k)\int_{x_1}^{x_2} \sigma\phi_i\phi_k\, dx. \tag{45}$$

HINT: Compare derivation of (26) of 7-4(a).

(b) Use (45) to prove (44) for all permissible combinations of the end-point conditions (6) and (7).

4. Prove the nondegeneracy of the Sturm-Liouville eigenvalues—that there exists only one linearly independent eigenfunction to each eigenvalue, that is. HINT: Compare exercise 3, Chap. 7, but show that $\tau w = $ constant $=$ zero.

5. In the manner of 7-4(c) develop a minimum characterization of the Sturm-Liouville eigenvalue-eigenfunction problem. Base the development upon the expansion theorem—valid also for Sturm-Liouville eigenfunctions, with $0 \leq x \leq L$ replaced, by $x_1 \leq x \leq x_2$—of 7-4(b).

6. (a) By means of the transformed Sturm-Liouville system (27), (28) of 8-2(d) and through suitable adaptation of the method of exercise 2, Chap. 7, give a geometric-intuitive proof of the existence of an infinite unbounded sequence of Sturm-Liouville eigenvalues. In particular show that λ_1 is greater than minus the maximum of (M/S) in $0 \leq z \leq \pi$.

(b) Give the existence proof also for the fixed- and free-fixed end-point problems.

7. (a) Use the method of 8-2(a) to obtain the self-adjoint forms of the equations

(i) $$x^2\phi'' + x\phi' + h(x)\phi = 0,$$
(ii) $$x\phi'' + (b - x)\phi' + \phi = 0.$$

(b) Write down the integral whose extremization leads to the (Hermite) differential equation

$$\phi'' - 2x\phi' + \lambda\phi = 0 \qquad (-\infty \leq x \leq \infty). \tag{46}$$

8. (a) *Without change of independent variable* use the method of 8-2(b) to rewrite the equation (i) of exercise 7(a) in the form $w'' + g(x)w = 0$, where $\phi = uw$. (Give u and $g(x)$ explicitly, the latter in terms of $h(x)$.) HINT: Note that $\dot{x} = 1$, and use (16,i,ii).

(b) Work out the details of the transformation of 8-2(c) as applied to the (Legendre) equation

$$\frac{d}{dx}[(1 - x^2)\phi'] + \lambda\phi = 0 \qquad (-1 \leq x \leq 1). \tag{47}$$

ANSWER:

$$\frac{d^2w}{dz^2} + \left[\frac{1}{4}(1 + \csc^2 z) + \lambda\right]w = 0, \tag{48}$$

where $x = -\cos z$, $w = \phi\sqrt{\sin z}$. NOTE: The boundedness of u is violated at $z = 0, \pi$ ($x = \pm 1$); this circumstance arises because of the singular condition $\tau = 0$ for $x = \pm 1$. The singular character of (47) is carried over into (48) as the unboundedness of the function $M(z)$.

9. Show that the orthonormality relationship—(44) plus the normalization (2)—is carried over directly into

$$\int_0^\pi Sw_iw_k\, dz = \delta_{ik}$$

by the general transformation of 8-2(b).

10. (a) List the eigenvalues and corresponding eigenfunctions of the system $x^4\phi'' + \lambda\phi = 0$, $\phi(1) = \phi(2) = 0$. HINT: Use (16,i to iii) to find a simplifying transformation. NOTE: $\tau = 1$ here, not x^4. ANSWER: $\lambda_n = 4\pi^2 n^2$,

$$\phi_n = 2x \sin[2\pi n(x - 1)/x],$$

for $n = 1, 2, 3, \ldots$.

(b) Show that the equation

$$\frac{d}{dx}(\tau\phi') + \lambda\sigma\phi = 0$$

can always be transformed into $\ddot{w} + \lambda K^2 w = 0$ if $\tau\sigma[\int(1/\tau)dx]^4 = $ constant.

CHAPTER 9

SEVERAL INDEPENDENT VARIABLES: THE VIBRATING MEMBRANE

9-1. Extremization of a Multiple Integral

(a) We fix our attention on the triple integral[1]

$$I = \iiint_R f(x,y,z,w,w_x,w_y,w_z)dx\,dy\,dz \tag{1}$$

carried out over a definite region R of xyz space. The integrand function f, supposed given explicitly as a function of the arguments indicated, is continuously twice differentiable with respect to any combination of them. The problem before us is to find the differential equation that must be satisfied by the function $w = w(x,y,z)$ which renders I an extremum with respect to twice-differentiable functions which assume prescribed values at all points of the boundary surface B of the region R.[2]

To effect the extremization of (1) we employ the technique of 3-3(b) and 7-1: We introduce a one-parameter family of comparison functions $W(x,y,z)$ as

$$W = w(x,y,z) + \epsilon\eta(x,y,z), \tag{2}$$

where it is assumed that $w(x,y,z)$ is the actual extremizing function and ϵ is the parameter of the family. Thus no matter what the choice of the function $\eta(x,y,z)$—arbitrary to within continuous differentiability and

$$\eta(x,y,z) = 0 \qquad \text{on } B \tag{3}$$

—the extremizing function w is a member of each comparison family for the parameter value $\epsilon = 0$. The condition (3) ensures that every comparison function assumes the same set of values on the boundary surface B.

In (1) we replace w by W—with

$$W_x = w_x + \epsilon\eta_x, \qquad W_y = w_y + \epsilon\eta_y, \qquad W_z = w_z + \epsilon\eta_z, \tag{4}$$

---

[1] In this chapter, as in 7, we often employ subscripts to indicate partial derivatives— w_x for $(\partial w/\partial x)$, etc.

[2] We use "extremum" here in the sense of 3-3(c), with obvious extension to the case of functions of several variables.

132

according to (2)—and so form

$$I(\epsilon) = \iiint\limits_{R} f(x,y,z,W,W_x,W_y,W_z)dx\,dy\,dz. \tag{5}$$

Since, for all permissible choices of η, $\epsilon = 0$ implies that W reduces to the actual extremizing function, we have that $I(0)$ is the extremum sought, or

$$I'(0) = 0. \tag{6}$$

We differentiate (5) with respect to ϵ, using (2) and (4) to obtain $(\partial W/\partial\epsilon) = \eta$, $(\partial W_x/\partial\epsilon) = \eta_x$, etc.; we set $\epsilon = 0$, whereby W, W_x, etc., become w, w_x, etc.; and finally we use (6) to achieve

$$\iiint\limits_{R}\left(\frac{\partial f}{\partial w}\,\eta + \frac{\partial f}{\partial w_x}\,\eta_x + \frac{\partial f}{\partial w_y}\,\eta_y + \frac{\partial f}{\partial w_z}\,\eta_z\right)dx\,dy\,dz = 0. \tag{7}$$

Applying Green's theorem (29) of 2-14 to the final three terms of (7) we get, since $\eta = 0$ on B,

$$\iiint\limits_{R}\eta\left[\frac{\partial f}{\partial w} - \frac{\partial}{\partial x}\left(\frac{\partial f}{\partial w_x}\right) - \frac{\partial}{\partial y}\left(\frac{\partial f}{\partial w_y}\right) - \frac{\partial}{\partial z}\left(\frac{\partial f}{\partial w_z}\right)\right]dx\,dy\,dz = 0. \tag{8}$$

Since (8) holds for arbitrary η vanishing on B, we use the basic lemma of 3-1(c), as extended to triple integrals, to conclude that

$$\frac{\partial f}{\partial w} - \frac{\partial}{\partial x}\left(\frac{\partial f}{\partial w_x}\right) - \frac{\partial}{\partial y}\left(\frac{\partial f}{\partial w_y}\right) - \frac{\partial}{\partial z}\left(\frac{\partial f}{\partial w_z}\right) = 0. \tag{9}$$

This we call the *Euler-Lagrange differential equation* generated upon the integrand f of (1).

(b) A generalization of (9) comes from considering an n-tuple integral over a fixed region of an n-dimensional space. If the integrand function is $f = f(x,y,\,\ldots\,,z,w,w_x,w_y,\,\ldots\,,w_z)$, the function w of the n independent variables x, y, $\ldots$, z which extremizes the integral in question must satisfy the Euler-Lagrange equation

$$\frac{\partial f}{\partial w} - \frac{\partial}{\partial x}\left(\frac{\partial f}{\partial w_x}\right) - \frac{\partial}{\partial y}\left(\frac{\partial f}{\partial w_y}\right) - \cdots - \frac{\partial}{\partial z}\left(\frac{\partial f}{\partial w_z}\right) = 0. \tag{10}$$

The derivation of (10) may be accomplished either by a generalization of Green's theorem to n dimensions, or by the method called for in exercise 1 at the end of this chapter.

(c) Extension of the above results to isoperimetric problems is achieved in the manner of 4-1 and 7-1(c): If w is a function which extremizes (1) with respect to functions which satisfy the subsidiary conditions

$$\iiint_R g_j(x,y,z,w,w_x,w_y,w_z)dx\,dy\,dz = C_j \qquad (j = 1,2, \ldots ,s),$$

where C_1, C_2, . . . , C_s are constants, then w satisfies the Euler-Lagrange equation

$$\frac{\partial f^*}{\partial w} - \frac{\partial}{\partial x}\left(\frac{\partial f^*}{\partial w_x}\right) - \frac{\partial}{\partial y}\left(\frac{\partial f^*}{\partial w_y}\right) - \frac{\partial}{\partial z}\left(\frac{\partial f^*}{\partial w_z}\right) = 0,$$

where

$$f^* = f - \lambda_1 g_1 - \lambda_2 g_2 - \cdots - \lambda_s g_s. \tag{11}$$

The constants λ_1, λ_2, . . . , λ_s are undetermined Lagrange multipliers. If n independent variables are involved, so that we deal with n-tuple integrals, we are led to (10), with f replaced by f^*, given by (11).

(d) It follows from Green's theorem (28) of 2-14 that, if P, Q, R are three arbitrary continuously differentiable functions of x, y, z, w and if

$$f = \frac{\partial P}{\partial x} + \frac{\partial Q}{\partial y} + \frac{\partial R}{\partial z}, \tag{12}$$

the integral I, given by (1), depends only on the values assumed by P, Q, R on the boundary surface B. Since w is supposed prescribed on B, it follows that I achieves the same value for all permissible choices of w and the extremization problem is meaningless. We now show, in fact, that the Euler-Lagrange equation (9) reduces to an identity in case (12) holds.

We rewrite (12) through the relations

$$\frac{\partial P}{\partial x} = P_x + P_w w_x, \qquad \frac{\partial Q}{\partial y} = Q_y + Q_w w_y, \qquad \frac{\partial R}{\partial z} = R_z + R_w w_z, \tag{13}$$

where P_x represents the derivative of P with respect to x when y, z, *and* w are held constant—and similarly for P_y, P_z—while $(\partial P/\partial x)$ is meant to take into account the fact that P varies with x by dint of the fact that w varies with x, as well as through the explicit dependence of P on x— and similarly for $(\partial Q/\partial y)$, $(\partial R/\partial z)$.[1] With (13), (12) becomes

$$f = P_x + Q_y + R_z + P_w w_x + Q_w w_y + R_w w_z,$$

[1] At best, any notation for a partial derivative is inadequate when it is not clear what is being held constant during the process of differentiation. Here $(\partial P/\partial x)$ implies merely that y and z are held constant, while P_x calls for holding constant y, z, *and* w during the differentiation. The reader unfamiliar with such distinctions may be straightened out by the following example: If $P = x^2 yzw^3$, we have, according to the notation adopted here, $P_x = 2xyzw^3$, while $(\partial P/\partial x) = 2xyzw^3 + 3x^2 yzw^2 w_x$.

and we have

$$\frac{\partial f}{\partial w} = P_{xw} + Q_{yw} + R_{zw} + P_{ww}w_x + Q_{ww}w_y + R_{ww}w_z; \tag{14}$$

and since $(\partial f/\partial w_x) = P_w$,

$$\frac{\partial}{\partial x}\left(\frac{\partial f}{\partial w_x}\right) = P_{wx} + P_{ww}w_x, \tag{15}$$

which is analogous to the first of (13) with P replaced by P_w. With (14), (15), and corresponding expressions with (P,x) replaced by (Q,y) and then by (R,z), it is directly noted that (9) reduces to an identity, since $P_{xw} = P_{wx}$, etc.

An immediate corollary of the above result is the fact that we may add any expression of the form (12)—a so-called "divergence" expression —to the integrand f of (1) without altering the Euler-Lagrange equation (9) generated upon f. This follows because every term of (9) depends upon f linearly.

9-2. Change of Independent Variables. Transformation of the Laplacian

It is clear from the form of f in (1) that (9) is in general a partial differential equation of second order—a relationship, that is, which involves derivatives of the unknown w no higher than the second. We usually arrive at such a differential equation with x, y, z representing cartesian (rectangular) coordinates, but the geometrical configuration of the problem at hand quite often happens to be better suited to some other coordinate system. For example, if the region R in which the Euler-Lagrange equation is to be solved is bounded by a sphere, or by concentric spheres, spherical coordinates are the most suitable, etc.

Although it is possible—often with a tremendous amount of tedious computation—to transform the Euler-Lagrange equation from cartesian coordinates to some other coordinate system by direct substitution in the differential equation itself, we have in the technique of the calculus of variations a means for significantly reducing the amount of labor required to effect the transformation. The method we derive directly below is analogous to the technique employed in 8-2 for altering the form of the Sturm-Liouville differential equation. The advantage of this method lies in the fact that we need go no further than the transformation of *first* partial derivatives; direct transformation of the differential equation, on the other hand, entails the much more complicated computations of transformed *second* partial derivatives.

(a) For convenience we write (x_1,x_2,x_3) instead of (x,y,z) and let

$$x_1 = x_1(r_1,r_2,r_3), \qquad x_2 = x_2(r_1,r_2,r_3), \qquad x_3 = x_3(r_1,r_2,r_3) \qquad (16)$$

be the equations of transformation from the cartesian system (x_1,x_2,x_3) to a general system of coordinates (r_1,r_2,r_3). We suppose that the equations in (16) are twice continuously differentiable in the region R on which we fix our attention below and that the jacobian of the transformation[1]—namely,

$$D = \frac{\partial(x_1,x_2,x_3)}{\partial(r_1,r_2,r_3)} = \begin{vmatrix} \dfrac{\partial x_1}{\partial r_1} & \dfrac{\partial x_2}{\partial r_1} & \dfrac{\partial x_3}{\partial r_1} \\[2mm] \dfrac{\partial x_1}{\partial r_2} & \dfrac{\partial x_2}{\partial r_2} & \dfrac{\partial x_3}{\partial r_2} \\[2mm] \dfrac{\partial x_1}{\partial r_3} & \dfrac{\partial x_2}{\partial r_3} & \dfrac{\partial x_3}{\partial r_3} \end{vmatrix} \qquad (17)$$

—does not change sign, although it may vanish at certain exceptional points or along certain exceptional curves, in R.

We consider the triple integral

$$I = \iiint_R f(x_1,x_2,x_3,w,w_{x_1},w_{x_2},w_{x_3})dx_1\,dx_2\,dx_3, \qquad (18)$$

which we transform according to (16). By (16) of 2-8(f), (18) becomes, with the aid of (16) and (17),

$$I = \iiint_{R'} F(r_1,r_2,r_3,w,w_{r_1},w_{r_2},w_{r_3})|D|dr_1\,dr_2\,dr_3, \qquad (19)$$

where R' is the region R, but described by the variables (r_1,r_2,r_3), and

$$F(r_1,r_2,r_3w,w_{r_1},w_{r_2},w_{r_3}) = f(x_1,x_2,x_3,w,w_{x_1},w_{x_2},w_{x_3}).$$

In f, that is, we substitute for (x_1,x_2,x_3) and express the derivatives $(w_{x_1},w_{x_2},w_{x_3})$ explicitly in terms of $(w_{r_1},w_{r_2},w_{r_3})$ and (r_1,r_2,r_3), through (16), and so form the function F.

The transformation of the first partial derivatives required in the formation of F from f is accomplished most easily by solving the linear system

$$w_{r_j} = \sum_{i=1}^{3} w_{x_i} \frac{\partial x_i}{\partial r_j} \qquad (j = 1,2,3) \qquad (20)$$

for $(w_{x_1},w_{x_2},w_{x_3})$. The coefficients $(\partial x_i/\partial r_j)$ are computed directly from the transformation equations (16). A specific example of this computation is carried out in (c) below.

[1] See 2-8(f).

(b) Based upon w, which is *any* continuously twice-differentiable function of position—of (x_1,x_2,x_3) or of (r_1,r_2,r_3), with the correspondence established through (16)—we form the auxiliary one-parameter family of functions

$$W = w + \epsilon\eta,$$

where η is an arbitrary continuously differentiable function of position which vanishes on the boundary of R, and therefore on the boundary of R'; and ϵ is the parameter of the family.   In both (18) and (19) we replace w by W (in terms of the appropriate variables), w_{x_i} by $W_{x_i} = w_{x_i} + \epsilon\eta_{x_i}$ in (18), w_{r_i} by $W_{r_i} = w_{r_i} + \epsilon\eta_{r_i}$ in (19), and so form two distinct expressions for the same quantity $I(\epsilon)$.

The process of forming $I'(0)$—that is, differentiating $I(\epsilon)$ with respect to ϵ and setting $\epsilon = 0$—and applying Green's theorem is identical with the process carried out in 9-1(a) up to equation (9) inclusive.   Guided by this fact, we may directly write down and equate the two forms of $I'(0)$ computed from the equal integrals $I(\epsilon)$ defined in the preceding paragraph:

$$I'(0) = \iiint_{R} \eta \left[\frac{\partial f}{\partial w} - \sum_{i=1}^{3} \frac{\partial}{\partial x_i}\left(\frac{\partial f}{\partial w_{x_i}}\right) \right] dx_1\, dx_2\, dx_3$$

$$= \iiint_{R'} \eta \left[\frac{\partial F|D|}{\partial w} - \sum_{i=1}^{3} \frac{\partial}{\partial r_i}\left(\frac{\partial F|D|}{\partial w_{r_i}}\right) \right] dr_1\, dr_2\, dr_3. \qquad (21)$$

The final stage of the development is reached by transforming the middle member of (21) according to (16).   In the same way that (18) is evolved into (19), that is, the middle member of (21) may be made to read

$$I'(0) = \iiint_{R'} \eta \left[\frac{\partial f}{\partial w} - \sum_{i=1}^{3} \frac{\partial}{\partial x_i}\left(\frac{\partial f}{\partial w_{x_i}}\right) \right] |D|\, dr_1\, dr_2\, dr_3.$$

It therefore follows from (21) that

$$\iiint_{R'} \eta \left\{ \left[\frac{\partial f}{\partial w} - \sum_{i=1}^{3} \frac{\partial}{\partial x_i}\left(\frac{\partial f}{\partial w_{x_i}}\right) \right] |D| \right. $$
$$\left. - \left[\frac{\partial F|D|}{\partial w} - \sum_{i=1}^{3} \frac{\partial}{\partial r_i}\left(\frac{\partial F|D|}{\partial w_{r_i}}\right) \right] \right\} dr_1\, dr_2\, dr_3 = 0.$$

Since η is arbitrary in $R,'$ we apply the basic lemma of 3-1(c), extended to three independent variables, and so obtain, on dividing through by $|D|$ and dropping the no-longer-needed absolute-value symbol,

$$\frac{\partial f}{\partial w} - \sum_{i=1}^{3} \frac{\partial}{\partial x_i}\left(\frac{\partial f}{\partial w_{x_i}}\right) = D^{-1}\left[\frac{\partial fD}{\partial w} - \sum_{i=1}^{3} \frac{\partial}{\partial r_i}\left(\frac{\partial fD}{\partial w_{r_i}}\right)\right], \qquad (22)$$

where f replaces F along with the understanding that in the left-hand member f is expressed in terms of $(x_1, x_2, x_3, w, w_{x_1}, w_{x_2}, w_{x_3})$, while in the right-hand member f is expressed in terms of $(r_1, r_2, r_3, w, w_{r_1}, w_{r_2}, w_{r_3})$. Since w is completely arbitrary to within twice continuous differentiability, (22) constitutes an identity in w and its derivatives: The transformation (16) carries the Euler-Lagrange expression given by the left-hand member of (22) into the transformed expression given by the equal right-hand member.

(c) For an illustration of the use of (22) we employ

$$f = w_x^2 + w_y^2 + w_z^2, \qquad (23)$$

where we replace (x_1, x_2, x_3) by (x, y, z) and effect the transformation to spherical coordinates[1]

$$x = r \sin\theta \cos\phi, \qquad y = r \sin\theta \sin\phi, \qquad z = r \cos\theta, \qquad (24)$$

with

$$0 \le r < \infty, \qquad 0 \le \theta \le \pi, \qquad 0 \le \phi < 2\pi. \qquad (25)$$

Here we have $r_1 = r$, $r_2 = \theta$, $r_3 = \phi$. The jacobian of the transformation (24) is readily calculated to be, according to (17),

$$D = r^2 \sin\theta. \qquad (26)$$

From (20), with the aid of (24), we compute

$$w_r = w_x \sin\theta \cos\phi + w_y \sin\theta \sin\phi + w_z \cos\theta,$$
$$w_\theta = w_x r \cos\theta \cos\phi + w_y r \cos\theta \sin\phi - w_z r \sin\theta,$$
$$w_\phi = -w_x r \sin\theta \sin\phi + w_y r \sin\theta \cos\phi.$$

Solving this system, we have the required transformation of the first partial derivatives

$$w_x = w_r \sin\theta \cos\phi + w_\theta \frac{\cos\theta \cos\phi}{r} - w_\phi \frac{\sin\phi}{r \sin\theta},$$

$$w_y = w_r \sin\theta \sin\phi + w_\theta \frac{\cos\theta \sin\phi}{r} + w_\phi \frac{\cos\phi}{r \sin\theta}, \qquad (27)$$

$$w_z = w_r \cos\theta - w_\theta \frac{\sin\theta}{r}.$$

[1] See exercise 5 at end of chapter.

Squaring and adding, we obtain from (23) and (26)

$$fD = w_r^2 r^2 \sin \theta + w_\theta^2 \sin \theta + w_\phi^2 \frac{1}{\sin \theta}. \tag{28}$$

Finally, we use (23), (26), and (28) for substitution into (22), whence we achieve the identity

$$\Delta^2 w = \frac{1}{r^2} \left[\frac{\partial}{\partial r} \left(r^2 \frac{\partial w}{\partial r} \right) + \frac{1}{\sin \theta} \frac{\partial}{\partial \theta} \left(\sin \theta \frac{\partial w}{\partial \theta} \right) + \frac{1}{\sin^2 \theta} \frac{\partial^2 w}{\partial \phi^2} \right], \tag{29}$$

according to the definition[1] $\nabla^2 w = w_{xx} + w_{yy} + w_{zz}$ of the *laplacian* of w.

(*d*) The transformation (29) of the laplacian from cartesian to spherical coordinates is a special case of its transformation to a general system of curvilinear orthogonal coordinates. A system of coordinates designated by the variables (r_1, r_2, r_3) is said to be orthogonal if, through any point, the surfaces $r_1 =$ constant, $r_2 =$ constant, $r_3 =$ constant intersect at right angles. A necessary and sufficient condition that a given system be orthogonal is that the equations (16) of transformation from cartesian coordinates lead to the relationship

$$dx_1^2 + dx_2^2 + dx_3^2 = h_1^2 \, dr_1^2 + h_2^2 \, dr_2^2 + h_3^2 \, dr_3^2, \tag{30}$$

where we find, on direct computation from (16), that

$$h_j^2 = \sum_{i=1}^{3} \left(\frac{\partial x_i}{\partial r_j} \right)^2 \qquad (j = 1, 2, 3) \tag{31}$$

and that (30) implies

$$\sum_{i=1}^{3} \frac{\partial x_i}{\partial r_j} \frac{\partial x_i}{\partial r_k} = 0 \qquad \text{for } j \neq k. \tag{32}$$

In fact we may write (31) and (32) in combination as

$$\sum_{i=1}^{3} \frac{\partial x_i}{\partial r_j} \frac{\partial x_i}{\partial r_k} = h_j^2 \delta_{jk} \qquad (j, k = 1, 2, 3, \text{ independently}), \tag{33}$$

where δ_{jk} is the Kronecker delta, equal to zero for $j \neq k$ and equal to unity for $j = k$. The quantities h_1, h_2, h_3 are nonnegative functions of r_1, r_2, r_3.

It is our aim to express the transformation of the laplacian from car-

[1] See 2-12(*c*).

tesian to general curvilinear orthogonal coordinates in terms of h_1, h_2, h_3. We first obtain an auxiliary formula useful in achieving this aim.

On multiplying both sides of (33) by $(\partial r_j/\partial x_m)$ and summing with respect to j, we obtain, with the aid of the definition of δ_{jk},

$$\sum_{i=1}^{3} \frac{\partial x_i}{\partial r_k}\left[\sum_{j=1}^{3} \frac{\partial x_i}{\partial r_j}\frac{\partial r_j}{\partial x_m}\right] = h_k^2 \frac{\partial r_k}{\partial x_m}. \tag{34}$$

The sum in square brackets is clearly $(\partial x_i/\partial x_m)$, and is therefore zero if $i \neq m$ and unity if $i = m$—is equal, that is, to δ_{im}. Thus the only non-vanishing term in the sum over i in (34) is the one for which $i = m$, with the coefficient of $(\partial x_m/\partial r_k)$ equal to unity, so that (34) reads

$$\frac{\partial x_m}{\partial r_k} = h_k^2 \frac{\partial r_k}{\partial x_m} \qquad (k,m = 1,2,3, \text{ independently}). \tag{35}$$

Replacing (m,k) by (i,j), we use (35) to make (20) read

$$w_{x_i} = \sum_{j=1}^{3} \frac{w_{r_j}}{h_j^2}\frac{\partial x_i}{\partial r_j}. \tag{36}$$

It is seen in (c) above that the left-hand member of (22) becomes a constant times the laplacian if we choose

$$f = \sum_{i=1}^{3} w_{x_i}^2 \tag{37}$$

as in (23). With (36) equation (37) becomes

$$f = \sum_{i=1}^{3}\sum_{j=1}^{3}\sum_{k=1}^{3} \frac{w_{r_j}w_{r_k}}{h_j^2 h_k^2}\frac{\partial x_i}{\partial r_j}\frac{\partial x_i}{\partial r_k} = \sum_{j=1}^{3}\sum_{k=1}^{3} \frac{w_{r_j}w_{r_k}}{h_j^2 h_k^2}\left[\sum_{i=1}^{3} \frac{\partial x_i}{\partial r_j}\frac{\partial x_i}{\partial r_k}\right]$$

$$= \sum_{j=1}^{3}\sum_{k=1}^{3} \frac{w_{r_j}w_{r_k}}{h_j^2 h_k^2} h_j^2 \delta_{jk} = \sum_{j=1}^{3} \frac{w_{r_j}^2}{h_j^2}, \tag{38}$$

where the penultimate expression is obtained from its predecessor with the aid of (33) and the final form follows from the definition of δ_{jk}.

To evaluate the jacobian D in terms of (h_1, h_2, h_3) we multiply the determinant (17) by itself according to the rule given in 2-8(c). Thus D^2 is a three-by-three determinant given by

$$D^2 = \begin{vmatrix} \sum_{i=1}^{3} \dfrac{\partial x_i}{\partial r_1}\dfrac{\partial x_i}{\partial r_1} & \sum_{i=1}^{3} \dfrac{\partial x_i}{\partial r_2}\dfrac{\partial x_i}{\partial r_1} & \sum_{i=1}^{3} \dfrac{\partial x_i}{\partial r_3}\dfrac{\partial x_i}{\partial r_1} \\[2ex] \sum_{i=1}^{3} \dfrac{\partial x_i}{\partial r_1}\dfrac{\partial x_i}{\partial r_2} & \sum_{i=1}^{3} \dfrac{\partial x_i}{\partial r_2}\dfrac{\partial x_i}{\partial r_2} & \sum_{i=1}^{3} \dfrac{\partial x_i}{\partial r_3}\dfrac{\partial x_i}{\partial r_2} \\[2ex] \sum_{i=1}^{3} \dfrac{\partial x_i}{\partial r_1}\dfrac{\partial x_i}{\partial r_3} & \sum_{i=1}^{3} \dfrac{\partial x_i}{\partial r_2}\dfrac{\partial x_i}{\partial r_3} & \sum_{i=1}^{3} \dfrac{\partial x_i}{\partial r_3}\dfrac{\partial x_i}{\partial r_3} \end{vmatrix}$$

$$= \begin{vmatrix} h_1^2 & 0 & 0 \\ 0 & h_2^2 & 0 \\ 0 & 0 & h_3^2 \end{vmatrix} = h_1^2 h_2^2 h_3^2,$$

with the aid of (33). Thus, since h_1, h_2, h_3 are all positive, we have

$$D = h_1 h_2 h_3. \tag{39}$$

With (37), (38), and (39) the transformation equation (22) yields the result

$$\nabla^2 w = \frac{1}{h_1 h_2 h_3}\left[\frac{\partial}{\partial r_1}\left(\frac{h_2 h_3}{h_1}\frac{\partial w}{\partial r_1}\right) + \frac{\partial}{\partial r_2}\left(\frac{h_3 h_1}{h_2}\frac{\partial w}{\partial r_2}\right) + \frac{\partial}{\partial r_3}\left(\frac{h_1 h_2}{h_3}\frac{\partial w}{\partial r_3}\right)\right]. \tag{40}$$

To illustrate the use of (40) we use (24) of (c) above to compute $dx^2 + dy^2 + dz^2 = dr^2 + r^2\,d\theta^2 + r^2\sin^2\theta\,d\phi^2$, so that $h_1 = 1$, $h_2 = r$, $h_3 = r\sin\theta$. With $r_1 = r$, $r_2 = \theta$, $r_3 = \phi$, (40) leads directly to (29) of (c).

(e) In transforming Euler-Lagrange expressions which involve only two independent variables, we may use the result (22), but the sums over i run from $i = 1$ to $i = 2$, only. Similarly, we may use (40) to transform the two-dimensional laplacian $\nabla^2 w = w_{xx} + w_{yy}$ by suppressing the final term and by setting $h_3 = 1$ in the remainder of the formula; that is, we have

$$\nabla^2 w = \frac{1}{h_1 h_2}\left[\frac{\partial}{\partial r_1}\left(\frac{h_2}{h_1}\frac{\partial w}{\partial r_1}\right) + \frac{\partial}{\partial r_2}\left(\frac{h_1}{h_2}\frac{\partial w}{\partial r_2}\right)\right], \tag{41}$$

where (r_1, r_2) are plane curvilinear orthogonal coordinates, with

$$dx_1^2 + dx_2^2 = h_1^2\,dr_1^2 + h_2^2\,dr_2^2.$$

For example, in the transformation from cartesian to plane polar coordinates, we have

$$x_1 = r\cos\theta, \qquad x_2 = r\sin\theta, \tag{42}$$

where $r_1 = r, r_2 = \theta$. From (42) it follows that $dx_1^2 + dx_2^2 = dr^2 + r^2\, d\theta^2$, so that $h_1 = 1, h_2 = r$. Thus (41) becomes

$$\nabla^2 w = \frac{1}{r} \frac{\partial}{\partial r} \left(r \frac{\partial w}{\partial r} \right) + \frac{1}{r^2} \frac{\partial^2 w}{\partial \theta^2}. \tag{43}$$

9-3. The Vibrating Membrane

To derive the partial differential equation which describes the motion of a vibrating elastic membrane we appeal to Hamilton's principle, introduced in 6-2, so extended as to apply to a mechanical system in which the mass is distributed continuously. This is done, as in the case of the vibrating string (7-2), by considering a continuous mass distribution as a limiting case of a system composed of discrete mass particles— the case to which Hamilton's principle is initially held applicable. Mathematically, the limiting process is effected in a natural way by replacing sums over the particles of the discrete system by integrals over the mass distribution of the continuous system.

(a) We consider a thin elastic membrane extended over a given non-self-intersecting closed curve C in the xy plane. The boundary curve C is supposed to consist of a finite number of arcs along each of which the curvature is continuous. The plane domain D enclosed by C coincides with the equilibrium configuration of the membrane.

We suppose the membrane to be in a state of vibration in which each of its points undergoes a motion in a direction perpendicular to the xy plane. For the present we confine our attention to the case in which the boundary edge of the membrane is constrained to coincide with the curve C. Aside from those which hold the boundary edge in place, the only forces which influence the membrane motion are the elastic forces which arise from the deformation of the membrane relative to its plane equilibrium configuration.

The displacement from equilibrium at time t of a given membrane point whose motion occurs along a line characterized by particular values of x and y is denoted by $w(x,y,t)$. Thus the configuration of the membrane as a whole is described, at any instant t, by the function $w = w(x,y,t)$ of the three independent variables indicated; w may assume, of course, both positive and negative values, with $w = 0$ indicating a point of the membrane instantaneously in the xy plane. In particular we have

$$w(x,y,t) = 0 \text{ along } C \text{ (all } t), \tag{44}$$

in view of the above-imposed condition which fixes the boundary edge in the xy plane. Since the membrane is supposed free of slits, and since mechanical motion cannot occur discontinuously, w is a continuous function of x and y in D and of t, for all t. Moreover, since a discontinuity of velocity with respect to position would induce a tear in the membrane, we have that the time rate of displacement—designated by $(\partial w/\partial t) = \dot{w}(x,y,t)$—is continuous in D as well as being a continuous function of t. Finally, we suppose that the first and second partial derivatives of w with respect to the position coordinates—w_x, w_y, w_{xx}, w_{yy}, w_{xy}—are continuous in D.[1]

(b) We denote the mass per unit area of the membrane by the continuous positive density function $\sigma = \sigma(x,y)$. We confine the motion to amplitudes of vibration so small that σ is effectively independent of w. Since the velocity of the point at (x,y) is $\dot{w}(x,y,t)$ at any instant t, the kinetic energy per unit area is accordingly $\frac{1}{2}\sigma(x,y)[\dot{w}(x,y,t)]^2$. Integrating over D, we obtain

$$T = \frac{1}{2} \iint_D \sigma \dot{w}^2 \, dx \, dy \tag{15}$$

for the total kinetic energy of the membrane as a function of time.

The elastic potential energy of the membrane in any configuration is equal to the amount of work which is required in order to bring the membrane from equilibrium to the given configuration. Since the membrane is assumed to be so flexible as to give no resistance to bending, the work of deformation must be entirely owing to the increase of the membrane's area relative to the equilibrium area of the domain D. We proceed to compute this quantity of work under certain simplifying assumptions characteristic of the usual membrane theory:

Across any arc drawn in the surface of a stretched membrane, the portion of the membrane on one side of the arc exerts a normal stretch-resisting force on the portion lying on the other side. If there is no lateral motion of any point of the membrane, and if the elastic properties of the membrane are isotropic (independent of direction)—both of which assumptions are appropriate to our theory—the stretch-resisting force per unit arc length is a constant with respect to position; and if we deal with only small deformations, it is likewise constant with respect to time. This positive constant force per unit length—the so-called surface tension—we denote by the symbol τ. An elementary physical analysis shows that the quantity of work required to increase the area of the membrane by a small amount ΔA is given by $(\tau \Delta A)$:

[1] This restriction is partially removed below in 9-7(a) to provide for the possible exception of a finite number of isolated points at which, and a finite number of smooth arcs across which, finite discontinuities of the derivatives may occur.

In Fig. 9-1 we suppose that the membrane is initially plane and is bounded by C; after deformation, it is still plane and is bounded by C'. Let any line normal to C have intercepted on it by C and C' a small segment of length $\delta = \delta(s)$, where s denotes the arc coordinate on C measured from some fixed point on C. The quantity of work done by the stretching agency is clearly given, if δ is everywhere small compared with the linear dimensions of D, by

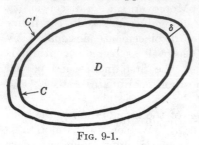

Fig. 9-1.

$$\int_0^{L_C} \tau\delta(s)ds = \tau \int_0^{L_C} \delta(s)ds = \tau\Delta A,$$

where L_C is the total length of C.

In any given configuration described by $w = w(x,y,t)$ the total area of the membrane is given[1] by

$$\iint_D \sqrt{1 + w_x^2 + w_y^2}\, dx\, dy.$$

Thus the potential energy of deformation is

$$V = \tau\left(\iint_D \sqrt{1 + w_x^2 + w_y^2}\, dx\, dy - \iint_D dx\, dy\right), \qquad (46)$$

where the second integral is the equilibrium area—the area of D.[2] We assume the deformation such that at every instant t the quantities w_x and w_y are so small that we may expand

$$\sqrt{1 + w_x^2 + w_y^2} = 1 + \tfrac{1}{2}(w_x^2 + w_y^2) + \cdots$$

and neglect without error the higher powers indicated by dots. With this assumption—which is the requirement of the usual theory that the deviation of the membrane from a plane figure is always slight—(46) becomes the working formula[3]

[1] See 2-9.

[2] Although the derivation upon which (46) is based takes into account no bending of the deformed membrane out of its original plane configuration, the fact that there is required no expenditure of work to bend the completely flexible membrane justifies the use of (46) here.

[3] The forces exerted by the external agency that keeps the boundary edge fixed involve no motion and so contribute nothing to the potential energy of the system.

$$V = \tfrac{1}{2}\tau \iint_D (w_x^2 + w_y^2)dx\,dy. \tag{47}$$

According to 6-1(d) we have from (45) and (47) that the lagrangian of the system constituted by the vibrating membrane is

$$L = T - V = \tfrac{1}{2}\iint_D [\sigma\dot{w}^2 - \tau(w_x^2 + w_y^2)]dx\,dy.$$

According to Hamilton's principle of 6-2 the integral

$$I = \int_{t_1}^{t_2} L\,dt = \tfrac{1}{2}\int_{t_1}^{t_2}\iint_D [\sigma\dot{w}^2 - \tau(w_x^2 + w_y^2)]dx\,dy\,dt \tag{48}$$

is extremized by the function $w(x,y,t)$ which describes the actual motion of the membrane; the extremization is effected with respect to functions w which vanish on C for all t, which describe the *actual* membrane configurations at $t = t_1$ and $t = t_2$, and which satisfy the regularity conditions set forth in (a) above.   The limits of integration t_1, t_2 are completely arbitrary.

To find the differential equation satisfied by the function $w(x,y,t)$ which extremizes (48) we may use the results of 9-1(a) if we replace z by t and, accordingly, w_z by $\dot{w}$.   The region R of 9-1(a) here becomes the cylindrical region generated by moving the domain D parallel to the "t direction" from $t = t_1$ to $t = t_2$.   The condition that w be prescribed on the boundary B of R is fulfilled here: By the imposed condition $w = 0$ on C we have that w vanishes on the cylindrical portion of B; since Hamilton's principle requires that w be prescribed at $t = t_1$ and $t = t_2$, the plane faces of B are taken care of.   Thus, with

$$f = \tfrac{1}{2}[\sigma\dot{w}^2 - \tau(w_x^2 + w_y^2)],$$

according to (48), the result (9) of 9-1(a) reads

$$\tau\nabla^2 w = \sigma\frac{\partial^2 w}{\partial t^2}, \tag{49}$$

where $\nabla^2 w = w_{xx} + w_{yy}$ is the two-dimensional laplacian of w.   We refer to (49) as the vibrating-membrane equation.

9-4.   Eigenvalue-Eigenfunction Problem for the Membrane

(a) The initial assault upon the membrane equation (49) consists of seeking a solution of the form

$$w = \phi(x,y)q(t), \tag{50}$$

where ϕ satisfies the boundary condition

$$\phi(x,y) = 0 \qquad \text{on } C, \tag{51}$$

—so that (44) of 9-3(a) is satisfied by (50)—and the normalization condition

$$\iint_D \sigma\phi^2 \, dx \, dy = 1. \tag{52}$$

From (50) it follows that

$$\nabla^2 w = q\nabla^2\phi, \qquad \frac{\partial^2 w}{\partial t^2} = \phi\ddot{q},$$

so that (49) becomes, on division by $\sigma w = \sigma\phi q$,

$$\frac{\tau\nabla^2\phi}{\sigma\phi} = \frac{\ddot{q}}{q}. \tag{53}$$

Since the left-hand member of (53) is independent of t, and since the right-hand member depends upon t alone, it follows that each member is a constant, which we denote by $-\lambda$. Thus we have inherent in (53) the pair of equations

$$\ddot{q} + \lambda q = 0 \tag{54}$$

and

$$\tau\nabla^2\phi + \lambda\sigma\phi = 0. \tag{55}$$

The values of λ are as yet undetermined.

With the fact that λ is positive, proved in (b) below, the general solution of (54) is directly found to be

$$q = A \cos \sqrt{\lambda}\, t + B \sin \sqrt{\lambda}\, t, \tag{56}$$

where A and B are arbitrary constants. We note that q is a periodic function of circular frequency (2π times frequency)

$$\omega = \sqrt{\lambda}, \tag{57}$$

so that (50) represents a membrane vibration which is periodic in time. Thus the determination of permissible values of λ constitutes the determination of the list of frequencies ($\omega/2\pi$) with which a given membrane is capable of executing periodic vibrations.

(b) Since (51) and (55) are homogeneous in ϕ, and because σ is a *positive* function, any nontrivial solution may be multiplied by a suitable constant in order that the normalization condition (52) be fulfilled.

The solution of (55) in the domain D, in conjunction with the boundary condition (51) on C, constitutes an eigenvalue-eigenfunction problem of the sort encountered in Chaps. 7 and 8.   Each value of λ for which (55) possesses a solution satisfying (51) is an eigenvalue of λ.

To prove that every eigenvalue of λ is positive—a fact used in the solution of (54) in (a) above—we multiply (55) by ϕ and integrate the resulting equation over D:

$$\lambda \iint_D \sigma\phi^2 \, dx \, dy = -\tau \iint_D \phi \nabla^2\phi \, dx \, dy. \qquad (58)$$

Because of the normalization (52) the left-hand member of (58) is simply λ.   Applying Green's theorem (23) of 2-13 to the right-hand member, we obtain from (58)

$$\lambda = \tau \iint_D (\phi_x^2 + \phi_y^2) dx \, dy - \tau \int_C \phi \frac{\partial \phi}{\partial n} \, ds, \qquad (59)$$

where the second integral on the right is a line integral taken along the boundary C of D and $(\partial\phi/\partial n)$ is the normal derivative of ϕ taken with respect to the direction exterior from D.   Because of the fact that $\phi = 0$ on C, however, the line integral vanishes and (59) shows that $\lambda \geqq 0$. For the equality to hold we must have $\phi_x = \phi_y = 0$, which implies that $\phi = $ constant in D; but since $\phi = 0$ on C, this means that $\phi = 0$ in D— a trivial solution, and therefore not an eigenfunction.   Thus we have $\lambda > 0$.

For purposes below we point out that (59) follows merely from the fact that ϕ satisfies (55) for the given value of λ; (59) in no way depends on the boundary condition placed upon ϕ.

(c) As in the case of the vibrating string, the eigenvalue-eigenfunction problem for the membrane may be characterized as an isoperimetric problem.   If we seek to extremize the integral

$$I = \tau \iint_D (\phi_x^2 + \phi_y^2) dx \, dy \qquad (60)$$

with respect to twice-differentiable functions ϕ which vanish on C and which satisfy the normalization condition

$$\iint_D \sigma\phi^2 \, dx \, dy = 1,$$

it follows directly from 7-1(c) that ϕ must satisfy the Euler-Lagrange equation

$$\tau \nabla^2 \phi + \lambda \sigma \phi = 0,$$

where λ plays the role of undetermined Lagrange multiplier. The identity with (55) is observed.

Comparison of (60) with (59), with $\phi = 0$ on C, shows that each extremized value of I is one of the eigenvalues of λ. In 9-9 below, this fact takes on a more precise meaning.

9-5. Membrane with Boundary Held Elastically. The Free Membrane

(a) We consider here the vibrating membrane which possesses all the characteristics of the membrane described in 9-3, with the exception that its boundary edge is not held in fixed position along the closed curve C in the xy plane. Instead, we suppose that the membrane edge is bound elastically to the curve C in such fashion that each point of the edge is free to move in a line through C perpendicular to the xy plane. The nature of the binding agency is such that it pulls each point of the boundary edge toward the point of C through which the point is free to move with a force proportional to the displacement of the point from the xy plane. Thus the equilibrium position of the membrane edge is the curve C.

With the arc length s measured along C from some fixed point on C we consider the binding force acting upon an arbitrary element of the membrane edge having the length ds in the equilibrium configuration. The binding force experienced by this element is $-p(s)w\,ds$, where the *positive* function $p(s)$ measures the strength of the binding along C and the minus sign indicates that the force opposes displacement from equilibrium. The potential energy associated with this force is

$$p(s)ds\textstyle\int w\,dw = \tfrac{1}{2}p(s)w^2\,ds, \tag{61}$$

where the arbitrary constant of integration is chosen so as to make the potential energy zero in equilibrium. Finally, the total binding potential energy, obtained by integrating (61) along C, is

$$V_B = \tfrac{1}{2}\int_C p(s)w^2\,ds, \tag{62}$$

where of course w is a function of s along C.

We suppose that the binding agency is such as to contribute negligibly to the kinetic energy of the system.

(b) If the integral (48) of 9-3(b) is to apply to the membrane under discussion here, the potential-energy term must be augmented to include the binding energy V_B given by (62). That is, the integral which is extremized by the function $w = w(x,y,t)$ that describes the actual membrane motion is

$$I = \tfrac{1}{2}\int_{t_1}^{t_2}\left\{\iint_D [\sigma\dot{w}^2 - \tau(w_x^2 + w_y^2)]dx\,dy - \int_C pw^2\,ds\right\}dt, \tag{63}$$

where the extremization is effected with respect to functions w which describe the actual membrane configurations at $t = t_1$ and $t = t_2$ and which satisfy the regularity conditions set forth in 9-3(a). *There is nothing in the mechanical problem under consideration which requires the imposition of any specific condition to be satisfied on the boundary curve C by the functions w eligible for the extremization.*

As a preliminary to the process of extremizing (63) we transform the line integral of (63) according to Green's theorem (21) of 2-13 as follows: We define the functions $P = P(x,y,t)$ and $Q = Q(x,y,t)$† as

$$P = \frac{1}{2}\,pw^2\,\frac{ds}{dy}, \qquad Q = -\frac{1}{2}\,pw^2\,\frac{ds}{dx} \qquad \text{on } C, \qquad (64)$$

and otherwise arbitrary—to within continuous twice differentiability—in the domain D. With (64), we have

$$\int_C pw^2\,ds = \frac{1}{2}\int_C \left(pw^2\,\frac{ds}{dy}\,dy + pw^2\,\frac{ds}{dx}\,dx\right) = \int_C (P\,dy - Q\,dx)$$

$$= \iint_D \left(\frac{\partial P}{\partial x} + \frac{\partial Q}{\partial y}\right)dx\,dy, \qquad (65)$$

according to (21) of 2-13. With (65) we may rewrite (63) as

$$I = \frac{1}{2}\int_{t_1}^{t_2}\iint_D \left[\sigma\dot{w}^2 - \tau(w_x^2 + w_y^2) - \left(\frac{\partial P}{\partial x} + \frac{\partial Q}{\partial y}\right)\right]dx\,dy\,dt, \qquad (66)$$

an integral carried out over the cylindrical region R of xyt space described in 9-3(b) just below (48).

Since there is a portion of the boundary B of R—namely, the cylindrical surface generated by the motion of C in the t direction from $t = t_1$ to $t = t_2$—on which the functions w eligible for the minimization are unrestricted, we cannot without alteration apply the result of 9-1(a); in that section it is required that eligible functions w be prescribed everywhere on B. We may, however, use (7) of 9-1(a), which is achieved without special assumption concerning the values of the arbitrary function η on B; we must, of course, replace z by t and w_z by $\dot{w}$ in (7). Thus we have

$$\int_{t_1}^{t_2}\iint_D \left(\frac{\partial f}{\partial w}\,\eta + \frac{\partial f}{\partial w_x}\,\eta_x + \frac{\partial f}{\partial w_y}\,\eta_y + \frac{\partial f}{\partial \dot{w}}\,\dot{\eta}\right)dx\,dy\,dt = 0, \qquad (67)$$

where

$$f = \frac{1}{2}\left[\sigma\dot{w}^2 - \tau(w_x^2 + w_y^2) - \left(\frac{\partial P}{\partial x} + \frac{\partial Q}{\partial y}\right)\right], \qquad (68)$$

† The derivatives (ds/dy) and (ds/dx) have reference to the curve C.

according to (66); η is arbitrary to within continuous differentiability and

$$\eta(x,y,t_1) = \eta(x,y,t_2) = 0, \tag{69}$$

according to the requirement of Hamilton's principle that the eligible functions w be prescribed at $t = t_1$ and $t = t_2$.

In the final term of (67) we carry out the integration over t first; integrating by parts, we get

$$\int_{t_1}^{t_2} \iint_D \frac{\partial f}{\partial \dot{w}} \dot{\eta} \, dx \, dy \, dt = \iint_D \int_{t_1}^{t_2} \frac{\partial f}{\partial \dot{w}} \dot{\eta} \, dt \, dx \, dy$$

$$= \iint_D \left\{ \left[\frac{\partial f}{\partial \dot{w}} \eta \right]_{t=t_1}^{t=t_2} - \int_{t_1}^{t_2} \eta \frac{\partial}{\partial t} \left(\frac{\partial f}{\partial \dot{w}} \right) dt \right\} dx \, dy$$

$$= - \iint_D \int_{t_1}^{t_2} \eta \frac{\partial}{\partial t} \left(\frac{\partial f}{\partial \dot{w}} \right) dt \, dx \, dy, \tag{70}$$

because of (69).

The second and third terms of (67) are transformed by means of (22) of 2-13 to give

$$\int_{t_1}^{t_2} \iint_D \left(\frac{\partial f}{\partial w_x} \eta_x + \frac{\partial f}{\partial w_y} \eta_y \right) dx \, dy \, dt$$

$$= - \int_{t_1}^{t_2} \iint_D \eta \left[\frac{\partial}{\partial x} \left(\frac{\partial f}{\partial w_x} \right) + \frac{\partial}{\partial y} \left(\frac{\partial f}{\partial w_y} \right) \right] dx \, dy \, dt$$

$$+ \int_{t_1}^{t_2} \int_C \eta \left[\frac{\partial f}{\partial w_x} \frac{dy}{ds} - \frac{\partial f}{\partial w_y} \frac{dx}{ds} \right] ds \, dt. \tag{71}$$

With (70) and (71) equation (67) becomes

$$\int_{t_1}^{t_2} \left\{ \iint_D \eta \left[\frac{\partial f}{\partial w} - \frac{\partial}{\partial x} \left(\frac{\partial f}{\partial w_x} \right) - \frac{\partial}{\partial y} \left(\frac{\partial f}{\partial w_y} \right) - \frac{\partial}{\partial t} \left(\frac{\partial f}{\partial \dot{w}} \right) \right] dx \, dy \right.$$

$$\left. + \int_C \eta \left[\frac{\partial f}{\partial w_x} \frac{dy}{ds} - \frac{\partial f}{\partial w_y} \frac{dx}{ds} \right] ds \right\} dt = 0 \tag{72}$$

for all η satisfying (69). In particular (72) holds for those η which vanish on C; for such η, only the triple integral remains, and we may apply the basic lemma of 3-1(c), extended to multiple integrals, to conclude that

$$\frac{\partial f}{\partial w} - \frac{\partial}{\partial x} \left(\frac{\partial f}{\partial w_x} \right) - \frac{\partial}{\partial y} \left(\frac{\partial f}{\partial w_y} \right) - \frac{\partial}{\partial t} \left(\frac{\partial f}{\partial \dot{w}} \right) = 0 \qquad \text{in } D, \tag{73}$$

for all t. With (73), and with η once again arbitrary on C, examination of (72) in light of 3-1(c) yields the further result

$$\frac{\partial f}{\partial w_x}\frac{dy}{ds} - \frac{\partial f}{\partial w_y}\frac{dx}{ds} = 0 \qquad \text{on } C. \tag{74}$$

(c) The membrane equation (49) is obtained in 9-3(b) by substituting

$$f = \tfrac{1}{2}[\sigma \dot{w}^2 - \tau(w_x^2 + w_y^2)] \tag{75}$$

into (9) of 9-1(a), with appropriate changes of notation.   If we now substitute (68) into (73), *which is identical with* (9), we must get precisely the same equation of motion as (49)—namely,

$$\tau \nabla^2 w = \sigma \frac{\partial^2 w}{\partial t^2}. \tag{76}$$

For (68) differs from (75) merely by an expression of the form $\tfrac{1}{2}[(\partial P/\partial x) + (\partial Q/\partial y)]$; according to 9-1($d$), with $R = 0$, the addition of such an expression to (68) can have no effect on the resulting Euler-Lagrange equation. The equation (76) describes the motion of the membrane of this section as well as that of the membrane with fixed boundary edge.

The influence of the final two terms of (68) is expressed when f is substituted into (74) for the derivation of the boundary condition which must be satisfied by the function w that describes the membrane motion. Since we have[1]

$$\frac{\partial P}{\partial x} = P_x + P_w w_x, \qquad \frac{\partial Q}{\partial y} = Q_y + Q_w w_y,$$

(68) becomes

$$f = \tfrac{1}{2}[\sigma \dot{w}^2 - \tau(w_x^2 + w_y^2) - (P_x + Q_y) - (P_w w_x + Q_w w_y)],$$

so that

$$\frac{\partial f}{\partial w_x} = -\tau w_x - \frac{1}{2}P_w, \qquad \frac{\partial f}{\partial w_y} = -\tau w_y - \frac{1}{2}Q_w,$$

or

$$\frac{\partial f}{\partial w_x} = -\tau w_x - \frac{1}{2}pw\frac{ds}{dy}, \qquad \frac{\partial f}{\partial w_y} = -\tau w_y + \frac{1}{2}pw\frac{ds}{dx} \qquad \text{on } C, \tag{77}$$

as we find with the aid of (64).   Substitution of (77) into (74) yields

$$\tau\left(w_x\frac{dy}{ds} - w_y\frac{dx}{ds}\right) + pw = 0 \qquad \text{on } C,$$

[1] See (13) of 9-1(d), with the accompanying footnote.

or, with the use of (13) of 2-7(d),

$$\tau \frac{\partial w}{\partial n} + pw = 0 \qquad \text{on } C; \tag{78}$$

($\partial w/\partial n$) is the normal derivative of w taken with respect to the direction exterior from D.

By way of summary, we have that the function $w = w(x,y,t)$ which describes the motion of the membrane whose edge is held elastically satisfies the equation (76) in D and the condition (78) on the boundary. The positive function $p = p(s)$, introduced in (a) above, is a measure of the strength of the force distribution which binds the boundary edge.

(d) We attack the equation (76), with the condition (78), in much the same way as we handle the corresponding problem of the membrane with fixed boundary edge in 9-4(a): We seek a solution of the form

$$w = \phi(x,y)q(t). \tag{79}$$

As in 9-4(a), it follows directly that

$$q = A \cos \sqrt{\lambda}\, t + B \sin \sqrt{\lambda}\, t, \tag{80}$$

with A and B arbitrary constants, and that ϕ satisfies

$$\tau \nabla^2 \phi + \lambda \sigma \phi = 0 \qquad \text{in } D, \tag{81}$$

with the boundary condition—derived directly from (78) and (79)—

$$\tau \frac{\partial \phi}{\partial n} + p\phi = 0 \qquad \text{on } C. \tag{82}$$

The eigenvalues of λ are the values of λ for which (81) possesses a solution which satisfies (82); such a solution is an eigenfunction, upon which we impose the normalization condition

$$\iint_D \sigma \phi^2 \, dx \, dy = 1. \tag{83}$$

To prove that λ is positive—a fact upon which the form (80) of $q(t)$ depends—we use (59) of 9-4(b), which is valid for any ϕ that satisfies (81) and (83). If, further, ϕ satisfies (82), then (59) reads

$$\lambda = \tau \iint_D (\phi_x^2 + \phi_y^2) dx \, dy + \int_C p\phi^2 \, ds, \tag{84}$$

which is clearly positive, since $\tau > 0$ and $p > 0$.

As in 9-4(c), we can show that the eigenvalue-eigenfunction problem under discussion may be set up as an isoperimetric problem. It is left for exercise 6(b) at the end of this chapter to prove that the functions which extremize

$$I = \tau \iint_D (\phi_x^2 + \phi_y^2)dx\,dy + \int_C p\phi^2\,ds$$

with respect to sufficiently regular ϕ for which the normalization (83) holds, *but upon which no boundary conditions are imposed*, must satisfy (81) and (82). A fuller significance of this fact is brought out in 9-9 below.

(e) Upon introduction of the binding-force-distribution function $p(s)$ in (a) above, it is assumed that p is everywhere positive along C. We may, however, remove this restriction by supposing that $p(s)$ may vanish[1] over any portion, or over all, of C. At those points of C at which $p(s) = 0$ the membrane edge is completely free of external constraint with regard to its motion perpendicular to the xy plane.[2] In particular if $p(s) = 0$ *identically* along C, the membrane edge is completely free and we speak of the *free membrane*. The physical realization of the free membrane, of course, is somewhat doubtful, but we consider it here for its mathematical interest alone.

With $p = 0$ identically on C the boundary condition for the free-membrane eigenfunctions becomes, according to (82),

$$\frac{\partial \phi}{\partial n} = 0 \qquad \text{on } C, \tag{85}$$

and from (81) it follows that $\phi = $ constant is an eigenfunction corresponding to the eigenvalue $\lambda = 0$. It is easily shown,[3] however, that the time-dependent factor $q(t)$ is not periodic when $\lambda = 0$, so that this eigenvalue is of no interest from the standpoint of vibrations; we cannot, however, ignore it completely, as we see below in 9-9.

(f) A membrane may have portions of its boundary edge held fixed to the equilibrium curve C, while the remaining portions are bound elastically in the manner described in (a) above. A slight modification of the analysis of (b) shows that the eigenfunctions are required to satisfy $\phi = 0$ on those portions of C to which the edge is held fixed and to satisfy $\tau(\partial\phi/\partial n) + p\phi = 0$ on the remaining portions of C, with $p \geqq 0$. The differential equation satisfied by the eigenfunctions is (81) in any case.

(g) Boundary conditions of the type considered in this chapter—either $\phi = 0$ on C, or (82), with $p \geqq 0$, or a mixture of both—are called homo-

---

[1] Nowhere, however, is $p(s)$ negative.

[2] It is always supposed, however, that each point of the edge is constrained to move only in a straight line through C perpendicular to the xy plane.

[3] See end-chapter exercise 10.

geneous in that any ϕ which satisfies them may be multiplied by an arbitrary constant without violating the boundary conditions. Since the same is true of any function which satisfies the associated differential equation $\tau \nabla^2 \phi + \lambda \sigma \phi = 0$, the membrane problem as we consider it is a linear homogeneous problem. It is this fact, of course, which makes possible the imposition of the convenient normalization condition (83).

9-6. Orthogonality of the Eigenfunctions. Expansion of Arbitrary Functions

(a) In the Sturm-Liouville eigenvalue-eigenfunction problem—including the vibrating-string problem as a special case—which is handled in Chap. 8, there corresponds to each eigenvalue of the parameter λ one and only one linearly independent[1] eigenfunction. In dealing with eigenvalue-eigenfunction problems involving two or more independent variables—such as we have in the membrane problem—we find, however, that to each eigenvalue of λ there may correspond one or more than one linearly independent eigenfunction. An eigenvalue to which there correspond two or more such eigenfunctions is called a multiple, or *degenerate*, eigenvalue; otherwise it is said to be simple, or *nondegenerate*.

The fact of possible degeneracy requires special attention in the discussion of the orthogonality of the membrane eigenfunctions in (b) and (c) below. Specific examples of degeneracy are observed in 9-8 and end-chapter exercise 13, in which we treat respectively the rectangular and circular membranes of uniform density.

(b) We prove, first, that the membrane eigenfunctions which correspond to distinct eigenvalues are orthogonal in the domain D with respect to the positive weight function $\sigma = \sigma(x,y)$. Next, in (c) below, we show that the independent membrane eigenfunctions which correspond to the same degenerate eigenvalue are always capable of being chosen so as to satisfy the same orthogonality relationship. We prove, that is, that

$$\iint_D \sigma \phi_j \phi_k \, dx \, dy = 0 \qquad (j \neq k),\tag{86}$$

where ϕ_j and ϕ_k are any two linearly independent membrane eigenfunctions which correspond to the same eigenvalue or to distinct eigenvalues.[2]

We have, according to (55) of 9-4(a), or the identical (81) of 9-5(d), that

$$\tau \nabla^2 \phi_j + \lambda_j \sigma \phi_j = 0, \qquad \tau \nabla^2 \phi_k + \lambda_k \sigma \phi_k = 0,\tag{87}$$

---

[1] See 2-8(e) for the definitions of linear independence and linear dependence.

[2] We deal, of course, with eigenfunctions of a *particular* membrane, for which σ, D (with boundary C), τ, and the binding function $p(s)$ are all given. Moreover, the boundary conditions are the same for each eigenfunction.

where λ_j and λ_k are the eigenvalues of λ which correspond respectively to ϕ_j and ϕ_k. If the membrane boundary edge is held fixed, we have $\phi_j = \phi_k = 0$ on C; if the boundary edge is held elastically (or is free, whereby $p = 0$), we have

$$\tau \frac{\partial \phi_j}{\partial n} + p\phi_j = 0, \qquad \tau \frac{\partial \phi_k}{\partial n} + p\phi_k = 0 \qquad \text{on } C, \tag{88}$$

according to (82) of 9-5(d). We suppose that $\lambda_j \neq \lambda_k$.

We multiply the first of (87) by ϕ_k and the second by ϕ_j, subtract, then integrate the result over D, to obtain

$$(\lambda_j - \lambda_k) \iint\limits_{D} \sigma \phi_j \phi_k \, dx \, dy = \tau \iint\limits_{D} (\phi_j \nabla^2 \phi_k - \phi_k \nabla^2 \phi_j) dx \, dy$$

$$= \tau \int_C \left(\phi_j \frac{\partial \phi_k}{\partial n} - \phi_k \frac{\partial \phi_j}{\partial n} \right) ds, \tag{89}$$

by virtue of Green's formula (24) of 2-13(d). In the case $\phi_j = \phi_k = 0$ on C the final member of (89) vanishes; the same fact holds if (88) is applicable, as we find on substituting from it for the normal derivatives. In either event the orthogonality (86) follows from (89), since $\lambda_j \neq \lambda_k$.

(c) We consider the linearly independent eigenfunctions $u_1, u_2, \ldots, u_N$ associated with a single degenerate eigenvalue λ_k; there is no further eigenfunction corresponding to λ_k which is linearly independent of the N functions listed.[1] Because, as it is pointed out in 9-5(g), the eigenvalue-eigenfunction problem for the membrane is linearly homogeneous, any arbitrary linear combination of $u_1, u_2, \ldots, u_N$ is also an eigenfunction which corresponds to the eigenvalue λ_k. We now show that it is always possible to construct a set of N linear combinations of $u_1, u_2, \ldots, u_N$ which form an orthogonal set in D with respect to the weight function σ:

We consider the functions $v_1, v_2, \ldots, v_N$, defined successively in terms of the given $u_1, u_2, \ldots, u_N$, through the relations

$$v_1 = u_1, \qquad v_2 = a_{21}v_1 + u_2, \qquad v_3 = a_{31}v_1 + a_{32}v_2 + u_3, \qquad \ldots,$$
$$v_m = a_{m1}v_1 + a_{m2}v_2 + \cdots + u_m, \qquad \ldots, \tag{90}$$
$$v_N = a_{N1}v_1 + a_{N2}v_2 + \cdots + u_N,$$

with the coefficients a_{mi} ($i = 1, 2, \ldots, m - 1; m = 2, 3, \ldots, N$) determined according to the needs of the orthogonality as follows: We have

$$\iint\limits_{D} \sigma v_1 v_2 \, dx \, dy = a_{21} \iint\limits_{D} \sigma v_1^2 \, dx \, dy + \iint\limits_{D} \sigma v_1 u_2 \, dx \, dy,$$

[1] That N must always be finite is shown in 9-12(d) below to be a consequence of the asymptotic formula for the membrane eigenvalues.

which vanishes, as required, if

$$a_{21} = -\frac{\iint_D \sigma v_1 u_2 \, dx \, dy}{\iint_D \sigma v_1^{\,2} \, dx \, dy}.$$

With v_2 thus determined as orthogonal to v_1, we evaluate the coefficients a_{31} and a_{32} so that

$$\iint_D \sigma v_1 v_3 \, dx \, dy = 0, \qquad \iint_D \sigma v_2 v_3 \, dx \, dy = 0 \tag{91}$$

are satisfied.   The first of (91) yields

$$a_{31} \iint_D \sigma v_1^2 \, dx \, dy + \iint_D \sigma v_1 u_3 \, dx \, dy = 0,$$

while the second of (91) gives

$$a_{32} \iint_D \sigma v_2^2 \, dx \, dy + \iint_D \sigma v_2 u_3 \, dx \, dy = 0.$$

The process is continued; with $v_1, v_2, \ldots, v_{m-1}$ thus determined as an orthogonal set, the $(m-1)$ coefficients in (90) are evaluated so as to make v_m orthogonal to each of $v_1, v_2, \ldots, v_{m-1}$.   It is left for exercise 11 at the end of this chapter to show that the required evaluation is

$$a_{mi} = -\frac{\iint_D \sigma v_i u_m \, dx \, dy}{\iint_D \sigma v_i^2 \, dx \, dy} \qquad (i = 1, 2, \ldots, m-1). \tag{92}$$

The process of orthogonalization—the so-called Schmidt process—is completed with $m = N$.

Each of the orthogonal set of functions $v_1, v_2, \ldots, v_m$ thus determined may be multiplied by a suitable constant in order that the normalization condition (83) of 9-5(d) be satisfied by every member of the set. With the result of (b) above we are therefore justified in assuming that the totality of eigenfunctions associated with a given membrane problem —$\phi_1, \phi_2, \phi_3, \ldots$—constitutes an orthonormal set in D with respect to σ as weight function:

$$\iint_D \sigma \phi_j \phi_k \, dx \, dy = \delta_{jk}. \tag{93}$$

We note that the Schmidt orthogonalization process does not depend upon the fact that u_1, u_2, . . . , u_N are membrane eigenfunctions, but merely on the fact that the functions listed are linearly independent in D. The same process can be carried out for any set, finite or infinite, of linearly independent functions of any number of variables if the domain— or interval, in the case of one independent variable—and the positive weight function are given.

(*d*) Let ϕ_1, ϕ_2, . . . , ϕ_m, . . . be the totality of the orthonormal eigenfunctions associated with a given membrane problem. Let $g(x,y)$ be an arbitrary continuous function defined in the membrane domain D; g is such that D may be divided into a finite number of subdomains by a finite set of smooth arcs such that the first partial derivatives g_x and g_y are continuous in each subdomain. Then, if we write

$$g(x,y) = \sum_{m=1}^{\infty} c_m \phi_m(x,y), \qquad c_m = \iint_D \sigma \phi_m g \, dx \, dy, \qquad (94)$$

the series converges uniformly to $g(x,y)$ in every subdomain of D in which $g(x,y)$ is continuous. The formula of (94) for the coefficients c_m follows directly[1] from the orthonormality (93).

Further, in every subdomain of D in which g_x and g_y are continuous, the series

$$g_x = \sum_{m=1}^{\infty} c_m \frac{\partial \phi_m}{\partial x}, \qquad g_y = \sum_{m=1}^{\infty} c_m \frac{\partial \phi_m}{\partial y} \qquad (95)$$

converge uniformly to g_x and g_y, respectively.

(To avoid going beyond the scope of our study, the above results are stated without proof.)

If the eigenfunctions $\phi_m(x,y)$ used in the expansion (94) all vanish on the boundary C of D, the series for g converges to zero on C. Thus at those points of C at which $g \neq 0$ the representation breaks down and the function represented by the series is discontinuous, and the convergence is nonuniform in the neighborhood. If, however, $g = 0$ everywhere on C, no such difficulty is incurred.

If the eigenfunctions $\phi_m(x,y)$ used in the expansion (94) satisfy the boundary condition $\tau(\partial\phi_m/\partial n) + p\phi_m = 0$ on C, the series converges to $g(x,y)$ on C, but the derivative series (95) do not in general converge respectively to g_x and g_y on C unless also $\tau(\partial g/\partial n) + pg = 0$ on C. Difficulties such as those mentioned in this and the preceding paragraph play no role in our study.

[1] See exercise 15 at end of chapter.

9-7. General Solution of the Membrane Equation

(a) Through the validity of the expansion theorem stated in 9-6(d) we are led to a solution of the membrane equation

$$\tau\nabla^2 w = \sigma\,\frac{\partial^2 w}{\partial t^2},\tag{96}$$

which is sufficiently general to embrace at least all cases which are of physical interest. Leaving the problem of the membrane with boundary edge held fixed for exercise 12(a) at the end of this chapter, we consider here the problem of the membrane whose boundary edge is held elastically to the curve C in the xy plane. We have, accordingly,

$$\tau\,\frac{\partial w}{\partial n} + pw = 0 \qquad\text{on } C,\tag{97}$$

with $p = p(s)$ a given positive function, by (78) of 9-5(c).

As in the vibrating-string problem handled in 7-5, determination of the general solution is based upon the prior complete solution of the associated eigenvalue-eigenfunction problem; that is, we have at our disposal the totality of the eigenvalues and orthonormal eigenfunctions of the system

$$\tau\nabla^2\phi_m + \lambda_m\sigma\phi_m = 0 \qquad\text{in } D,\tag{98}$$

$$\tau\,\frac{\partial\phi_m}{\partial n} + p\phi_m = 0 \qquad\text{on } C,\tag{99}$$

for $m = 1, 2, 3, \ldots\ldots$ (If any given eigenvalue is degenerate, it appears consecutively in the list $\lambda_1, \lambda_2, \ldots, \lambda_m, \ldots$ a number of times equal to the number of independent eigenfunctions associated with it. We suppose that the eigenvalues are numbered in ascending order, so that $\lambda_m \leqq \lambda_{m+1}$ for all m.)

At this point we weaken the restrictions upon the solution $w = w(x,y,t)$ of (96) which are set forth in 9-3(a). It is not necessary for further purposes to keep the restriction of continuity of the partial derivatives w_x and w_y everywhere in D: We allow a finite number of isolated points at which, and a finite number of smooth arcs across which, finite discontinuities of w_x and w_y may occur. Since the conditions for expanding w and $\dot w$, for any t, as infinite series of eigenfunctions, according to 9-6(d), are clearly met, we may write

$$w(x,y,t) = \sum_{m=1}^{\infty} c_m\phi_m(x,y), \qquad \dot w(x,y,t) = \sum_{m=1}^{\infty} d_m\phi_m(x,y),\tag{100}$$

where, according to (94),

$$c_m = c_m(t) = \iint_D \sigma \phi_m w \, dx \, dy, \qquad d_m = d_m(t) = \iint_D \sigma \phi_m \dot{w} \, dx \, dy. \quad (101)$$

From (101), with the rule for differentiating an integral, it follows that $d_m = \dot{c}_m$; the second of (100) therefore reads

$$\dot{w}(x,y,t) = \sum_{m=1}^{\infty} \dot{c}_m \phi_m(x,y) \quad (102)$$

—the term-by-term time derivative of the first of (100). Further, we have

$$w_x = \sum_{m=1}^{\infty} c_m \frac{\partial \phi_m}{\partial x}, \qquad w_y = \sum_{m=1}^{\infty} c_m \frac{\partial \phi_m}{\partial y}. \quad (103)$$

According to (63) of 9-5(b) Hamilton's principle calls for the extremization of the integral

$$I = \tfrac{1}{2} \int_{t_1}^{t_2} \left\{ \iint_D [\sigma \dot{w}^2 - \tau(w_x^2 + w_y^2)] dx \, dy - \int_C p w^2 \, ds \right\} dt \quad (104)$$

by the function $w(x,y,t)$ which describes the actual membrane motion, with respect to functions w which fit the actual membrane configurations at $t = t_1$ and $t = t_2$. Into one factor of each term of (104) we substitute from (102), (103), and (100), as appropriate:

$$I = \tfrac{1}{2} \sum_{m=1}^{\infty} \int_{t_1}^{t_2} \left\{ \dot{c}_m \iint_D \sigma \phi_m \dot{w} \, dx \, dy \right.$$

$$\left. - c_m \left[\tau \iint_D \left(w_x \frac{\partial \phi_m}{\partial x} + w_y \frac{\partial \phi_m}{\partial y} \right) dx \, dy + \int_C p w \phi_m \, ds \right] \right\} dt. \quad (105)$$

We transform the second integral over D according to Green's theorem (23) of 2-13 to obtain

$$\iint_D \left(w_x \frac{\partial \phi_m}{\partial x} + w_y \frac{\partial \phi_m}{\partial y} \right) dx \, dy = \int_C w \frac{\partial \phi_m}{\partial n} \, ds - \iint_D w \nabla^2 \phi_m \, dx \, dy.$$

Thus (105) becomes, since the coefficient of $\dot{c}_m$ in (105) is $d_m = \dot{c}_m$, according to (101),

$$
I = \frac{1}{2} \sum_{m=1}^{\infty} \int_{t_1}^{t_2} \left\{ \dot{c}_m^2 + c_m \left[\tau \iint_D w \nabla^2 \phi_m \, dx \, dy - \int_C w \left(\tau \frac{\partial \phi_m}{\partial n} + p \phi_m \right) ds \right] \right\} dt
$$

$$
= \frac{1}{2} \sum_{m=1}^{\infty} \int_{t_2}^{t_2} \left\{ \dot{c}_m^2 - c_m \lambda_m \iint_D \sigma \phi_m w \, dx \, dy \right\} dt = \frac{1}{2} \sum_{m=1}^{\infty} \int_{t_1}^{t_2} (\dot{c}_m^2 - \lambda_m c_m^2) dt,
$$

$$
\tag{106}
$$

with the aid of (98), (99), and finally the first of (101).

Thus the extremization of (104) is reduced to the extremization of the final form of (106) with respect to the infinite set of quantities c_1, c_2, . . . , c_m, . . . , which are prescribed at $t = t_1$ and $t = t_2$, according to the requirements of Hamilton's principle. This extremization problem is worked out in 7-5(a), with the result

$$
c_m = A_m \cos \sqrt{\lambda_m}\, t + B_m \sin \sqrt{\lambda_m}\, t \qquad (m = 1,2,3, \ldots),
$$

where A_m and B_m are arbitrary constants. According to (100), therefore, we have for the general solution of the membrane problem characterized analytically by (96) and (97)

$$
w(x,y,t) = \sum_{m=1}^{\infty} (A_m \cos \sqrt{\lambda_m}\, t + B_m \sin \sqrt{\lambda_m}\, t) \phi_m(x,y). \tag{107}
$$

Evaluation of the coefficients A_m, B_m through the imposition of initial ($t = 0$) conditions is left for end-chapter exercise 12(c).

(b) The result (107) justifies the analysis of any given state of vibration of a membrane as a linear superposition of vibrations, each of which is characterized by a single frequency. Comparison with (79) and (80) of 9-5(d) shows that each term of (107) corresponds to one of the single-frequency modes of vibration which the membrane is capable of executing.

9-8. The Rectangular Membrane of Uniform Density

(a) The problem of determining the eigenvalues and corresponding eigenfunctions for a given membrane is in general tractable only if the boundary curve C is so shaped and the density function σ so constituted that the differential equation

$$
\tau \nabla^2 \phi + \lambda \sigma \phi = 0 \tag{108}
$$

lends iself to a separation of variables. First, we must choose a coordinate system in which the entire boundary curve is describable through constant values of the coordinate variables. In the case of a rectangle of sides a and b, for example, we use the cartesian system whose origin is at one of the corners and along whose positive axes two sides of the rectangle extend. In this case the boundary curve is describable by the sequence of constant values $y = 0$, $x = a$, $y = b$, $x = 0$. For a circular membrane of radius c, the equation of the boundary is simply $r = c$ in a system of polar coordinates whose origin is at the center of the circle.

Once a coordinate system is determined to suit the boundary needs, the next requirement is that the partial differential equation (108) separate into two *ordinary* differential equations through substitution into it of a product function—a function of one of the coordinate variables times a function of the other. Separation is effected when the equation is put into a form such that each of its two members depends respectively on one or the other, alone, of the independent variables. In this event the two members must equal an undetermined constant, whereby two ordinary differential equations result.

It may happen that the form of the function σ is such as to make a separation of variables impossible; but even if σ is a constant, there is only a limited number of coordinate systems in which separation may occur. In the event of inseparability the quest for a precise solution of the problem is in general hopeless, so that methods of approximation must be resorted to. Discussion of a method of approximation is found in 9-13 below.

(b) In this section we illustrate the precise solution of a membrane eigenvalue-eigenfunction problem with the rectangular membrane of uniform density. If the rectangle side lengths are a and b, we set up a cartesian coordinate system so that the membrane is bounded by $x = 0$, $x = a$, $y = 0$, $y = b$. For the fixed-edge membrane, which we consider first, the boundary conditions thus read

$$\phi(0,y) = \phi(a,y) = 0, \qquad \phi(x,0) = \phi(x,b) = 0. \tag{109}$$

Into (108), with $\sigma = \sigma_0$, a constant, we substitute the product function

$$\phi(x,y) = X(x)Y(y) \tag{110}$$

—whereby the conditions (109) become

$$X(0) = X(a) = 0, \qquad Y(0) = Y(b) = 0. \tag{111}$$

Since it follows from (110) that $\nabla^2\phi = X''Y + XY''$ (where primes indicate differentiation with respect to the appropriate independent variable), (108) becomes $\tau(X''Y + XY'') + \lambda\sigma_0 XY = 0$, or

$$\frac{X''}{X} + \frac{\lambda\sigma_0}{\tau} = -\frac{Y''}{Y}. \tag{112}$$

Since the left-hand member of (112) is a function of x alone while the right-hand is independent of x, it follows that each member is a constant, which we denote by β. Thus (112) results in the two ordinary equations

$$\frac{d^2X}{dx^2} + \alpha X = 0, \qquad \frac{d^2Y}{dy^2} + \beta Y = 0, \qquad (113)$$

where we introduce the constant $\alpha = (\lambda\sigma_0/\tau) - \beta$, so that

$$\lambda = \frac{\tau}{\sigma_0}(\alpha + \beta). \qquad (114)$$

The permissible values of α and β—and thereby the eigenvalues of λ— are determined directly.

For $\alpha < 0$ the solutions of the first of (113) are real exponentials, which cannot be combined so as to vanish both at $x = 0$ and $x = a$. Thus we have $\alpha \geqq 0$, with the solution

$$X(x) = C \cos \sqrt{\alpha}\, x + D \sin \sqrt{\alpha}\, x. \qquad (115)$$

Imposing the first pair of conditions (111), we obtain

$$X(0) = C = 0, \qquad X(a) = D \sin \sqrt{\alpha}\, a = 0,$$

whence it follows that $\sqrt{\alpha}\, a = m\pi$, with m an integer, or

$$\alpha = \frac{m^2\pi^2}{a^2} \qquad (m = 1,2,3, \ldots); \qquad (116)$$

we ignore the trivial solution which arises if $D = 0$ or $\alpha = 0$. The corresponding functions (115) are, for each m,

$$X_m = D_m \sin \frac{m\pi x}{a} \qquad (m = 1,2,3, \ldots).$$

In similar fashion we obtain for the functions $Y(y)$ which satisfy the second set of conditions (111)

$$Y_k = E_k \sin \frac{k\pi y}{b} \qquad (k = 1,2,3, \ldots),$$

with the corresponding values

$$\beta = \frac{k^2\pi^2}{b^2} \qquad (k = 1,2,3, \ldots). \qquad (117)$$

From (116), (117), and (114) we have for the eigenvalues of λ

$$\lambda_{mk} = \frac{\tau\pi^2}{\sigma_0}\left(\frac{m^2}{a^2} + \frac{k^2}{b^2}\right) \qquad (m,k = 1,2,3, \ldots, \text{independently}), \qquad (118)$$

with the corresponding eigenfunctions

$$\phi_{mk} = F_{mk} \sin \frac{m\pi x}{a} \sin \frac{k\pi y}{b}, \qquad (119)$$

for $m,k = 1, 2, 3, \ldots$, independently. The constants F_{mk} may be determined by the requirements of normalization.

(c) The attack upon the eigenvalue-eigenfunction problem for the *free* membrane is handled in much the same way. We seek a solution of the form (110), but the boundary conditions (111) are here replaced by

$$X'(0) = X'(a) = 0, \qquad Y'(0) = Y'(b) = 0. \qquad (120)$$

For, the condition[1] $(\partial\phi/\partial n) = 0$ on C reads, for the rectangle, $(\partial\phi/\partial x) = 0$ on $x = 0$, $x = a$, and $(\partial\phi/\partial y) = 0$ on $y = 0$, $y = b$; (120) follows directly.

Applying the first of (120) to (115), we obtain

$$X'(0) = \sqrt{\alpha}\, D - 0, \qquad X'(a) = -\sqrt{\alpha}\, C \sin \sqrt{\alpha}\, a = 0,$$

whence $\sqrt{\alpha}\, a = m\pi$, with m an integer, or

$$\alpha = \frac{m^2\pi^2}{a^2} \qquad (m = 0,1,2, \ldots).$$

Similarly, we find $\beta = (k^2\pi^2/b^2)$, so that the eigenvalues of λ are, according to (114),

$$\lambda_{mk} = \frac{\tau\pi^2}{\sigma_0}\left(\frac{m^2}{a^2} + \frac{k^2}{b^2}\right) \qquad (m,k = 0,1,2, \ldots, \text{independently}), \qquad (121)$$

with the corresponding eigenfunctions

$$\phi_{mk} = F_{mk} \cos \frac{m\pi x}{a} \cos \frac{k\pi y}{b}. \qquad (122)$$

It is significant that we may choose the values $m = 0$ and $k = 0$ in the free-edge problem, whereas these values must be ignored in the fixed-edge case. The reason for the difference is made clear on comparison of the eigenfunctions (119) and (122): Setting m or k equal to zero in (119) yields the trivial solution $\phi = 0$, while setting m, k, or both, equal to zero in (122) leaves us with a nontrivial solution—a constant in the extreme case $m = k = 0$.

(d) It is obvious from (119) and (122) that an eigenvalue exhibits degeneracy, defined above in 9-6(a), if there exist four integers m, k, m', k' $(m \neq m', k \neq k')$ such that $\lambda_{mk} = \lambda_{m'k'}$. It is immediately seen

[1] See 9-5(e).

from the form of (118) and (121) that a membrane for which the ratio (a/b) is rational possesses an infinite number of degenerate eigenvalues. The most apparent type of degeneracy arises in the case of the square membrane, $a = b$. Mere interchange of (unequal) values of m and k leads to a new eigenfunction without altering the eigenvalue.

If $(a/b)^2$ is irrational, every eigenvalue for the rectangular membrane is nondegenerate.

9-9. The Minimum Characterization of the Membrane Eigenvalues

(a) We proceed to prove the following theorem concerning the eigenvalues of the vibrating-membrane problem associated with the domain D, the density function σ, the elastic constant τ, and having its boundary edge either held fixed, or elastically bound,[1] with the binding function p, to the boundary curve C:

We arrange the totality of the eigenvalues in the ascending order $\lambda_1 \leqq \lambda_2 \leqq \cdots \leqq \lambda_k \leqq \cdots$, with each degenerate eigenvalue appearing consecutively in the list a number of times equal to the number of independent eigenfunctions associated with it. *The kth eigenvalue λ_k is the minimum of the quantity*

$$I = \tau \iint_D (\phi_x^2 + \phi_y^2)dx\,dy + \int_C p\phi^2\,ds \tag{123}$$

with respect to those functions ϕ which satisfy the normalization condition

$$\iint_D \sigma\phi^2\,dx\,dy = 1 \tag{124}$$

and the $(k - 1)$ orthogonality relations

$$\iint_D \sigma\phi_m\phi\,dx\,dy = 0 \qquad (m = 1,2,\ldots,k-1), \tag{125}$$

where ϕ_m $(m = 1,2,3,\ldots)$ is the eigenfunction which satisfies

$$\tau\nabla^2\phi_m + \lambda_m\sigma\phi_m = 0 \qquad \text{in } D \qquad (m = 1,2,3,\ldots) \tag{126}$$

and

$$\tau\frac{\partial\phi_m}{\partial n} + p\phi_m = 0 \qquad \text{on } C \qquad (m = 1,2,3,\ldots). \tag{127}$$

Further, the functions ϕ eligible for the minimization must be continuous everywhere in D and have partial derivatives ϕ_x and ϕ_y which are continuous, except possibly for a finite number of isolated points at which, and a

---

[1] "Elastically bound" includes the case of the free edge, with the binding function p identically zero on C. The theorem of this section also includes the mixed case in which part of the boundary edge is held fixed, part held elastically (or free).

finite number of smooth arcs across which, finite discontinuities of the derivatives may occur.

For the membrane with fixed boundary edge the *additional* restriction $\phi = 0$ on C must be imposed on the eligible functions; in this case the boundary line integral of (123) clearly drops out. Also, the condition (127) upon the eigenfunctions ϕ_m appearing in (125) is replaced by $\phi_m = 0$ on C. (*No* boundary restrictions are imposed upon the eligible functions ϕ if the membrane edge is held elastically.)

The minimum λ_k of I under the stated restrictions is achieved when $\phi = \phi_k$.

(*b*) The proof of the theorem stated in (*a*) runs along the line of the corresponding proof of the minimum character of the vibrating-string eigenvalues carried out in 7-4(*c*). We use the eigenfunctions of the system (126) and (127) to expand the eligible functions ϕ according to the expansion theorem of 9-6(*d*):

$$\phi = \sum_{m=1}^{\infty} c_m \phi_m(x,y) \qquad \left(c_m = \iint_D \sigma \phi_m \phi \, dx \, dy \right), \tag{128}$$

$$\phi_x = \sum_{m=1}^{\infty} c_m \frac{\partial \phi_m}{\partial x}, \qquad \phi_y = \sum_{m=1}^{\infty} c_m \frac{\partial \phi_m}{\partial y}; \tag{129}$$

(127) is replaced by $\phi_m = 0$ on C if the condition $\phi = 0$ on C is imposed (the fixed-edge case). According to the parenthetic part of (128) the orthogonality conditions (125) take the form

$$c_1 = c_2 = \cdots = c_{k-1} = 0. \tag{130}$$

Substituting (128) for one factor of (124), we have for the normalization condition

$$\sum_{m=1}^{\infty} c_m \iint_D \sigma \phi_m \phi \, dx \, dy = \sum_{m=1}^{\infty} c_m^2 = 1, \tag{131}$$

with the aid of the parenthetic part of (128).

We substitute the appropriate member of (128), (129) for one factor of each term of (123) to obtain

$$I = \sum_{m=1}^{\infty} c_m \left[\tau \iint_D \left(\phi_x \frac{\partial \phi_m}{\partial x} + \phi_y \frac{\partial \phi_m}{\partial y} \right) dx \, dy + \int_C p \phi \phi_m \, ds \right]$$

$$= \sum_{m=1}^{\infty} c_m \left[-\tau \iint_D \phi \nabla^2 \phi_m \, dx \, dy + \int_C \phi \left(\tau \frac{\partial \phi_m}{\partial n} + p \phi_m \right) ds \right], \tag{132}$$

with the aid of Green's theorem (23) of 2-13. From (127)—or, in the fixed-edge case, since $\phi = 0$ on C—it follows that the line integral of (132) vanishes for each m. In the double integral of (132) we use (126) and so obtain

$$I = \sum_{m=1}^{\infty} c_m \lambda_m \iint_D \sigma \phi_m \phi \, dx \, dy = \sum_{m=1}^{\infty} c_m^2 \lambda_m, \qquad (133)$$

according to the parenthetic part of (128).

With the aid of (130) and (131) we rewrite (133) as

$$I = \lambda_k + \sum_{m=k+1}^{\infty} c_m^2 (\lambda_m - \lambda_k). \qquad (134)$$

Since, for $m > k$, $\lambda_m \geqq \lambda_k$, this result implies

$$I \geqq \lambda_k. \qquad (135)$$

Equality is achieved if—but not necessarily only if—$c_k = 1$ and $c_m = 0$ for $m \neq k$, whereby the imposed conditions (130) and (131) are fulfilled. But assignment of this set of coefficients implies $\phi = \phi_k$, according to (128), so that the final part of the theorem of (a) is proved.

Aside from the trivial alternative $c_k = -1$ the "not necessarily only if" of the preceding paragraph is added because of the possibility of degeneracy. If it happens that $\lambda_k = \lambda_{k+1} = \cdots = \lambda_{k+N}$, it follows from (134) that the equality sign prevails in (135) if any one of the coefficients $c_k, c_{k+1}, \ldots, c_{k+N}$ is chosen equal to unity while every other c_m is set equal to zero.

(c) The theorem proved in (b) above provides us with a fresh statement of the membrane eigenvalue-eigenfunction problem:

Given the expression I of (123) and the density function σ, we consider the class of functions ϕ defined in D with the regularity properties stated and which satisfy the normalization condition (124); only if we deal with the fixed-edge membrane do we exclude from the class those functions for which $\phi \neq 0$ on C. The class of functions so defined we call K_1—the class of functions eligible for the first minimization of I. The minimum of I with respect to K_1 is the lowest eigenvalue λ_1 associated with the membrane; a function in K_1 which renders I equal to λ_1 is the associated eigenfunction ϕ_1.

We define the class K_2—the class of functions ϕ eligible for the second minimization—by removing from K_1 all functions ϕ which do not satisfy the condition (125) for $m = 1$; that is, K_2 includes only functions which are in K_1 and which are orthogonal to ϕ_1 in D with respect to σ. The minimum of I with respect to K_2 is the second eigenvalue λ_2; a function in K_2 which renders I equal to λ_2 is ϕ_2, the associated eigenfunction.

The process continues indefinitely: The class K_k—the class of functions ϕ eligible for the kth minimization of I—includes all functions in K_{k-1} (and therefore in K_1, K_2, . . . , K_{k-2}) which are orthogonal to ϕ_1, ϕ_2, . . . , ϕ_{k-1}. The minimum of I with respect to K_k is the kth eigenvalue λ_k; a function[1] in K_k which renders I equal to λ_k is the associated eigenfunction ϕ_k.

9-10. Consequences of the Minimum Characterization of the Membrane Eigenvalues

As a preliminary to the maximum-minimum characterization of the membrane eigenvalues presented in 9-11 following, we draw some simple inferences from the minimum characterization given in 9-9(c); and, of possibly greater importance, we point out, with regard to the minimum characterization, a fundamental limitation which is overcome only through the vastly more powerful maximum-minimum principle. Since every consequence of the minimum principle can also be derived from the maximum-minimum, only a few results are treated in this section.

(a) It is useful here and in following sections of this chapter to introduce the concept of a *membrane system*, which we define as follows:

A membrane system consists of a membrane eigenvalue-eigenfunction problem in which we are given the domain D (and thus its boundary C), the tension constant τ, the binding function p, the density function σ, and the class of functions K_1 which includes those and only those functions ϕ eligible for competition in the *first* minimization of the quantity I of (123). (The orthogonality conditions (125) are clearly not included in the determination of a membrane system.) Examples follow:

One membrane system S_X is defined, say, by a rectangular region of given dimensions, with a definite tension constant and a definite constant density, a binding function identically zero, and a class $K_1^{(X)}$ of normalized functions ϕ which satisfy no special conditions other than the standard regularity conditions stated in (a) above. This system, clearly, is associated with the rectangular free membrane of constant density considered above in 9-8(c). If we now take a membrane which has the identical physical characteristics of the foregoing, with the exception that we hold its boundary edge fixed in the equilibrium plane, we have to deal with a second membrane system S_Y: rectangular region of given dimensions,

---

[1] We say "*a* function" rather than "*the* function" because of the possibility of degeneracy [see the closing paragraph of (b)].

definite tension constant, definite constant density, binding function[1] identically zero, $K_1^{(Y)}$ including regular normalized functions ϕ which vanish on rectangle perimeter. A third system S_Z may be formed from the second by removing from the class $K_1^{(Y)}$ all functions ϕ for which $\phi \neq 0$ at the intersection of the diagonals of the rectangle. Physically, the third system is associated with the fixed-boundary membrane of the second, but with an additional constraint which holds the center point at rest in the equilibrium plane.

A membrane system S_A is defined as *narrower* than a system S_B if S_A and S_B have in common the first four of the defining characteristics of a system—namely, D, τ, p, and σ—and if every function of the class $K_1^{(A)}$ (the class K_1 for S_A) is included in the class $K_1^{(B)}$ (the class K_1 for S_B), *but not vice versa*. For example, of the systems of the preceding paragraph, S_Z is narrower than S_Y, S_Y is narrower than S_X, and S_Z is narrower than S_X. (This is a simple instance of the obvious property of transitivity for the "narrower" relation: If S_A is narrower than S_B and S_B is narrower than S_C, then S_A is narrower than S_C.)

(i) We prove directly that *if S_A is narrower than S_B, the lowest eigenvalue $\lambda_1^{(A)}$ of S_A is not less than $\lambda_1^{(B)}$, the lowest eigenvalue of S_B:* For any given ϕ the quantity I of (123), whose minimum is sought, is the same for both systems. A function $\phi_1^{(A)}$ which minimizes I with respect to $K_1^{(A)}$ is also in $K_1^{(B)}$; hence the minimum $\lambda_1^{(B)}$ of I with respect to $K_1^{(B)}$ is less than or equal to its minimum $\lambda_1^{(A)}$ with respect to $K_1^{(A)}$.

Since, according to 9-4(a) and 9-5(d), the successive frequencies of the single-frequency modes of vibration of which a membrane is capable[2] are proportional to the respective square roots of the eigenvalues of λ, the theorem (i) may be reworded: If S_A is narrower than S_B, the fundamental frequency of the membrane associated with S_A is no lower than that of the membrane associated with S_B. Applied to the three systems S_X, S_Y, S_Z defined above, (i) implies that the fundamental frequency of the free membrane is no higher than that of the same membrane with fixed boundary, which is no higher than that of the same fixed-boundary membrane having its center point constrained to remain in the equilibrium plane. This result is a special case of a more general consequence of (i):

---

[1] To this point the binding function $p(s)$ is not defined for the membrane whose boundary edge is held fixed in the equilibrium plane. From the physical point of view we could regard the fixed boundary to be the limiting case of the elastically held boundary as $p \to \infty$. It is more convenient for our purposes, however, to define p as completely arbitrary in the fixed-boundary eigenvalue-eigenfunction problem. This definition is valid because the line integral of (123), the only quantity in which p appears, vanishes if $\phi = 0$ on C. In the case at hand, $p = 0$ is the most useful choice.

[2] The so-called natural vibration frequencies, the lowest of which is called the fundamental.

The addition of constraints to a given membrane (such as holding it fixed at certain points or along certain arcs, etc.) cannot lower its fundamental vibration frequency.

Conversely, the removal of constraints from a given membrane cannot raise its fundamental frequency. An important application of this fact lies in considering the effect of introducing a slit into a membrane surface. Analytically, a slit along a given arc admits into the class K_1 functions which are discontinuous across the arc, thus giving rise to a system less narrow than that associated with the unmutilated membrane: The introduction of a slit cannot increase the fundamental frequency of a membrane.

(*b*) We suppose that two membrane systems S_A and S_B have in common the characteristics D, τ, σ, and K_1, but the respective binding functions p_A and p_B are not necessarily equal. We prove the following theorem:

(ii) *If* $p_A \geqq p_B$, *the lowest eigenvalue* $\lambda_1^{(A)}$ *of* S_A *is not less than* $\lambda_1^{(B)}$, *the lowest eigenvalue of* S_B.

If, in (123) of 9-9(*a*), we write $p = p_A$, I is written I_A; if we write $p = p_B$, I is written I_B. Since $p_A \geqq p_B$, it follows that $I_A \geqq I_B$ for any given function ϕ in K_1 (common to the two systems). If the minimum $\lambda_1^{(A)}$ of I_A with respect to K_1 is achieved through $\phi = \phi_1^{(A)}$, and if I_B^* is the value of I_B when $\phi = \phi_1^{(A)}$, we have $\lambda_1^{(A)} \geqq I_B^* \geqq \lambda_1^{(B)}$.

In physical terms (ii) implies that the tightening of the agency which binds the boundary edge elastically (*i.e.*, the increase of p) may raise, but cannot lower, the fundamental frequency of a given membrane.

(*c*) We consider the two membrane systems S_A and S_B which have in common the characteristics D, τ and p, but for which the respective density functions σ_A and σ_B are not identically equal in D. The members of the eligible classes $K_1^{(A)}$ and $K_1^{(B)}$, we suppose, are required to satisfy the same conditions of regularity and the same set of special conditions (such as vanishing at certain points, or along certain arcs, etc.) which may be imposed. The classes $K_1^{(A)}$ and $K_1^{(B)}$, clearly, are not in general identical, because of the difference in the normalization which springs from the nonidentity of σ_A and σ_B. Any function ϕ_A in $K_1^{(A)}$ can be converted, however, to membership in $K_1^{(B)}$ simply through being multiplied by a suitable constant. For if

$$\iint\limits_D \sigma_B \phi_A^2 \, dx \, dy = c^2,$$

(ϕ_A/c) is clearly a member of $K_1^{(B)}$. Similarly, any function ϕ_B in $K_1^{(B)}$ can be converted to membership in $K_1^{(A)}$ in the same manner.

We prove the following theorem concerning two systems satisfying the relationship stated in the preceding paragraph:

(iii) *If $\sigma_A \leqq \sigma_B$ in D, the lowest eigenvalue $\lambda_1^{(A)}$ of S_A is not less than $\lambda_1^{(B)}$, the lowest eigenvalue of S_B.* It follows from the hypothesis that

$$c^2 = \iint\limits_D \sigma_B \phi_A^2 \, dx \, dy \geqq \iint\limits_D \sigma_A \phi_A^2 \, dx \, dy = 1, \qquad (136)$$

where ϕ_A is any member of $K_1^{(A)}$ and c is a positive constant, in general different for different members of $K_1^{(A)}$, defined by the left-hand equation of (136). We therefore conclude from (136) and the result above that for any member ϕ_A of $K_1^{(A)}$ there is a corresponding member (ϕ_A/c) of $K_1^{(B)}$, with $c \geqq 1$. It accordingly follows that, if any member ϕ_A of $K_1^{(A)}$ renders I of (123) equal to I_A, there exists a member (ϕ_A/c) of $K_1^{(B)}$ which renders I equal to I_B, where

$$I_B = \frac{1}{c^2} I_A \leqq I_A,$$

according to (136). From this we have that the *minimum* of I with respect to $K_1^{(B)}$ is less than or equal to its minimum with respect to $K_1^{(A)}$, and theorem (iii) is proved.

The physical implication of (iii) is that an increase in the density of a membrane, without other change, may lower, but cannot raise, the fundamental vibration frequency.

(d) In the statements of the three theorems of this section no mention whatever is made concerning any but the lowest eigenvalue of a given system. What about the higher eigenvalues? We may ask, for example: If S_A is narrower than S_B, what can we say concerning the relative magnitudes of the kth eigenvalue $\lambda_k^{(A)}$ of S_A and the kth eigenvalue $\lambda_k^{(B)}$ of S_B, for $k \geqq 2$? Does there exist a relation $\lambda_k^{(B)} \leqq \lambda_k^{(A)}$ which holds, according to theorem (i), for $k = 1$? The answer is affirmative, but it is not given by the minimum principle employed in the proof of theorem (i); it does, however, follow from the maximum-minimum principle of 9-11 below.

It is not difficult to see why the minimum characterization of the eigenvalues fails to provide information concerning the relative magnitudes of the higher eigenvalues of different systems. Reference to 9-9(c) reveals the source of the limitation in the following way:

We suppose S_A narrower than S_B so that $K_1^{(B)}$ includes every function ϕ in $K_1^{(A)}$. The minimum $\lambda_1^{(A)}$ of I of (123) with respect to $K_1^{(A)}$ is achieved with $\phi = \phi_1^{(A)}$; the minimum $\lambda_1^{(B)}$ of I with respect to $K_1^{(B)}$ is achieved with $\phi = \phi_1^{(B)}$, which is not in general the same as $\phi_1^{(A)}$. The class $K_2^{(A)}$, with respect to which $\lambda_2^{(A)}$ is the minimum of I, is formed by removing from $K_1^{(A)}$ all functions ϕ which do not satisfy

$$\iint_D \sigma \phi_1^{(A)} \phi \, dx \, dy = 0;$$

the corresponding class $K_2^{(B)}$ is formed by removing from $K_1^{(B)}$ all functions ϕ which do not satisfy

$$\iint_D \sigma \phi_1^{(B)} \phi \, dx \, dy = 0.$$

From this limited information concerning the formation of $K_2^{(A)}$ and $K_2^{(B)}$, we are in no position to know whether every ϕ of $K_2^{(A)}$ is a member of $K_2^{(B)}$, or vice versa. In fact, neither instance of all-inclusion is generally exhibited.

Since the proof that $\lambda_1^{(B)} \leqq \lambda_1^{(A)}$ depends upon the fact that every ϕ in $K_1^{(A)}$ is a member of $K_1^{(B)}$, lack of corresponding information concerning $K_2^{(A)}$ and $K_2^{(B)}$ makes it impossible to infer from the minimum principle a similar relationship between $\lambda_2^{(A)}$ and $\lambda_2^{(B)}$. It is clear that this limitation extends to the higher eigenvalues $\lambda_k^{(A)}$ and $\lambda_k^{(B)}$, for $k > 2$.

9-11. The Maximum-Minimum Characterization of the Membrane Eigenvalues[1]

(a) We fix our attention upon a single membrane system S, characterized by the domain D (with boundary C), the constant τ, the functions $\sigma(x,y)$ and $p(s)$, and the class K_1 of functions ϕ eligible for the first minimization of

$$I = \tau \iint_D (\phi_x^2 + \phi_y^2) dx \, dy + \int_C p\phi^2 \, ds; \tag{137}$$

all ϕ in K_1 satisfy the normalization condition

$$\iint_D \sigma \phi^2 \, dx \, dy = 1. \tag{138}$$

We form the class $K(U_k)$ of functions ϕ $(k \geqq 2)$ by removing from K_1 all members which do not satisfy the $(k-1)$ orthogonality relationships

$$\iint_D \sigma u_j \phi \, dx \, dy = 0 \qquad (j = 1,2, \ldots ,k-1), \tag{139}$$

---

[1] The important ideas in this and the following sections apparently originate with Courant (2). See also Courant-Hilbert, Chap. 6.

where the functions $u_1(x,y)$, $u_2(x,y)$, . . . , $u_{k-1}(x,y)$ are *completely arbitrary* in D to within continuity, except for a finite number of smooth arcs across which finite discontinuities may occur. We denote the set of arbitrary functions u_1, u_2, . . . , u_{k-1}, in aggregate, by the single symbol U_k.

We now prove that the minimum of I with respect to functions ϕ in $K(U_k)$ is less than or equal to the kth eigenvalue λ_k of the system S; to do this we merely show that there is at least one ϕ in $K(U_k)$ which renders I less than or equal to λ_k: The linear combination

$$\phi = \sum_{m=1}^{k} c_m \phi_m \qquad \left(c_m = \iint_D \sigma \phi_m \phi \, dx \, dy \right) \tag{140}$$

of the first k orthonormal eigenfunctions of the system S is a member of $K(U_k)$ if the coefficients c_1, c_2, . . . , c_k satisfy the $(k-1)$ conditions imposed by (139),

$$\sum_{m=1}^{k} c_m \iint_D \sigma u_j \phi_m \, dx \, dy = 0 \qquad (j = 1, 2, \ldots, k-1), \tag{141}$$

as well as the condition

$$\sum_{m=1}^{k} c_m^2 = 1, \tag{142}$$

which results directly from substituting (140) into (138). It is easily shown that a system of $(k-1)$ linear homogeneous equations, such as (141), in k unknowns subject to a normalization condition, such as (142), always possesses a solution.[1] We may assume, therefore, that there is a function ϕ of the form (140) which is a member of $K(U_k)$.

We substitute (140) into (137) and, in the manner of achieving (133) in 9-9(*b*), we show that

$$I = \sum_{m=1}^{k} c_m^2 \lambda_m = \lambda_k - \sum_{m=1}^{k-1} c_m^2 (\lambda_k - \lambda_m), \tag{143}$$

with the aid of (142). Since $\lambda_m \leqq \lambda_k$ for $m < k$, it follows from (143) that $I \leqq \lambda_k$. Thus the assertion that the *minimum* of I with respect to $K(U_k)$ is less than or equal to λ_k is proved.

Furthermore, from the statement of the minimum principle in 9-9(*c*) it is clear that the minimum of I with respect to $K(U_k)$ is *precisely* λ_k if the set U_k consists of the first $(k-1)$ eigenfunctions ϕ_1, ϕ_2, . . . , ϕ_{k-1}—that is, if $u_j = \phi_j$ for $j = 1, 2, \ldots, k-1$. For, in this event,

[1] See exercise 15 at end of chapter.

$K(U_k)$ is identical with the class K_k defined in 9-9(c), with respect to which the minimum of I is λ_k.

Therefore, if we let $I(U_k)$ denote the minimum of I with respect to $K(U_k)$, we have $I(U_k) \leqq \lambda_k$, and *the maximum of $I(U_k)$ with respect to all sets U_k is λ_k, the kth eigenvalue of the system S. This maximum minimum of I is achieved when U_k is the set of the first $(k-1)$ eigenfunctions* $\phi_1, \phi_2, \ldots, \phi_{k-1}$ *of S.*

Thus we may reformulate the membrane eigenvalue-eigenfunction problem as follows: Given a set U_k of arbitrary functions $u_1, u_2, \ldots, u_{k-1}$, we first minimize I of (137) with respect to those functions ϕ in K_1 which further satisfy the $(k-1)$ orthogonality relations (139)—with respect to the functions ϕ in $K(U_k)$, that is. The minimum so achieved, namely $I(U_k)$, depends in general upon the particular set U_k which is employed. We next proceed to maximize $I(U_k)$ with respect to U_k; that is, we form the minimum $I(U_k)$ for each of all possible choices of sets U_k, and we select from among these minima the largest. The maximum of $I(U_k)$ so achieved is the kth eigenvalue λ_k of the system S, according to the maximum-minimum principle proved directly above.

(It is clear that the maximum-minimum formulation of the membrane eigenvalue-eigenfunction problem reduces to the minimum formulation of 9-9(c) when $k = 1$—that is, in the quest for the lowest eigenvalue λ_1. For the set U_1 is an empty set, containing no functions whatever.)

In the paragraphs following, we derive several consequences of the maximum-minimum principle, some of which are generalizations of the theorems proved in 9-10.

(b) We prove:

Theorem (i). *If the membrane system S_A is narrower[1] than the system S_B, the kth eigenvalue $\lambda_k^{(A)}$ of S_A is no less than $\lambda_k^{(B)}$, the kth eigenvalue of S_B.*

By definition every ϕ in $K_1^{(A)}$ is a member of $K_1^{(B)}$. With the use of any given set U_k the formation of $K_A(U_k)$ involves removing from $K_1^{(A)}$ a subset of functions ϕ—those which do not satisfy the $(k-1)$ orthogonality conditions (139). In the formation of $K_B(U_k)$ the same subset is removed from $K_1^{(B)}$; any additional functions removed from $K_1^{(B)}$ to form $K_B(U_k)$ are not in $K_1^{(A)}$ to begin with, and therefore not in $K_A(U_k)$. Thus every function in $K_A(U_k)$ is a member of $K_B(U_k)$. It thus follows that the minimum $I_A(U_k)$ of I of (137) with respect to $K_A(U_k)$ is not less than $I_B(U_k)$, the minimum of I with respect to $K_B(U_k)$. If U_k' is the set—the first $(k-1)$ eigenfunctions of S_B—which maximizes $I_B(U_k)$, we have, according to the maximum-minimum principle,

$$\lambda_k^{(B)} = I_B(U_k') \leqq I_A(U_k') \leqq \lambda_k^{(A)}.$$

Theorem (i) is thus proved.

[1] See the definition in 9-10(a).

We may express (i) in terms of the numbers $N_A(\lambda)$ and $N_B(\lambda)$, where $N_A(\lambda)$ is defined as the number of eigenvalues of S_A which are less than or equal to the number λ, and $N_B(\lambda)$ has the same meaning for S_B:

Theorem (i'). *If S_A is narrower than S_B, it follows that $N_A(\lambda) \leqq N_B(\lambda)$.*

(c) We prove:

Theorem (ii). *If the membrane systems S_A and S_B have in common the characteristics D, τ, σ, and K_1, but the respective binding functions satisfy the relation $p_A \geqq p_B$, then $\lambda_k^{(A)} \geqq \lambda_k^{(B)}$.* Or,

Theorem (ii'). $N_A(\lambda) \leqq N_B(\lambda)$.

If in (137) above we write $p = p_A$, I is written I_A; if we write $p = p_B$, I is written I_B. Since $p_A \geqq p_B$, it follows that $I_A \geqq I_B$ for any given function ϕ in $K(U_k)$—common to both systems—for any U_k employed. By the argument of 9-10(b), it also follows that the respective minima of I with respect to $K(U_k)$ satisfy the relation $I_A(U_k) \geqq I_B(U_k)$. By the argument of (b) above it follows that $\lambda_k^{(A)} \geqq \lambda_k^{(B)}$ and therefore that $N_A(\lambda) \leqq N_B(\lambda)$.

(d) We prove a third direct consequence of the maximum-minimum principle:

Theorem (iii). *The systems S_A and S_B have in common the characteristics D, τ, and p; and the eligible classes $K_1^{(A)}$ and $K_1^{(B)}$ are required to satisfy the same conditions of regularity and the same set of special conditions which may be imposed. If $\sigma_A \leqq \sigma_B$ in D, then $\lambda_k^{(A)} \geqq \lambda_k^{(B)}$.* Or,

Theorem (iii'). $N_A(\lambda) \leqq N_B(\lambda)$.

We proceed to form the class of functions $K_A(U_k)$ by removing from $K_1^{(A)}$ all functions ϕ which do not satisfy

$$\iint_D \sigma_A u_j \phi \, dx \, dy = 0 \qquad (j = 1,2, \ldots ,k-1). \qquad (144)$$

With U_k given we form the set V_k such that its members $v_1, v_2, \ldots , v_{k-1}$ satisfy the relations

$$\sigma_B v_j = \sigma_A u_j \quad \text{in } D \qquad (j = 1,2, \ldots ,k-1). \qquad (145)$$

With V_k thus established we form the class $K_B(V_k)$ by removing from $K_1^{(B)}$ all functions ϕ which fail to satisfy

$$\iint_D \sigma_B v_j \phi \, dx \, dy = 0 \qquad (j = 1,2, \ldots ,k-1), \qquad (146)$$

which is identical with (144) because of (145).

In 9-10(c) it is shown that, if ϕ is any member of $K_1^{(A)}$, there is a corresponding member (ϕ/c) in $K_1^{(B)}$, where c is a constant which may differ for different ϕ, but where always

$$c^2 \geqq 1. \tag{147}$$

If and only if the formation of $K_A(U_k)$ involves the removal of a given ϕ from $K_1^{(A)}$, the formation of $K_B(V_k)$ involves the removal of the corresponding (ϕ/c) from $K_1^{(B)}$, because of the identity of (146) with (144). We therefore conclude that for each ϕ in $K_A(U_k)$ there is a corresponding (ϕ/c) in $K_B(V_k)$.

Hence if the minimum $I_A(U_k)$ of I with respect $K_A(U_k)$ is achieved when $\phi = \phi_A$, it follows from (137) that the function (ϕ_A/c) in $K_B(V_k)$ renders I equal to

$$I_B = \frac{1}{c^2} I_A(U_k) \leqq I_A(U_k), \tag{148}$$

because of (147). But since I_B is not less than $I_B(V_k)$, the minimum of I with respect to $K_B(V_k)$, we conclude from (148) that

$$I_B(V_k) \leqq I_A(U_k). \tag{149}$$

We now proceed to maximize $I_B(V_k)$ with respect to V_k and suppose that the maximum $\lambda_k^{(B)}$ is attained when $V_k = V_k'$—the set of the first $(k - 1)$ eigenfunctions of S_B. If U_k' corresponds to V_k' through (145), it follows from (149) that

$$\lambda_k^{(B)} = I_B(V_k') \leqq I_A(U_k') \leqq \lambda_k^{(A)},$$

according to the maximum-minimum principle. Theorem (iii) is thus proved, and (iii′) follows directly.

The physical implications in the theorems (i) to (iii) above are obvious generalizations of the inferences drawn in 9-10 from the corresponding theorems (i) to (iii) of that section. Fuller discussion is left for exercise 18 at the end of this chapter.

(e) We consider the membrane system S_B which is characterized by the domain D_B (with boundary C_B), the density function σ_B, the tension constant τ, the binding function $p = 0$, and the class $K_1^{(B)}$ of functions ϕ which satisfy the standard regularity conditions[1] and $\phi = 0$ on C_B. S_B is associated, that is, with a given fixed-edge membrane.

A second system S_A is characterized by the domain D_A whose boundary C_A lies entirely in D_B or is, at most, in partial coincidence with C_B; the density function σ_A defined so that $\sigma_A = \sigma_B$ in D_A; the same tension constant τ as for S_B; the binding function $p = 0$; and the class $K_1^{(A)}$ of functions ϕ which satisfy the standard regularity conditions and $\phi = 0$

[1] See 9-9(a), just following equation (127).

on C_A. Clearly, the membrane associated with S_A may be constructed from the membrane of S_B by holding the latter fixed along an internal closed non-self-intersecting curve C_A and ignoring whatever of D_B is left over (see Fig. 9-2).

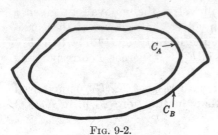

FIG. 9-2.

We characterize a third system S'_A by the domain D_B (with boundary C_B), the density function σ_B, the tension constant τ (same as above), the binding function $p = 0$, and the class $K_1^{(A')}$ of standardly regular functions ϕ which satisfy $\phi = 0$ on C_A, $\phi = 0$ on C_B, $\phi = \phi_x = \phi_y = 0$ throughout the portion of D_B exterior to C_A. It is evident that the sequence of eigenvalues $\lambda_1^{(A')}$, $\lambda_2^{(A')}$, . . . , $\lambda_k^{(A')}$, . . . of S'_A is identical with the sequence $\lambda_1^{(A)}$, $\lambda_2^{(A)}$, . . . , $\lambda_k^{(A)}$, . . . of S_A. For we have from (137) of (a) above, if ϕ is any member of $K_1^{(A')}$,

$$I = \tau \iint_{D_B} (\phi_x^2 + \phi_y^2)dx\,dy = \tau \iint_{D_A} (\phi_x^2 + \phi_y^2)dx\,dy;$$

also

$$\iint_{D_B} \sigma_B \phi^2\,dx\,dy = \iint_{D_A} \sigma_A \phi^2\,dx\,dy,$$

because $\phi = \phi_x = \phi_y = 0$ between C_A and C_B. Since the functions in $K_1^{(A')}$ satisfy the same requirements in D_A as do the members of $K_1^{(A)}$, it follows that the S_A and S'_A eigenvalue-eigenfunction problems are identical.

Comparison of the characteristics defining S_B and S'_A shows, according to the definition in 9-10(a), that S'_A is narrower than S_B. From theorem (i) of (b) above, it therefore follows that $\lambda_k^{(B)} \leqq \lambda_k^{(A')} = \lambda_k^{(A)}$. We thus have a proof of:

Theorem (iv). *The shrinking of the boundary, without any other change,*[1] *of a fixed-edge membrane may increase, but cannot lower, each eigenvalue* λ_k $(k = 1,2,3, . . .)$. Or,

Theorem (iv'). *The shrinking may decrease, but cannot increase, the number* $N(\lambda)$ *of the fixed-edge-membrane eigenvalues which are less than or equal to any number* λ.

---

[1] The "shrinking" is effected in the way the system S_A is created above, on the basis of S_B.

9-12. The Asymptotic Distribution of the Membrane Eigenvalues

In this section we derive, as a consequence of the maximum-minimum principle, an asymptotic formula for the kth eigenvalue of a vibrating membrane. The results we achieve are applicable to any membrane of which the domain D is divisible into a finite number of congruent squares. Finally, with the aid of theorem (iv) of the preceding section, we extend our results to include the fixed-edge membrane of arbitrary shape. To avoid an uninteresting morass of tedious detail, however, we merely state without proof the corresponding extension to the elastically bound (or free) membrane of arbitrary shape.

(a) We consider the membrane system S_A characterized by the domain D, the tension constant τ, the binding function $p = 0$, the density function σ, and the eligible class $K_1^{(A)}$ of standardly regular[1] functions ϕ which satisfy $\phi = 0$ on the boundary C of D. A second system $S_{A'}$ has the first four characteristics—D, τ, p, and σ—in common with S_A, but the functions ϕ in its eligible class $K_1^{(A')}$ are required to vanish not only on C but also along a given network of piecewise smooth arcs which subdivide D, without gap or overlap, into a finite set of r subdomains $D_1, D_2, \ldots,$ D_r (see Fig. 9-3).   Clearly $S_{A'}$ is narrower than S_A, so that the kth eigenvalue $\lambda_k^{(A')}$ of $S_{A'}$ is not less than $\lambda_k^{(A)}$, the kth eigenvalue of S_A, according to theorem (i) of 9-11(b). Or by the equivalent theorem (i') of 9-11(b),

$$N_{A'}(\lambda) \leqq N_A(\lambda), \tag{150}$$

where the two members of (150) are respectively the numbers of eigenvalues of $S_{A'}$ and S_A less than or equal to a given number λ.

We seek a second characterization of the eigenvalue $\lambda_k^{(A')}$ in terms of the eigenvalues of the r systems $S_{A_1}, S_{A_2}, \ldots, S_{A_r}$ defined as follows: The system S_{A_j} ($j = 1, 2, \ldots, r$) is characterized by the domain D_j (jth subdomain of D), the tension constant τ (same as for S_A, $S_{A'}$), the density function σ which coincides with σ of S_A and $S_{A'}$ in D_j, arbitrary binding function p, and the eligible class $K_1^{(A_j)}$ of standardly regular functions ϕ which vanish on the boundary C_j of D_j. We prove now that every eigenvalue of $S_{A'}$ is an eigenvalue of one of the systems S_{A_j} and, conversely, that every eigenvalue of each S_{A_j} is an eigenvalue of $S_{A'}$:

If $\lambda_k^{(A')}$ is an eigenvalue of $S_{A'}$ and ϕ_k is the corresponding eigenfunction, we have, according to (126) of 9-9(a),

$$\tau\nabla^2\phi_k + \lambda_k^{(A')}\sigma\phi = 0 \qquad \text{in } D, \tag{151}$$

and therefore in each D_j ($j = 1, 2, \ldots, r$). Since ϕ_k is not identically

[1] See 9-9(a), just following equation (127).

zero, there is at least one subdomain—say D_i—in which it is not identically zero; and since ϕ_k is a member of $K_1^{(A')}$, ϕ_k vanishes on the boundary C_i of D_i. The function $\phi_k^{(i)}$, defined as identical with ϕ_k in D_i and identically zero outside D_i, is therefore the eigenfunction associated with the eigenvalue $\lambda_k^{(A')}$ of S_{A_i}. For, according to (151), we have

$$\tau\nabla^2\phi_k^{(i)} + \lambda_k^{(A')}\sigma\phi_k^{(i)} = 0 \qquad \text{in } D_i$$

and $\phi_k^{(i)} = 0$ on C_i. That is, $\lambda_k^{(A')}$ is an eigenvalue (in general not the kth) of S_{A_i}.

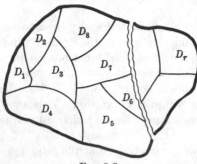

If λ^* is any eigenvalue of S_{A_i} and ϕ^* is the corresponding eigenfunction, we have that (λ^*,ϕ^*) also constitutes an eigenvalue-eigenfunction pair of the system $S_{A'}$, provided we extend the definition of ϕ^* by means of $\phi^* = 0$, identically, outside D_i. For we have

$$\tau\nabla^2\phi^* + \lambda^*\sigma\phi^* = 0 \qquad (152)$$

Fig. 9-3.

in D_i; and, with the extended definition of ϕ^*, equation (152) holds also in D, with $\phi^* = 0$ on C.

If no two systems S_{A_i}, S_{A_j} $(i \neq j)$ have an eigenvalue in common, we conclude from the preceding two paragraphs that *the list of eigenvalues* $\lambda_1^{(A')}$, $\lambda_2^{(A')}$, . . . , $\lambda_k^{(A')}$, . . . *of $S_{A'}$ may be formed by writing down the aggregate of the eigenvalues of all the systems S_{A_1}, S_{A_2}, . . . , S_{A_r}, and arranging them in ascending order.* We now show that the same statement holds even if several of the systems S_{A_i} have in common any number of eigenvalues: Let s systems of the S_{A_i}—which, for the sake of simplicity, we suppose to be S_{A_1}, S_{A_2}, . . . , S_{A_s} $(s \leq r)$—have in common any eigenvalue λ^*, and let the corresponding eigenfunction in S_{A_j} be ϕ_j^* $(j = 1,2, . . . ,s)$. We extend the definition of ϕ_j^* by means of $\phi_j^* = 0$, identically, outside D_j. According to the preceding paragraph, λ^* is also an eigenvalue of $S_{A'}$ corresponding to the s eigenfunctions ϕ_1^*, ϕ_2^*, . . . , ϕ_s^* (with the extended definitions). Since the extended ϕ_1^*, ϕ_2^*, . . . , ϕ_s^* are linearly independent,[1] the eigenvalue λ^* is at least[2] s-fold degenerate in $S_{A'}$ and thus appears consecutively at least s times in the list $\lambda_1^{(A')}$, $\lambda_2^{(A')}$, . . . , $\lambda_k^{(A')}$, . . . We therefore conclude that the italicized statement above holds in all cases.

If $N_{A_j}(\lambda)$ is the number of eigenvalues of S_{A_j} $(j = 1,2, . . . ,r)$ less

[1] For no two are different from zero at any point of D.

[2] "At least" because λ^* may be degenerate in any or all of S_{A_1}, S_{A_2}, . . . , S_{A_s}.

than or equal to λ, and $N_{A'}(\lambda)$ is the corresponding number for $S_{A'}$, the above result is equivalent to writing

$$N_{A'}(\lambda) = N_{A_1}(\lambda) + N_{A_2}(\lambda) + \cdots + N_{A_r}(\lambda). \tag{153}$$

From (150) it therefore follows that

$$N_{A_1}(\lambda) + N_{A_2}(\lambda) + \cdots + N_{A_r}(\lambda) \leqq N_A(\lambda), \tag{154}$$

where $N_A(\lambda)$ applies to the original fixed-edge membrane bounded by C.

(b) Holding in attention the membrane systems S_A, S_{A_1}, S_{A_2}, . . . , S_{A_r} defined in (a) above, we consider also the system S_B characterized by the same D, τ, σ, and p ($= 0$ identically) as S_A but having a class $K_1^{(B)}$ of eligible functions ϕ which satisfy no restrictions in D or on C except for standard regularity. It is clear that S_A is narrower than S_B and, therefore, by theorem (i') of 9-11(b), that

$$N_A(\lambda) \leqq N_B(\lambda), \tag{155}$$

where $N_B(\lambda)$ is the number of eigenvalues of S_B less than or equal to λ, and $N_A(\lambda)$ is similarly defined in (a).

We next define the system $S_{B'}$, characterized by the same D, τ, σ, and p as in S_A and S_B but which has the eligible class $K_1^{(B')}$ of functions ϕ which are permitted to exhibit finite discontinuities across each arc of the network which subdivides D into the subdomains $D_1, D_2, \ldots, D_r$— the same set of subdivisions used in the definition of $S_{A'}$ in (a). (See Fig. 9-3.)[1] Clearly, S_B is narrower than $S_{B'}$, since every ϕ in $K_1^{(B)}$ is a member of $K_1^{(B')}$, while the latter contains discontinuous functions as well. We therefore have, from theorem (i') of 9-11(b), that

$$N_B(\lambda) \leqq N_{B'}(\lambda). \tag{156}$$

Finally, we define the r systems S_{B_1}, S_{B_2}, . . . , S_{B_r} the free-edge counterparts of S_{A_1}, S_{A_2}, . . . , S_{A_r} of (a)—as follows: The system S_{B_j} ($j = 1, 2, \ldots, r$) is characterized by the domain D_j, the tension constant τ (same as for $S_{B'}$, etc.), the density function σ which coincides with σ of $S_{A'}$, etc., in D_j, binding function $p = 0$, identically, and the eligible class $K_1^{(B_j)}$ of functions ϕ which satisfy standard regularity conditions and no other special requirements. With point-by-point correspondence of details we may use the method of deriving (153) in (a) above to prove also that

$$N_{B'}(\lambda) = N_{B_1}(\lambda) + N_{B_2}(\lambda) + \cdots + N_{B_r}(\lambda) \tag{157}$$

---

[1] In barely realizable physical terms, $S_{B'}$ has to do with the free-edge membrane associated with S_B after it has been sliced into r free-edge membranes in the domains $D_1, D_2, \ldots, D_r$.

—with the single exception that we must replace the boundary condition $\phi = 0$ in the proof above with the boundary condition $(\partial\phi/\partial n) = 0$ for the present proof. This fact requires special discussion:

From (127) of 9-9(a) it follows that every eigenfunction ϕ of S_{Bj} must satisfy $(\partial\phi/\partial n) = 0$ on the boundary C_j of D_j since $p = 0$ in the definition of S_{Bj} ($j = 1, 2, \ldots, r$). Thus a preliminary requirement for the proof of (157) is that the eigenfunction ϕ_k of $S_{B'}$ also satisfy $(\partial\phi_k/\partial n) = 0$ on each of $C_1, C_2, \ldots, C_r$, for every $k = 1, 2, 3, \ldots$. The proof of this fact depends essentially upon the "$\epsilon\eta$ process" of extremizing

$$I = \tau \iint_D (\phi_x^2 + \phi_y^2)dx\,dy$$

with respect to those functions ϕ in $K_1^{(B')}$—which may exhibit finite discontinuities across those portions of $C_1, C_2, \ldots, C_r$ which do not coincide with the boundary C of the whole of D—which satisfy the orthogonality conditions

$$\iint_D \sigma\phi_m\phi\,dx\,dy = 0 \qquad (m = 1, 2, \ldots, k - 1).$$

The proof, which is quite straightforward, is left for end-chapter exercise 20, which is presented with an ample supply of guiding hints.

We may therefore accept the validity of (157), which we combine with (156), (155), and (154) to achieve the important result

$$\sum_{j=1}^{r} N_{Aj}(\lambda) \leqq N_A(\lambda) \leqq N_B(\lambda) \leqq \sum_{j=1}^{r} N_{Bj}(\lambda). \tag{158}$$

(c) In (118) of 9-8(b) we have, by setting $a = b$, the explicit formula

$$\lambda_{mk} = \frac{\tau\pi^2}{\sigma_0 b^2}(m^2 + k^2) \qquad (m, k = 1, 2, 3, \ldots, \text{independently}) \tag{159}$$

for the eigenvalues of a fixed-edge square membrane of side length b and of uniform density σ_0. If we write

$$R^2 = \frac{\sigma_0 b^2 \lambda}{\tau\pi^2}, \tag{160}$$

with R positive, it follows from (159) that the number $N_{A_1}(\lambda)$ of eigenvalues of the system less than or equal to λ is the number of pairs of positive integers (m,k) which satisfy the inequality

$$m^2 + k^2 \leqq R^2 \qquad (mk \neq 0). \tag{161}$$

In the language of analytic geometry $N_{A_1}(\lambda)$ is thus the number of lattice points[1] lying within and on the first quadrant of the circle

$$x^2 + y^2 = R^2, \tag{162}$$

exclusive of points lying on the x and y axes (see Fig. 9-4).

We associate with each lattice point $x = m$, $y = k$ for which (161) is satisfied the unit square of which it is the *upper right-hand* corner. With this, it is clear that $N_{A_1}(\lambda)$ is the total area of those unit squares which lie completely within the quarter-circle under consideration. To obtain an upper bound to the difference Δ_1 between $N_{A_1}(\lambda)$ and $(\pi R^2/4)$, the quarter-circle area, we note that a concentric circle of radius $(R - \sqrt{2})$ excludes all the partial squares whose total area is Δ_1. It therefore follows that Δ_1 is less than the first-quadrant area of the annular ring between the circles having the respective radii R and $(R - \sqrt{2})$ (see Fig. 9-4); that is,

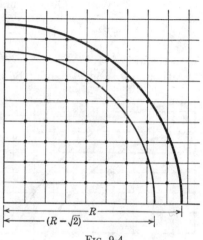

Fig. 9-4.

$$\Delta_1 = \tfrac{1}{4}\pi R^2 - N_{A_1}(\lambda) < \tfrac{1}{4}\pi R^2 - \tfrac{1}{4}\pi(R - \sqrt{2})^2$$
$$= \tfrac{1}{4}\pi \sqrt{2}\,(2R - \sqrt{2}) < \tfrac{1}{2}\sqrt{2}\,\pi R.$$

Thus we may write

$$N_{A_1}(\lambda) = \tfrac{1}{4}\pi R^2 - \tfrac{1}{2}\sqrt{2}\,\pi\theta_A R,$$

where $0 < \theta_A < 1$; or, with the aid of (160), we have

$$N_{A_1}(\lambda) = \frac{\sigma_0 b^2 \lambda}{4\pi\tau} - \theta_A b \sqrt{\frac{\lambda\sigma_0}{2\tau}} \qquad (0 < \theta_A < 1). \tag{163}$$

In similar fashion we obtain an analogous expression for $N_{B_1}(\lambda)$, the corresponding quantity for the free-edge square membrane of side b and uniform density σ_0. The only difference, according to (121) of 9-8(c), between the fixed- and free-edge cases is that we may admit zero values

[1] A lattice point in the xy plane is any point both of whose coordinates are integers—positive, negative, or zero.

of m and k. We therefore associate, in Fig. 9-5, each lattice point which represents a free-edge eigenvalue with the unit square of which it is the *lower left-hand* corner. With this, it is clear that $N_{B_1}(\lambda)$ is the total area of those unit squares which lie completely *or partially* within the first quadrant of the circle (162). The difference Δ_2 between $N_{B_1}(\lambda)$ and $(\pi R^2/4)$ clearly satisfies the relation

$$\Delta_2 = N_{B_1}(\lambda) - \tfrac{1}{4}\pi R^2 < \tfrac{1}{4}\pi(R + \sqrt{2})^2 - \tfrac{1}{4}\pi R^2$$
$$= \tfrac{1}{4}\pi \sqrt{2}\,(2R + \sqrt{2}) < \pi R,$$

if $R > (1 + \tfrac{1}{2}\sqrt{2})$. Thus we may write

$$N_{B_1}(\lambda) = \frac{1}{4}\pi R^2 + \theta_B \pi R = \frac{\sigma_0 b^2 \lambda}{4\pi\tau} + \theta_B b \sqrt{\frac{\lambda \sigma_0}{\tau}} \qquad (0 < \theta_B < 1), \quad (164)$$

with the aid of (160).

Although θ_A and θ_B cannot be evaluated in simple form as functions

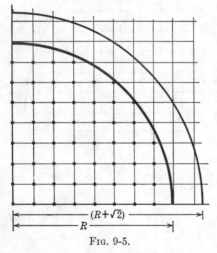

of λ, the results (163) and (164) are useful in that the unevaluated final term is, in each case, small compared with the term $(\sigma_0 b^2 \lambda/4\pi\tau)$, for sufficiently large λ. Both of (163) and (164) are used, in conjunction with (158) of (*b*) above, in deriving the asymptotic results below.

(*d*) We consider now a membrane of uniform density σ_0 and tension constant τ, for which the domain D may be subdivided, without gap or overlap, into a finite number r of congruent squares of side length b. Associated with this membrane we con-

$$\longleftarrow (R + \sqrt{2}) \longrightarrow$$
$$\longleftarrow R \longrightarrow$$

FIG. 9-5.

sider the systems S_A, S_B, S_{A_j}, S_{B_j} $(j = 1,2, \ldots ,r)$, which correspond to the systems defined in (*a*) and (*b*) above:

S_A: Entire membrane, boundary edge held fixed

S_B: Entire membrane, boundary edge free

S_{A_j}: Membrane of jth square of side length b, boundary edge held fixed $(j = 1,2, \ldots ,r)$

S_{B_j}: Membrane of jth square of side length b, boundary edge free $(j = 1,2, \ldots ,r)$.

The symbols $N_A(\lambda)$, $N_B(\lambda)$, $N_{A_j}(\lambda)$, $N_{B_j}(\lambda)$ have the meanings assigned to them in (*a*) and (*b*).

For each j, we may apply to the system S_{A_j} the result (163) of (c)— namely,

$$N_{A_j}(\lambda) = \frac{\sigma_0 b^2 \lambda}{4\pi\tau} - \theta_A b \sqrt{\frac{\lambda\sigma_0}{2\tau}} \qquad (0 < \theta_A < 1). \qquad (165)$$

To the system S_{B_j} we similarly apply (164), for each j:

$$N_{B_j}(\lambda) = \frac{\sigma_0 b^2 \lambda}{4\pi\tau} + \theta_B b \sqrt{\frac{\lambda\sigma_0}{\tau}} \qquad (0 < \theta_B < 1). \qquad (166)$$

Since the expressions for both $N_{A_j}(\lambda)$ and $N_{B_j}(\lambda)$ are independent of the index j, summation of each over j from $j = 1$ to $j = r$ merely involves multiplication by r. Thus, with (165) and (166), the relations (158) of (b) above read

$$r\left(\frac{\sigma_0 b^2 \lambda}{4\pi\tau} - \theta_A b \sqrt{\frac{\lambda\sigma_0}{2\tau}}\right) \leqq N_A(\lambda) \leqq N_B(\lambda) \leqq r\left(\frac{\sigma_0 b^2 \lambda}{4\pi\tau} + \theta_B b \sqrt{\frac{\lambda\sigma_0}{\tau}}\right). \qquad (167)$$

We use the fact that rb^2 is the total area, and therefore

$$M = \sigma_0 r b^2$$

the total mass, of the membrane to conclude from (167) that

$$N_A(\lambda) = \frac{M\lambda}{4\pi\tau} + \theta'_A rb \sqrt{\frac{\lambda\sigma_0}{2\tau}} \qquad (-1 < \theta'_A < \sqrt{2}) \qquad (168)$$

and

$$N_B(\lambda) = \frac{M\lambda}{4\pi\tau} + \theta'_B rb \sqrt{\frac{\lambda\sigma_0}{2\tau}} \qquad (-1 < \theta'_B < \sqrt{2}). \qquad (169)$$

For λ large compared with $(8\pi^2 r^2 b^2 \sigma_0 \tau / M^2) = (8\pi^2\tau/\sigma_0 b^2)$ the second term of each of (168) and (169) is negligible compared with the first. For such large λ we may therefore write

$$N_A(\lambda) \sim N_B(\lambda) \sim \frac{M\lambda}{4\pi\tau}, \qquad (170)$$

or, equivalently,

$$\lim_{\lambda \to \infty} \frac{N_A(\lambda)}{\lambda} = \lim_{\lambda \to \infty} \frac{N_B(\lambda)}{\lambda} = \frac{M}{4\pi\tau}. \qquad (171)$$

We let $\lambda_k^{(A)}$ represent the kth eigenvalue of the fixed-edge membrane (S_A) and $\lambda_k^{(B)}$ the kth eigenvalue of the free-edge membrane (S_B). If[1]

---

[1] This means that λ_k (either superscript) is nondegenerate or that in the ascending sequence of the eigenvalues of the system involved λ_k denotes the final listing of a given degenerate eigenvalue.

both $\lambda_k^{(A)} < \lambda_{k+1}^{(A)}$ and $\lambda_k^{(B)} < \lambda_{k+1}^{(B)}$, we have, by definition of $N_A(\lambda)$ and $N_B(\lambda)$,

$$N_A(\lambda_k^{(A)}) = k, \qquad N_B(\lambda_k^{(B)}) = k. \tag{172}$$

Thus the asymptotic results (170) imply

$$\lambda_k^{(A)} \sim \lambda_k^{(B)} \sim \frac{4\pi\tau k}{M} \qquad \text{(large } k\text{)}. \tag{173}$$

Asymptotically, that is, the kth eigenvalue of a fixed-edge membrane is equal to the kth eigenvalue of the corresponding free-edge membrane, with the common asymptotic value depending only on the total mass M, the tension constant τ, and the index k.   Although this result is proved here only for the membrane of uniform density whose shape is such that it may be subdivided into a finite number of congruent squares, it is extended below so as to justify the italicized statement in its full generality.[1]

It is to be pointed out that the final member of (173) is not an approximation to $\lambda_k^{(A)}$ or $\lambda_k^{(B)}$ in the sense that the error of approximation can be made arbitrarily small by taking k sufficiently large.   The mode of derivation—in particular, the neglect of terms involving the undetermined θ_A' and θ_B' in (168) and (169), respectively—makes it clear that the reverse is in general true.   But although the error increases with increasing k, the *relative* error approaches zero with increasing k.   The approximation given by (173) is good only in the *asymptotic* sense; i.e., the ratio of any pair of members of (173) approaches unity as $k \to \infty$.

We may draw two important conclusions from the results of the preceding paragraphs.   First, *the multiplicity of any given (degenerate) eigenvalue is necessarily finite.*   For it follows from (168) and (169) that below any given number there lie only a finite number of eigenvalues.   Second, *the eigenvalues of a given membrane form an infinite unbounded sequence.*   Our ability to prove these facts depends upon the explicit solubility of the eigenvalue-eigenfunction problems for the uniform square membranes of fixed and free boundary edges.

The assumptions $\lambda_k^{(A)} < \lambda_{k+1}^{(A)}$ and $\lambda_k^{(B)} < \lambda_{k+1}^{(B)}$ upon which (172) is founded are unessential to the results (173).   If $\lambda_k = \lambda_{k+j} < \lambda_{k+j+1}$ (either superscript), the number k in the final member of (173) should be replaced by $(k + j)$.   Since, according to the preceding paragraph, j is necessarily finite, failure to make this replacement incurs a *relative* error which approaches zero as k increases indefinitely.

(e) We consider next a membrane of tension constant τ, for which the domain D may be subdivided, without gap or overlap, into a finite number r of congruent squares of side length b, but whose density $\sigma = \sigma(x,y)$

---

[1] In end-chapter exercise 20 the general result is extended to the membrane whose boundary edge is held elastically.

varies continuously over D. We suppose that the maximum and minimum of σ in the jth subdomain D_j are given respectively by† σ_{M_j} and σ_{m_j}, so that

$$\sigma_{m_j} \leqq \sigma \leqq \sigma_{M_j} \quad \text{in } D_j \quad (j = 1,2, \ldots ,r). \quad (174)$$

In addition to the $(2r + 2)$ membrane systems defined at the opening of (d) above we consider also the $2r$ systems

$S_{A'_j}$: Membrane of jth square of side length b, boundary edge held fixed, σ replaced by the constant σ_{m_j} $(j = 1,2, \ldots ,r)$
$S_{B'_j}$: Membrane of jth square of side length b, boundary edge free, σ replaced by the constant σ_{M_j} $(j = 1,2, \ldots ,r)$.

The symbols $N_{A'_j}(\lambda)$, $N_{B'_j}(\lambda)$ have the same meanings with respect to $S_{A'_j}$ and $S_{B'_j}$ as $N_{A_j}(\lambda)$ and $N_{B_j}(\lambda)$ have with respect to S_{A_j} and S_{B_j}, for each j.

It follows directly from theorem (iii') of 9-11(d) that

$$N_{A'_j}(\lambda) \leqq N_{A_j}(\lambda) \quad \text{and} \quad N_{B_j}(\lambda) \leqq N_{B'_j}(\lambda), \quad (175)$$

by virtue of (174). Since $S_{A'_j}$ is associated with a fixed-edge square membrane of side b and *constant* density σ_{m_j}, the result (165) of (d) above, with σ_0 replaced by σ_{m_j}, may be taken over to read

$$N_{A'_j}(\lambda) = \frac{\sigma_{m_j} b^2 \lambda}{4\pi\tau} - \theta_A b \sqrt{\frac{\lambda\sigma_{m_j}}{2\tau}} \quad (0 < \theta_A < 1).$$

With the first of (175), therefore, we have

$$\frac{\sigma_{m_j} b^2 \lambda}{4\pi\tau} - \theta_A b \sqrt{\frac{\lambda\sigma_{m_j}}{2\tau}} \leqq N_{A_j}(\lambda) \quad (0 < \theta_A < 1). \quad (176)$$

With similar application of (166) of (d), with σ_0 replaced by σ_M, we have, further,

$$N_{B_j}(\lambda) \leqq \frac{\sigma_{M_j} b^2 \lambda}{4\pi\tau} + \theta_B b \sqrt{\frac{\lambda\sigma_{M_j}}{\tau}} \quad (0 < \theta_B < 1). \quad (177)$$

Because of (176) and (177) the general result (158) of (b) above implies

$$\frac{\lambda}{4\pi\tau} \sum_{j=1}^{r} \sigma_{m_j} b^2 - \theta_A b \sqrt{\frac{\lambda}{2\tau}} \sum_{j=1}^{r} \sqrt{\sigma_{m_j}} \leqq N_A(\lambda)$$

$$\leqq N_B(\lambda) \leqq \frac{\lambda}{4\pi\tau} \sum_{j=1}^{r} \sigma_{M_j} b^2 + \theta_B b \sqrt{\frac{\lambda}{\tau}} \sum_{j=1}^{r} \sqrt{\sigma_{M_j}} \quad (0 < \theta_A, \theta_B < 1). \quad (178)$$

† The use of the subscript M to denote maximum should not be confused with the use of the same symbol (but never as subscript) to denote the total mass of the membrane.

From the definition of the double integral we have

$$\sum_{j=1}^{r} \sigma_{m_j} b^2 = \iint_D \sigma \, dx \, dy - \delta_1 = M - \delta_1 \qquad (179)$$

and

$$\sum_{j=1}^{r} \sigma_{M_j} b^2 = \iint_D \sigma \, dx \, dy + \delta_2 = M + \delta_2, \qquad (180)$$

where M is the total membrane mass and δ_1, δ_2 are *positive* numbers which can be made arbitrarily close to zero by taking b sufficiently small (or, since rb^2 is the membrane area, by taking r sufficiently large). Further, if we denote the maximum of σ in D by σ_M, we have

$$\sum_{j=1}^{r} \sqrt{\sigma_{m_j}} \leqq \sum_{j=1}^{r} \sqrt{\sigma_{M_j}} \leqq \sum_{j=1}^{r} \sqrt{\sigma_M} = r \sqrt{\sigma_M}. \qquad (181)$$

With (179), (180), and (181), the inequalities (178) imply

$$\frac{\lambda}{4\pi\tau} (M - \delta_1) - \theta_A r b \sqrt{\frac{\lambda\sigma_M}{2\tau}} \leqq N_A(\lambda)$$

$$\leqq N_B(\lambda) \leqq \frac{\lambda}{4\pi\tau} (M + \delta_2) + \theta_B r b \sqrt{\frac{\lambda\sigma_M}{\tau}} \qquad (0 < \theta_A, \theta_B < 1). \quad (182)$$

Ignoring, temporarily, the quantity $N_B(\lambda)$, we infer from (182) that

$$-\frac{\delta_1}{4\pi\tau} - \theta_A r b \sqrt{\frac{\sigma_M}{2\tau\lambda}} \leqq \frac{N_A(\lambda)}{\lambda} - \frac{M}{4\pi\tau}$$

$$\leqq \frac{\delta_2}{4\pi\tau} + \theta_B r b \sqrt{\frac{\sigma_M}{\tau\lambda}} \qquad (0 < \theta_A, \theta_B < 1). \quad (183)$$

From (183) it follows that

$$\left| \frac{N_A(\lambda)}{\lambda} - \frac{M}{4\pi\tau} \right| < \frac{\delta}{4\pi\tau} + r b \sqrt{\frac{\sigma_M}{\tau\lambda}}, \qquad (184)$$

where δ is the larger of the positive numbers δ_1, δ_2, both of which, according to (179) and (180), can be made arbitrarily small by taking b sufficiently small.

Letting λ increase indefinitely (with b held fixed), we see that the limit of the left-hand member of (184) as $\lambda \to \infty$ is less than ($\delta/4\pi\tau$). But, since this limit is independent of δ, and since δ can be made as close to zero as we please, it follows that the limit is zero, or

$$\lim_{\lambda \to \infty} \frac{N_A(\lambda)}{\lambda} = \frac{M}{4\pi\tau}. \qquad (185)$$

From (182) we may also derive the result (185), with $N_A(\lambda)$ replaced by $N_B(\lambda)$, in identical fashion.

Thus (185) and the equivalent statement for $N_B(\lambda)$ bring us to the result (171) of (d) above. We therefore conclude that all the asymptotic results enunciated in (d) as springing from (171) are applicable to the membrane of nonuniform density which is divisible into a finite number of congruent squares.

(f) With the aid of theorem (iv') of 9-11(e) we can extend the asymptotic results achieved above to the fixed-edge membrane of arbitrary

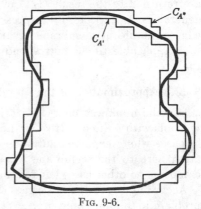

FIG. 9-6.

shape. We let S_A denote the system associated with a given fixed-edge membrane whose domain D is of arbitrary shape. The system $S_{A'}$ is characterized by the domain $D_{A'}$ which is divisible into a finite number of congruent squares and whose boundary $C_{A'}$ lies entirely in D, the same tension constant as for S_A, and a density function which coincides with that of S_A over $D_{A'}$. The system $S_{A''}$ is characterized by the domain $D_{A''}$ which is also divisible into a finite number of congruent squares and whose boundary $C_{A''}$ completely encloses D, the same tension constant as for S_A, $S_{A'}$, and a density function which coincides with that of S_A in D and is arbitrary outside D (see Fig. 9-6). All three systems are associated with fixed-edge membranes.

If the symbols $N_A(\lambda)$, $N_{A'}(\lambda)$, $N_{A''}(\lambda)$ have their usual meanings, it follows from theorem (iv') of 9-11(e) that

$$N_{A'}(\lambda) \leqq N_A(\lambda) \leqq N_{A''}(\lambda),$$

from which we have, on subtracting $(M/4\pi\tau)$ from each member,

$$\frac{N_{A'}(\lambda)}{\lambda} - \frac{M'}{4\pi\tau} + \frac{M'-M}{4\pi\tau} \leqq \frac{N_A(\lambda)}{\lambda} - \frac{M}{4\pi\tau}$$
$$\leqq \frac{N_{A''}(\lambda)}{\lambda} - \frac{M''}{4\pi\tau} + \frac{M''-M}{4\pi\tau}, \quad (186)$$

where M' is the membrane mass in $D_{A'}$ and M'' is the mass in $D_{A''}$. Because $D_{A'}$ and $D_{A''}$ are both divisible into a finite number of congruent squares, it follows from (e) above that

$$\lim_{\lambda \to \infty} \left| \frac{N_{A'}(\lambda)}{\lambda} - \frac{M'}{4\pi\tau} \right| = \lim_{\lambda \to \infty} \left| \frac{N_{A''}(\lambda)}{\lambda} - \frac{M''}{4\pi\tau} \right| = 0.$$

We therefore conclude from (186) that

$$\lim_{\lambda \to \infty} \left| \frac{N_A(\lambda)}{\lambda} - \frac{M}{4\pi\tau} \right| \leq \frac{\Delta M}{4\pi\tau}, \tag{187}$$

where ΔM is the larger of the differences $|M' - M|$, $|M'' - M|$.

If the boundary C of D is made up of a finite number of smooth arcs—which we assume to be the case—it is always possible to construct domains $D_{A'}$ and $D_{A''}$ in such fashion that ΔM is arbitrarily small. Since the limit on the left is independent of ΔM, it is therefore zero. Thus we are returned to the result (171) of (d) above, whence it follows that the general asymptotic results expressed in (d) are applicable to the fixed-edge nonuniform membrane of arbitrary shape. The results are likewise applicable to the corresponding free-edge membrane, but we omit the proof.

9-13. Approximation of the Membrane Eigenvalues

(a) The minimum characterization of the membrane eigenvalues provides us with a direct (Ritz) method for approximating these eigenvalues in cases where explicit solution cannot be effected. We limit consideration here to the membrane whose boundary edge is held fixed, with extension to other cases left for the end-chapter exercises.

According to 9-9 the kth eigenvalue λ_k of a given fixed-edge membrane problem is the minimum of

$$I = \tau \iint_D (\phi_x^2 + \phi_y^2) dx \, dy \tag{188}$$

with respect to those sufficiently regular functions ϕ which vanish on C and satisfy the normalization condition

$$\iint_D \sigma \phi^2 \, dx \, dy = 1 \tag{189}$$

and the $(k - 1)$ orthogonality relations

$$\iint_D \sigma \phi_m \phi \, dx \, dy = 0 \qquad (m = 1, 2, \ldots, k - 1),$$

where ϕ_m is the minimizing function which renders I equal to λ_m. We denote the system, in the language of 9-10(a), associated with this minimization problem by S, with K_1 the class of functions ϕ eligible for the first minimization of I.

For an approximation procedure we replace S by a system S', which has in common with it the system-defining characteristics D, τ, σ, p but for which the class K_1' of functions ϕ eligible for the first minimization is a certain subclass of K_1. According to the definition given in 9-10(a) the system S' is therefore *narrower* than S.   If $\Phi_1(x,y)$, $\Phi_2(x,y)$, . . . , $\Phi_s(x,y)$ are s conveniently given functions continuously differentiable in D, we let the class K_1' consist of all functions ψ which exhibit the form

$$\psi = c_1\Phi_1 + c_2\Phi_2 + \cdots + c_s\Phi_s, \tag{190}$$

where $c_1, c_2, \ldots , c_s$ are arbitrary constants consistent with the normalization condition (189), with ϕ replaced by ψ.

We denote by $\psi_1, \psi_2, \ldots , \psi_s$ the first s approximate eigenfunctions[1] sought and the corresponding approximate eigenvalues by $\Lambda_1, \Lambda_2, \ldots , \Lambda_s$.   In accordance with (190) we write

$$\psi_m = \sum_{j=1}^{s} c_j^{(m)}\Phi_j \qquad (m = 1,2, \ldots ,s), \tag{191}$$

so that the problem of finding each minimizing ψ_m is equivalent to that of determining the set of values $c_1^{(m)}$, $c_2^{(m)}$, . . . , $c_s^{(m)}$ for the coefficients $c_1, c_2, \ldots , c_s$, respectively, in (190), for each m.   Since the functions ψ eligible for the kth minimization of I must be orthogonal in D to the first $(k - 1)$ approximate eigenfunctions $\psi_1, \psi_2, \ldots , \psi_{k-1}$ with respect to $\sigma(x,y)$, we have, because of (190) and (191),

$$\iint_D \sigma\psi\psi_m \, dx \, dy = \sum_{i=1}^{s} \sum_{j=1}^{s} c_i c_j^{(m)}\sigma_{ij} = 0 \qquad (m = 1,2, \ldots ,k - 1), \tag{192}$$

where we define

$$\sigma_{ji} = \sigma_{ij} = \iint_D \sigma \Phi_i\Phi_j \, dx \, dy. \tag{193}$$

Substitution of (190) into the normalization condition (189) gives, further, the condition

$$\sum_{i=1}^{s} \sum_{j=1}^{s} c_i c_j\sigma_{ij} = 1. \tag{194}$$

Finally, if we define

$$\Gamma_{ji} = \Gamma_{ij} = \tau \iint_D \left(\frac{\partial \Phi_i}{\partial x} \frac{\partial \Phi_j}{\partial x} + \frac{\partial \Phi_i}{\partial y} \frac{\partial \Phi_j}{\partial y} \right) dx \, dy, \tag{195}$$

substitution of (190) for ϕ in (188) gives

$$I = \sum_{i=1}^{s} \sum_{j=1}^{s} c_i c_j\Gamma_{ij} \tag{196}$$

for the quantity whose successive minima we seek.

[1] See 7-6(b).

The problem of determining the minimum of (196) with respect to the s quantities $c_1, c_2, \ldots, c_s$ which satisfy the normalization condition (194) with the $(k - 1)$ subsidiary conditions (192) is readily seen to be identical with the corresponding problem worked out in 7-6(c). We may therefore state directly the following results:

The first s approximate eigenvalues $\Lambda_1, \Lambda_2, \ldots, \Lambda_s$ of the system S (the *precise* eigenvalues of the narrower system S') are given by the s roots of the equation in Λ

$$\begin{vmatrix} -\Gamma_{11} + \Lambda\sigma_{11} & -\Gamma_{12} + \Lambda\sigma_{12} & \cdots & -\Gamma_{1s} + \Lambda\sigma_{1s} \\ -\Gamma_{21} + \Lambda\sigma_{21} & -\Gamma_{22} + \Lambda\sigma_{22} & \cdots & -\Gamma_{2s} + \Lambda\sigma_{2s} \\ \cdots\cdots\cdots\cdots\cdots\cdots\cdots\cdots\cdots\cdots\cdots \\ -\Gamma_{s1} + \Lambda\sigma_{s1} & -\Gamma_{s2} + \Lambda\sigma_{s2} & \cdots & -\Gamma_{ss} + \Lambda\sigma_{ss} \end{vmatrix} = 0. \quad (197)$$

The coefficients $c_1^{(k)}, c_2^{(k)}, \ldots, c_s^{(k)}$, which—when k ranges over the values $1, 2, \ldots, s$—supply, through (191), the corresponding approximate eigenfunctions $\psi_1, \psi_2, \ldots, \psi_s$, are obtained by solving the system of s linear homogeneous equations

$$\sum_{j=1}^{s} (\Lambda_k\sigma_{ij} - \Gamma_{ij})c_j^{(k)} = 0 \qquad (i = 1, 2, \ldots, s)$$

in conjunction with the normalization requirement

$$\sum_{i=1}^{s} \sum_{j=1}^{s} c_i^{(k)} c_j^{(k)\prime}\sigma_{ij} = 1,$$

for each k. The constants σ_{ij} are computed by means of their definition (193); the Γ_{ij} are computed from (195).

From theorem (i) of 9-11(b) it follows directly that $\lambda_k \leqq \Lambda_k$ for all k, since S' is by definition narrower than S; that is, the approximation of *each* eigenvalue of the original system is an *approximation from above.*

(b) If the boundary curve C of D may be described by the equation $u(x,y) = 0$, a simple choice for the functions Φ_j ($j = 1, 2, \ldots, s$) introduced in (a) above is the following:

$$\Phi_1 = u, \quad \Phi_2 = ux, \quad \Phi_3 = uy, \quad \Phi_4 = ux^2, \quad \Phi_5 = uxy, \quad \Phi_6 = uy^2,$$

and, in general,

$$\Phi_j = ux^{p-q}y^q, \qquad \text{with } j = \tfrac{1}{2}p(p + 1) + (q + 1),$$

where $j = 1, 2, \ldots, s$; $q = 0, 1, \ldots, p$;

$$p = 1, 2, \ldots, \tfrac{1}{2}(\sqrt{8s + 1} - 3).$$

(Thus the choice of s is restricted to values for which $(8s + 1)$ is the square of an integer.) We employ this choice of the functions Φ_j to approximate the three lowest eigenvalues associated with the circular membrane of uniform density, with $s = 3$.

We suppose that the membrane density is the constant σ_0, with the radius equal to R. For the function $u(x,y)$ which vanishes on the boundary, we choose

$$u = R - \sqrt{x^2 + y^2}.$$

With $s = 3$ we have, according to the preceding paragraph,

$$\Phi_1 = R - \sqrt{x^2 + y^2}, \quad \Phi_2 = x(R - \sqrt{x^2 + y^2}), \quad \Phi_3 = y(R - \sqrt{x^2 + y^2}).$$

With the introduction of polar coordinates ($x = r \cos \theta$, $y = r \sin \theta$) so that (193) and (195) become respectively

$$\sigma_{ij} = \sigma_{ji} = \sigma_0 \int_0^R \int_0^{2\pi} \Phi_i \Phi_j r \, d\theta \, dr$$

and

$$\Gamma_{ij} = \Gamma_{ji} = \tau \int_0^R \int_0^{2\pi} \left(\frac{\partial \Phi_i}{\partial x} \frac{\partial \Phi_j}{\partial x} + \frac{\partial \Phi_i}{\partial y} \frac{\partial \Phi_j}{\partial y} \right) r \, d\theta \, dr,$$

we compute directly that

$$\Gamma_{11} = \pi \tau R^2, \qquad \Gamma_{22} = \Gamma_{33} = \tfrac{1}{4}\pi \tau R^4,$$
$$\Gamma_{12} = \Gamma_{21} = \Gamma_{13} = \Gamma_{31} = \Gamma_{23} = \Gamma_{32} = 0$$

and

$$\sigma_{11} = \tfrac{1}{6}\pi \sigma_0 R^4, \qquad \sigma_{22} = \sigma_{33} = \tfrac{1}{60}\pi \sigma_0 R^6,$$
$$\sigma_{12} = \sigma_{21} = \sigma_{13} = \sigma_{31} = \sigma_{23} = \sigma_{32} = 0.$$

With these results the determinantal equation (197) assumes the particularly simple form

$$(-\pi \tau R^2 + \tfrac{1}{6}\pi \sigma_0 R^4 \Lambda)(-\tfrac{1}{4}\pi \tau R^4 + \tfrac{1}{60}\pi \sigma_0 R^6 \Lambda)^2 = 0,$$

whence we obtain for the first three approximate eigenvalues of the uniform circular membrane

$$\Lambda_1 = \frac{6\tau}{\sigma_0 R^2}, \qquad \Lambda_2 = \Lambda_3 = \frac{15\tau}{\sigma_0 R^2}.$$

The precise results[1] are

$$\lambda_1 = \frac{j_{01}^2 \tau}{\sigma_0 R^2}, \qquad \lambda_2 = \lambda_3 = \frac{j_{11}^2 \tau}{\sigma_0 R^2}.$$

where $j_{01} = 2.40$, $j_{11} = 3.83$, to two decimal places.

[1] See exercise 13 at the end of this chapter.

Although the foregoing illustration is relatively free from the computational difficulties one generally encounters, it should supply an adequate guide to the application of the method to cases which involve greater complications.

EXERCISES

1. (a) Starting with equation (7), and without the aid of Green's theorem, derive (8) by carrying out a suitable integration by parts in each of the last three integrand terms.

(b) Generalize the procedure of part (a) to derive the general Euler-Lagrange equation (10).

2. The equations

$$x = x' \cos Q - y' \sin Q,$$
$$y = x' \sin Q + y' \cos Q, \tag{198}$$

define the transformation from a plane cartesian xy coordinate system to a second cartesian $x'y'$ system having the same origin, where Q (a constant) is the angle through which the axes of the former system must be rotated counterclockwise to bring them into coincidence with the axes of the latter.

(a) Use (22), reduced to two independent variables, with $f = \frac{1}{2}w_x^2$, to derive the relation

$$w_{xx} = w_{x'x'} \cos^2 Q + w_{y'y'} \sin^2 Q - 2w_{x'y'} \sin Q \cos Q.$$

(b) Derive analogous expressions for w_{yy} and w_{xy} with the aid of (22).

(c) Use the results of parts (a) and (b) to show that the expression $w_{xx}w_{yy} - w_{xy}^2$ is unaltered by a rotation of the coordinate axes.

(d) Show that the laplacian $w_{xx} + w_{yy}$ is unaltered by a rotation of coordinate axes.

3. Show that the solution of the system (20) for w_{x_1}, w_{x_2}, w_{x_3} is always possible if the jacobian of the transformation is nonvanishing, as required.

4. (a) Given the transformation from cartesian to paraboloidal coordinates (p, q, ϕ)

$$x = pq \cos \phi, \qquad y = pq \sin \phi, \qquad z = \frac{1}{2}(p^2 - q^2),$$

show that

$$\nabla^2 w = \frac{1}{p^2 + q^2}\left[\frac{1}{p}\frac{\partial}{\partial p}(pw_p) + \frac{1}{q}\frac{\partial}{\partial q}(qw_q) \right] + \frac{1}{p^2q^2} w_{\phi\phi}.$$

HINT: Show that (30) is fulfilled and that $h_p = h_q = \sqrt{p^2 + q^2}$, $h_\phi = pq$. Then use (40).

Describe the three families of surfaces $p = $ constant, $q = $ constant, $\phi = $ constant.

(b) Given the transformation from plane cartesian to plane elliptic coordinates (p, q)

$$x = \frac{\sqrt{pq}}{c}, \qquad y = \frac{\sqrt{(p - c^2)(c^2 - q)}}{c} \qquad (0 \leqq q \leqq c^2 \leqq p), \tag{199}$$

where c is a positive constant, show that

$$\nabla^2 w = \frac{4\sqrt{p(p - c^2)}}{p - q}\frac{\partial}{\partial p}(\sqrt{p(p - c^2)}\,w_p) + \frac{4\sqrt{q(c^2 - q)}}{p - q}\frac{\partial}{\partial q}(\sqrt{q(c^2 - q)}\,w_q).$$

Describe the families of curves $p = $ constant, $q = $ constant, as defined by (199).

5. By solving the equations of transformation to spherical coordinates

$$x = r \sin \theta \cos \phi, \qquad y = r \sin \theta \sin \phi, \qquad z = r \cos \theta$$

for r, θ, ϕ, show that r represents the distance from (x,y,z) to the origin, θ the angle between the positive z axis and the line drawn to (x,y,z) from the origin, ϕ the angle between the xz plane (positive x) and the half-plane bounded by the z axis and containing (x,y,z). Thus describe the families of surfaces $r =$ constant, $\theta =$ constant, $\phi =$ constant.

6. (a) Work out the details of the assertion made in 9-4(c).

(b) Prove the assertion made in the final paragraph of 9-5(d).

7. A membrane having all the characteristics of the membrane described in 9-3 is subjected, additionally, to a nonconservative transverse force per unit area given by the expression $F(x,y,t)$. (That is, an element of area $dx\,dy$ experiences the externally applied force $F(x,y,t)dx\,dy$ perpendicular to the xy plane.)

(a) Use the extended Hamilton's principle of 6-7 to show that the equation of motion of the membrane so influenced is

$$\sigma\,\frac{\partial^2 w}{\partial t^2} = \tau\nabla^2 w + F(x,y,t). \tag{200}$$

(b) Extend the method of 9-7 to show that the solution of (200), with $w = 0$ on C, is

$$w = \sum_{j=1}^{\infty} c_j(t)\phi_j(x,y),$$

where

$$\tau\nabla^2\phi_j + \lambda_j\sigma\phi_j = 0 \text{ in } D, \qquad \phi_j = 0 \text{ on } C,$$

and

$$\frac{d^2 c_j}{dt^2} + \lambda_j c_j = \iint_D F(x,y,t)\phi_j dx\,dy.$$

(Each ϕ_j is normalized in D with respect to σ.)

8. Given the inhomogeneous boundary condition $w = g(x,y)$ on C for the membrane equation

$$\sigma\,\frac{\partial^2 w}{\partial t^2} = \tau\nabla^2 w,$$

show that we may write $w = u(x,y,t) + v(x,y)$ where $u = 0$ and $v = g(x,y)$ on C, $\nabla^2 v = 0$ and $\sigma\ddot{u} = \tau\nabla^2 u$ in D. Thus we reduce a membrane problem with an inhomogeneous boundary condition to one having a homogeneous boundary condition, plus a solution of the two-dimensional Laplace's equation with boundary values prescribed. (The latter part of the problem is discussed in Chap. 12.)

9. Suppose that the tension constant τ, introduced in 9-3(b), is replaced by the continuously differentiable positive function $\tau = \tau(x,y)$.

(a) Derive the differential equation of motion (corresponding to (49)) for such a membrane.

(b) Derive the equation satisfied by ϕ if $w = \phi(x,y)q(t)$ is a solution of this equation of motion.

(c) Prove that the eigenfunctions of the equation derived in part (b) form an orthogonal set in D with respect to the weight function σ if they are required to satisfy on C any of the homogeneous boundary conditions discussed in this chapter—namely, $\phi = 0$, $\tau(\partial\phi/\partial n) + p\phi = 0$, or a mixture.

10. In 9-5(e) it is pointed out that $\lambda = 0$ is an eigenvalue of the free-edge membrane.

Return to 9-4(a) to show that the time-dependent factor corresponding to the associated eigenfunction (a constant) is $q = A + Bt$, where A and B are arbitrary constants.

11. (a) Derive the result (92) of 9-6(c).

(b) Use the Schmidt process to show that a set of three linear combinations of $1, x, y$ which form an orthogonal set with respect to the weight function $\sigma = 1$ in the domain $0 \leq x \leq 1, 0 \leq y \leq 1$ is $v_1 = 1, v_2 = x - \frac{1}{2}, v_3 = y - \frac{1}{2}$.

(c) Show that the Schmidt orthogonalization process is in general not unique in its determination of the N orthogonal linear combinations.

(d) Prove that a set $\phi_1, \phi_2, \ldots, \phi_N$ of orthogonal functions is linearly independent.

12. (a) Modify the analysis of 9-7 to show that the results achieved are equally valid in the case of the fixed-edge membrane ($w = 0$ on C).

(b) Show that in the case of the free-edge membrane the term $(A_0 + B_0 t)$ must be added to the right-hand member of (107) of 9-7(a) (see exercise 10 above).

(c) Given that $w(x,y,0) = g(x,y)$ and $\dot{w}(x,y,0) = h(x,y)$, show that the coefficients in (107) have the values

$$A_m = \iint\limits_D \sigma \phi_m g \, dx \, dy, \qquad B_m = \frac{1}{\sqrt{\lambda_m}} \iint\limits_D \sigma \phi_m h \, dx \, dy.$$

Give the values of A_0 and B_0 in the case mentioned in part (b).

13. We consider, throughout this exercise, a circular membrane of radius R. We use the polar coordinates (r,θ) with origin at the center of the circle, so that $r = R$ is the equation of the membrane boundary.

(a) Use (43) of 9-2(e) to show that the equation

$$\tau \nabla^2 \phi + \lambda \sigma \phi = 0 \tag{201}$$

is separable (in the sense of 9-8(a)) if and only if σ is independent of θ.

(b) We must require that $\phi = \phi(r,\theta)$ be a single-valued function of position in D, so that $\phi(r, \theta + 2\pi) = \phi(r,\theta)$. Show that, if $\sigma = \sigma(r)$, equation (201) has solutions of the form $H_n(r) \sin n\theta$ and $H_n(r) \cos n\theta$, where $n = 0, 1, 2, \ldots$.

(c) If $\sigma = \sigma_0$, a constant, show that

$$H_n(r) = J_n\left(r \sqrt{\frac{\lambda \sigma_0}{\tau}} \right),$$

where $J_n(z)$ is the nth-order Bessel function of the first kind, provided we impose the condition that ϕ be finite everywhere in D. HINT: Compare the r-dependent differential equation with (41) of 8-3(c).

(d) If we impose the boundary condition $\phi = 0$ for $r = R$, show that the eigenvalues are given by the scheme

$$\lambda_{nk} = \frac{\tau j_{nk}^2}{R^2 \sigma_0} \qquad (n = 0,1,2, \ldots ; k = 1,2,3, \ldots, \text{independently}),$$

where j_{nk} is the kth positive zero of $J_n(z)$. Thus show that the two (unnormalized) eigenfunctions which correspond to the eigenvalue λ_{nk} ($n = 1,2,3, \ldots$) are $J_n(rj_{nk}/R) \cos n\theta$ and $J_n(rj_{nk}/R) \sin n\theta$. How many eigenfunctions are associated with each of the eigenvalues λ_{0k}?

(e) Show that the eigenfunctions associated with λ_{nk} vanish on the circles whose radii are given by $(j_{ni}/j_{nk})R$, for $i = 1, 2, \ldots, k - 1$. (These are the so-called *nodal circles*.)

(f) Show that the vibration frequency associated with the eigenvalue λ_{nk} is $(1/2\pi)$ $\sqrt{\tau/\sigma_0}\,(j_{nk}/R)$.

14. Use the orthonormality of the eigenfunctions to derive the expression for c_m given in (94) of 9-6(d).

15. Show that $(k-1)$ linear homogeneous equations among the quantities c_1, c_2, . . . , c_k—subject to the condition

$$\sum_{j=1}^{k} c_j^2 = 1$$

—always possess at least one solution. HINT: Consider the system of k linear homogeneous equations consisting of the original $(k-1)$ equations and any one of these repeated once. Evaluate the determinant of this system (see 2-8(b)).

16. List the physical consequences of theorems (i), (ii), (iii) of 9-11 which are direct generalizations of the physical consequences drawn in the text from theorems (i), (ii), (iii) of 9-10.

17. Prove the following extension of theorem (iii) of 9-11(d):
If the maximum of σ_A is less than the minimum of σ_B in D, then $\lambda_k^{(A)} > \lambda_k^{(B)}$. HINT: First show, from the differential equation, that the kth eigenvalue of a constant-σ system is inversely proportional to σ.

18. It is required to extremize the integral

$$I = \iint_D f(\phi_x, \phi_y)\,dx\,dy$$

with respect to functions which are continuous, with their first derivatives, in D— except for a finite number of smooth arcs which subdivide D, without gap or overlap, into a finite number of subdomains; across these arcs, the eligible functions ϕ may exhibit finite discontinuities. Let the subdomains be denoted by D_1, D_2, . . . , D_r and the respective boundaries by C_1, C_2, . . . , C_r.

(a) Show that the "$\epsilon\eta$ process" of 9-5(b)—with the time integral suppressed, and with extension to take care of the allowable discontinuities—leads to the result

$$\sum_{i=1}^{r} \left\{ -\iint_{D_i} \eta \left[\frac{\partial}{\partial x}\left(\frac{\partial f}{\partial \phi_x}\right) + \frac{\partial}{\partial y}\left(\frac{\partial f}{\partial \phi_y}\right) \right] dx\,dy + \int_{C_i} \eta \left[\frac{\partial f}{\partial \phi_x}\frac{dy}{ds} - \frac{\partial f}{\partial \phi_y}\frac{dx}{ds} \right] ds \right\} = 0.$$

HINT: Apply Green's theorem (22) of 2-13 to each subdomain D_i separately.

(b) Show that the permissibility of discontinuities across each C_i of the eligible functions ϕ allows us to choose η arbitrarily in the line integral along each C_i. Hence, conclude that

$$\frac{\partial f}{\partial \phi_x}\frac{dy}{ds} - \frac{\partial f}{\partial \phi_y}\frac{dx}{ds} = 0 \qquad \text{on } C_i \qquad (i = 1, 2, \ldots, r).$$

But every point of the boundary C of D is a point of at least one of the C_i; thus we have the result that the boundary condition satisfied by the extremizing ϕ is the same at each subdivision boundary as it is at the exterior boundary of the whole domain.

(c) Generalize the final result of part (b) to include cases in which ϕ is required to satisfy normalization and orthogonality conditions in D. Establish the assertion made in 9-12(b) that any eigenfunction ϕ_k of $S_{B'}$—which satisfies $(\partial\phi_k/\partial n) = 0$ on C—must satisfy the same relation on each of C_1, C_2, . . . , C_r.

19. In 9-12(c) show that $(\theta_B + \frac{1}{2}\sqrt{2}\,\theta_A) < (2/\pi)$, and therefore that $\theta_B < (2/\pi)$, $\theta_A < (2\sqrt{2}/\pi)$. HINT: First show that $N_{B_1}(\lambda) \leqq N_{A_1}(\lambda) + (2R - 1)$.

20. We consider three membrane systems S, S', S'' which involve the same physical membrane; but S involves the boundary edge fixed, S' the boundary edge held elastically, and S'' the boundary edge free.

(a) If λ_k, λ_k', λ_k'' represent the kth eigenvalues of the respective systems, prove that $\lambda_k'' \leqq \lambda_k' \leqq \lambda_k$.

(b) Use the result of part (a) to extend the asymptotic results of 9-12 to the case of the membrane with boundary held elastically.

21. (a) Use the orthogonality property to prove that a given membrane system can possess at most one eigenfunction which does not vanish in the interior of D.

(b) If a membrane eigenfunction changes sign anywhere in D, continuity requires that it must do so either across an arc which has its end points on the boundary or across some closed curve in D. A curve across which an eigenfunction changes sign is called a *nodal line*.

(c) Prove that the eigenfunction associated with the lowest eigenvalue of a fixed-edge membrane system S can have no nodal lines in the interior of D.

PROOF: Assume the contrary: Suppose that the nodal line divides D into two or more subdomains in each of which the eigenfunction ϕ_1 has one sign, with a reversal of sign between adjacent subdomains; let D^* be one of these subdomains. Show that the function ϕ^*, defined as equal to ϕ_1 in D^*, is an eigenfunction of the system S^* associated with D^* (boundary edge fixed, same σ, τ as for S) which corresponds with the *lowest* (see (iv) of 9-11(e)) eigenvalue λ_1 of S^*, equal to the lowest eigenvalue of S. Let D' be any subdomain of D such that D' contains D^* as a subdomain. Prove, with the aid of theorem (iv) of 9-11(e), that the lowest eigenvalue of the system S' associated with D' (boundary edge fixed, same σ, τ as for S) is also λ_1; let ϕ be the corresponding eigenfunction. Show that the function which is equal to ϕ in D' and is identically zero outside D' is an eigenfunction of S associated with the lowest eigenvalue λ_1. Construct a sequence D_1', D_2', . . . , D_m' of subdomains of the type D', where D_{j-1}' is a subdomain of D_j' ($j = 2,3, . . . ,m - 1$). Thus show that it is possible to construct m *linearly independent* eigenfunctions $\phi^{(1)}$, $\phi^{(2)}$, . . . , $\phi^{(m)}$—with $\phi^{(j)}$ identically zero outside D_j'—of the system S which all correspond with the single eigenvalue λ_1. Since m can be chosen arbitrarily, this result contradicts the conclusion of 9-12(d) that the multiplicity of any given eigenvalue is necessarily finite.

(d) Show that the lowest eigenvalue of a fixed-edge membrane is nondegenerate.

(e) Illustrate parts (c) and (d) by means of the rectangular and circular fixed-edge membranes of uniform density.

22. (a) Return to Chap. 8 and derive, in the manner of 9-11, a maximum-minimum characterization of the Sturm-Liouville eigenvalues.

(b) On the basis of part (a) prove that an increase of the function $\tau = \tau(x)$ cannot decrease the nth Sturm-Liouville eigenvalue λ_n; that an increase of $\sigma = \sigma(x)$ cannot increase λ_n.

(c) Prove the result analogous to (158) of 9-12(b) for the Sturm-Liouville eigenvalues, where the interval $x_1 \leqq x \leqq x_2$ is subdivided into r sections in the fixed- and free-end-point cases.

(d) On the basis of parts (b) and (c) derive the asymptotic formula

$$\lambda_n \sim \frac{n^2\pi^2}{\left(\displaystyle\int_{x_1}^{x_2} \sqrt{\sigma/\tau}\,dx\right)^2}$$

in the case $\mu = 0$. HINT: Use the fact that $\lambda_n = [n^2\pi^2\tau/\sigma(x_2 - x_1)^2]$ if σ and τ are both constant, and compare 9-12(e)—with the use of the quantities $(\sigma/\tau)_{m_j}$ and $(\sigma/\tau)_{M_j}$ in place of (σ_{m_j}/τ) and (σ_{M_j}/τ), respectively; part (b) above is required.

23. Apply the final remark of 9-13(a) to the approximation of the vibrating-string eigenvalues developed in 7-6(c). HINT: Use exercise 22(a) above.

24. What change in the procedure of 9-13(a) is required if it is to be applicable to the free-edge membrane?

25. Extend the work of the foregoing chapter to the three-dimensional analogue of the vibrating membrane; that is, consider the case in which

$$T = \tfrac{1}{2} \iiint\limits_R \sigma \dot{w}^2 \, dx \, dy \, dz, \qquad V = \tfrac{1}{2}\tau \iiint\limits_R (w_x^2 + w_y^2 + w_z^2) dx \, dy \, dz,$$

where T and V are respectively the kinetic and potential energies of the given system which occupies the region R of three-dimensional space. Here $\sigma = \sigma(x,y,z)$ may be interpreted initially as mass per unit volume, τ as an elastic constant; w may be considered to measure some sort of displacement from equilibrium. We consider two cases: $w = 0$ on the boundary B of R, and w completely unspecified on B.

Work out the details of the following outline of procedure:

(a) Use Hamilton's principle to derive the differential equation

$$\tau\nabla^2 w = \sigma \frac{\partial^2 w}{\partial t^2}, \tag{202}$$

where $\nabla^2 w$ is here the *three*-dimensional laplacian. Show that the eigenfunctions of the problem satisfy

$$\tau\nabla^2\phi + \lambda\sigma\phi = 0 \qquad \text{in } R,$$

with either $\phi = 0$ on B or $(\partial\phi/\partial n) = 0$ on B.

(b) Assuming the validity of an expansion theorem analogous to that given in 9-9(d), prove a minimum, then a maximum-minimum, characterization of the eigenvalues of the system.

(c) If $(\tau/\sigma) = c^2$, a constant, solve the eigenvalue-eigenfunction problem for the cube of side length b in the case $\phi = 0$ on B. Show that the eigenvalues are given by

$$\lambda_{mkj} = \frac{c^2\pi^2}{b^2}(m^2 + k^2 + j^2) \qquad (m,k,j = 1,2,3, \ldots).$$

Show that for the case $(\partial\phi/\partial n) = 0$ on B (the free-boundary case) the eigenvalues are given by the same formula, except that m, k, j may each take on the value zero (cf. 9-8(b,c)).

(d) Let $N_{A_1}(\lambda)$ be the number of eigenvalues less than or equal to λ in the fixed-boundary case in part (c) and let $N_{B_1}(\lambda)$ be the corresponding quantity for the cube with boundary free. In the manner of 9-12(c) derive the expressions

$$N_{A_1}(\lambda) = \frac{b^3}{6\pi^2 c^3}\lambda^{\frac{3}{2}} - \frac{1}{2}\sqrt{3}\,\theta_A\frac{\lambda b^2}{\pi c^2} \qquad (0 < \theta_A < 1),$$

$$N_{B_1}(\lambda) = \frac{b^3}{6\pi^2 c^3}\lambda^{\frac{3}{2}} + \sqrt{3}\,\theta_B\frac{\lambda b^2}{\pi c^2} \qquad (0 < \theta_B < 1).$$

(e) Use these last results, together with the maximum-minimum principle of part (b) above, to derive the asymptotic formula (for the fixed-boundary case)

$$N(\lambda) \sim \frac{W}{6\pi^2 c^3} \lambda^{\frac{3}{2}}$$

for the number of eigenvalues less than or equal to λ, where W is the volume of the region R; it is required that the eigenfunctions vanish on the boundary B of R. (The proof requires merely a repetition of the steps carried out in the two-dimensional investigation of 9-12. All the intermediate results which are required can be derived from the maximum-minimum principle for the eigenvalues in the three-dimensional problem.)

(*f*) *Cavity (black-body) radiation.* Show, as an adjunct to part (*b*) above, that w is capable of varying periodically in time with frequency $\nu = (1/2\pi)\sqrt{\lambda}$, if λ is an eigenvalue of the problem; such values of ν are termed "natural vibration frequencies," as in the case of the membrane. Thus show that if $n(\nu)$ is the number of natural frequencies less than or equal to ν, we have, in the fixed-boundary case,

$$n(\nu) \sim \frac{4\pi W}{3c^3} \nu^3, \tag{203}$$

where W is again the volume of R.

This last result is of tremendous importance in the theory of thermal radiation in a cavity—so-called *black-body radiation*. In the theory of this radiation, which is described by the differential equation (202) of part (*a*) above, it is required to determine the asymptotic distribution of radiation frequencies. In physics texts the derivation is usually carried out for the cubical region and is followed by a statement of its provable validity—with b^3 replaced by the volume W—for volumes of arbitrary shape. The proof is embodied in this exercise. (For application to cavity radiation the right-hand member of (203) must be multiplied by the factor 2 because of the two possible polarization directions which are associated with each electromagnetic vibration. Here c is the velocity of light.)

The result (203) is also applied to the theory of vibrations of a crystalline solid.

CHAPTER 10

THEORY OF ELASTICITY

In the ensuing chapter we consider some phases of the mathematical theory of elasticity in its relationship to the calculus of variations. The first part of the chapter is devoted mainly to deriving the basic equations of elasticity theory as direct consequences of the extended Hamilton's principle (6-7). The latter portions focus attention on the problems of the vibrating rod and the vibrating plate.

While this chapter should be of importance mainly to those individuals who possess some background in the theory of elasticity, its subject matter is meant to be sufficiently self-contained to be of interest to persons not specifically studied in the field but who have mastered the content of the preceding chapters of this book. The introductory discussion of the basic notions involved is necessarily held to minimal compactness, however.

Throughout we limit consideration to the usual linear theory—*i.e.*, to the study of deformations so small that the generally useful Hooke's law $(10\text{-}1(d))$ is applicable.

10-1. Stress and Strain

(a) We consider a deformable solid body under the influence of two sets of force distribution: (i) so-called *body forces*, which in general act through the entire extent R of the body—whereby the force exerted upon any volume element in the neighborhood of a given point is proportional to the volume of the element; and (ii) so-called *surface forces*, which act only at the boundary surface B of the body—whereby the force exerted upon any element of surface area in the neighborhood of a given surface point is proportional to the area of the element.

The three cartesian (x_1, x_2, x_3) components of the body-force density are denoted by F_1, F_2, F_3, respectively, so that the total body force acting upon the given solid in the x_k direction is accordingly

$$\iiint_R F_k \, dx_1 \, dx_2 \, dx_3 \qquad (k = 1,2,3),$$

where the integration is carried out over the entire region R occupied by the body. In general F_1, F_2, F_3 are functions of position in R.

The three cartesian components of the surface-force density are respectively denoted by T_1, T_2, T_3, so that the total x_k component of the surface force acting upon the given solid is

$$\iint_B T_k \, dS \qquad (k = 1,2,3),$$

where the integration is carried out over the entire boundary surface B of the body. In general T_1, T_2, T_3 are functions of position on B.

The most usual example of a body-force distribution is the influence of a gravitational field. Surface forces are in operation whenever a body is subject to contact with external agencies at its surface.

(b) As the result of the application of body and surface forces to a deformable solid body there occur, in general—in addition to the accelerations considered in the study of rigid mechanics—displacements of the particles of the body relative to one another; *i.e.*, a state of strain exists within the body. If the body is elastic, the imposition of a state of strain calls into operation forces which behave in such fashion as to resist the deformation and so tend to restore the body to its unstrained state—*i.e.*, to the state which would obtain in the absence of all body and surface forces.

The elastic forces which tend to oppose deformation are described in terms of a system of stresses defined in the following way: At any given point of the body we draw an arbitrary plane element of area normal to a direction denoted by n; we consider the elastic force per unit area exerted across the element by the material on the positive (with respect to the n direction arbitrarily chosen) side of the element upon the material on the opposite side. The three cartesian components of this force per unit area—the so-called *stress vector*—are denoted respectively by T_{n1}, T_{n2}, T_{n3}. In general the values of these components depend upon the orientation of the element of area as well as the point of the body under consideration. In particular, if we choose the n direction to coincide with the cartesian x_k direction $(k = 1,2,3)$, the components of the stress vector are denoted by

$$T_{k1}, \ T_{k2}, \ T_{k3} \qquad (k = 1,2,3). \tag{1}$$

The nine quantities appearing in (1) are called the elements of the *stress tensor* evaluated at the point under consideration. They are assumed to be continuous, single-valued, continuously twice-differentiable functions of position within and on the boundary surface of the body. From the definition of the stress tensor it is clear that the "diagonal" elements T_{11}, T_{22}, T_{33} represent pure tensions or pressures normal to plane elements parallel to the cartesian coordinate planes—tensions if positive, pressures if negative. An "off-diagonal" element T_{kj} $(j \neq k)$ represents a shearing stress in the x_j direction and acting across a plane element normal to the x_k direction.

Elementary considerations[1] lead to the symmetry of the stress tensor—namely,

$$T_{kj} = T_{jk} \qquad (k,j = 1,2,3 \text{ independently}),$$

so that only six of the nine elements T_{jk} are independent. Further,[2] the stress vector across any elementary plane area of arbitrary normal direction n is related to the stress tensor at the point under consideration by the set of three equations

$$T_{nk} = T_{1k} \cos (n,1) + T_{2k} \cos (n,2) + T_{3k} \cos (n,3) \qquad (k = 1,2,3), \quad (2)$$

where $\cos (n,j)$ is the cosine of the angle between the x_j direction and the positive direction of the normal n.

(c) In the analysis of strain in a given solid body R we fix our attention upon a single point of R, whose cartesian coordinates in the unstrained state are x_1, x_2, x_3, and the close neighborhood of this point. We suppose that in the strained state the cartesian coordinates of the same point have become $(x_1 + u_1)$, $(x_2 + u_2)$, $(x_3 + u_3)$, with† $u_k = u_k(x_1,x_2,x_3)$. We consider also a close neighboring point whose coordinates before strain are x_1', x_2', x_3' and whose coordinates under strain are $(x_1' + u_1')$, $(x_2' + u_2')$, $(x_3' + u_3')$, with $u_k' = u_k(x_1',x_2',x_3')$. Thus the components of *relative* displacement of the two points in passing from the unstrained to the strained state are

$$u_k' - u_k = u_k(x_1',x_2',x_3') - u_k(x_1,x_2,x_3) \qquad (k = 1,2,3).$$

The definition of "close neighborhood" is such that the partial derivatives $(\partial u_1/\partial x_1)$, etc., which appear in the analysis following may be considered as constant over the neighborhood and that only terms linear in $(x_k' - x_k)$ need be kept in the expressions giving the relative displacements $(u_k' - u_k)$.

We develop the displacements $u_k' = u_k(x_1',x_2',x_3')$ as Taylor series[3] with neglect of quadratic and higher terms:

[1] See Sokolnikoff, pp. 41–43.
[2] Sokolnikoff, p. 39. An independent proof is called for in exercise 5 at the end of this chapter.
† In general $u_k = u_k(x_1,x_2,x_3,t)$ where t is the time variable. Since the analysis immediately following applies to a single instant of time, we ignore the fact that each u_k may vary with time.
[3] See 2-10.

$$u'_k = u_k + \frac{\partial u_k}{\partial x_1}(x'_1 - x_1) + \frac{\partial u_k}{\partial x_2}(x'_2 - x_2) + \frac{\partial u_k}{\partial x_3}(x'_3 - x_3)$$

$$(k = 1,2,3), \quad (3)$$

since $u_k(x_1,x_2,x_3) = u_k$, by definition. A useful rearrangement of the result (3) is

$$u'_k - u_k = \frac{1}{2}\left(\frac{\partial u_k}{\partial x_1} + \frac{\partial u_1}{\partial x_k}\right)(x'_1 - x_1) + \frac{1}{2}\left(\frac{\partial u_k}{\partial x_2} + \frac{\partial u_2}{\partial x_k}\right)(x'_2 - x_2)$$

$$+ \frac{1}{2}\left(\frac{\partial u_k}{\partial x_3} + \frac{\partial u_3}{\partial x_k}\right)(x'_3 - x_3) + \frac{1}{2}\left(\frac{\partial u_k}{\partial x_1} - \frac{\partial u_1}{\partial x_k}\right)(x'_1 - x_1)$$

$$+ \frac{1}{2}\left(\frac{\partial u_k}{\partial x_2} - \frac{\partial u_2}{\partial x_k}\right)(x'_2 - x_2) + \frac{1}{2}\left(\frac{\partial u_k}{\partial x_3} - \frac{\partial u_3}{\partial x_k}\right)(x'_3 - x_3), \quad (4)$$

for $k = 1, 2, 3$. With the definitions

$$e_{jk} = e_{kj} = \frac{1}{2}\left(\frac{\partial u_k}{\partial x_j} + \frac{\partial u_j}{\partial x_k}\right) \quad (j,k = 1,2,3, \text{ independently}) \quad (5)$$

and

$$\omega_1 = \frac{1}{2}\left(\frac{\partial u_3}{\partial x_2} - \frac{\partial u_2}{\partial x_3}\right), \quad \omega_2 = \frac{1}{2}\left(\frac{\partial u_1}{\partial x_3} - \frac{\partial u_3}{\partial x_1}\right), \quad \omega_3 = \frac{1}{2}\left(\frac{\partial u_2}{\partial x_1} - \frac{\partial u_1}{\partial x_2}\right) \quad (6)$$

the three equations (4) for the components of relative displacement may be rewritten

$$u'_1 - u_1 = e_{11}(x'_1 - x_1) + e_{12}(x'_2 - x_2) + e_{13}(x'_3 - x_3)$$
$$- \omega_3(x'_2 - x_2) + \omega_2(x'_3 - x_3),$$
$$u'_2 - u_2 = e_{21}(x'_1 - x_1) + e_{22}(x'_2 - x_2) + e_{23}(x'_3 - x_3)$$
$$+ \omega_3(x'_1 - x_1) - \omega_1(x'_3 - x_3), \quad (7)$$
$$u'_3 - u_3 = e_{31}(x'_1 - x_1) + e_{32}(x'_2 - x_2) + e_{33}(x'_3 - x_3)$$
$$- \omega_2(x'_1 - x_1) + \omega_1(x'_2 - x_2).$$

The quantities defined by (5) and (6) are assumed so small compared with unity that squares and products—for example, ω_1^2, $e_{12}\omega_3$, etc.—may be neglected in the linear theory to which we restrict our attention.

It is easily demonstrated[1] by means of the equations (7) that the change which the distance between two neighboring points (x_1,x_2,x_3) and (x'_1,x'_2,x'_3) undergoes as the result of strain is independent of the quantities ω_1, ω_2, ω_3, but depends only on the quantities e_{jk}—provided we ignore squares and products of these quantities, as stipulated in the preceding paragraph. For this reason we should expect elastic forces—*i.e.*, stresses —to develop only as a result of those relative displacements embodied in

---

[1] See exercise 1 at end of chapter.

the terms of (7) which involve the quantities e_{jk}. The quantities e_{jk}, defined by (5), are called the elements of the *strain tensor* evaluated at (x_1,x_2,x_3). According to (5) the strain tensor is symmetric—that is, $e_{jk} = e_{kj}$—so that in general only six of its nine elements are independent. These elements e_{jk} are assumed to be twice continuously differentiable functions of position in R and on B.

(d) The main body of the mathematical theory of elasticity rests upon the assumption of a linear homogeneous relation between the elements of the stress tensor on the one hand and the elements of the strain tensor on the other. This type of relationship, known as *Hooke's law*, is generally applicable, provided the strain elements involved lie below certain values characteristic of the material under consideration. The most general form of Hooke's law is embodied in the six (since $e_{jk} = e_{kj}$ and $T_{jk} = T_{kj}$) equations

$$T_{jk} = C_{11}^{jk}e_{11} + C_{22}^{jk}e_{22} + C_{33}^{jk}e_{33} + C_{12}^{jk}e_{12} + C_{23}^{jk}e_{23} + C_{31}^{jk}e_{31}$$
$$(j,k = 1,2,3, \text{ independently}), \quad (8)$$

where the quantities C_{rs}^{jk} are elastic constants of the material to which the law is applied.

If we limit our study to bodies which are elastically isotropic—*i.e.*, whose elastic properties at any given point are independent of direction—the number of independent elastic constants is reduced from 36 to 2, and the Hooke's law equations (8) read

$$T_{jk} = A\,\delta_{jk}(e_{11} + e_{22} + e_{33}) + 2Be_{jk} \quad (j,k = 1,2,3, \text{ independently}), \quad (9)$$

where A and B are experimentally determined positive elastic constants of the material, assumed homogeneous as well as isotropic. (As in earlier chapters, δ_{jk} is the Kronecker delta—zero for $j \neq k$, unity for $j = k$.) In the following sections of this chapter, wherever a relation between stress and strain is required, we assume the validity of (9)—Hooke's law for a homogeneous isotropic elastic solid.

Solving the six equations (9) for the strain elements, we obtain

$$e_{kk} = \frac{1}{E}[T_{kk} - \sigma(T_{jj} + T_{ii})], \; e_{kj} = \frac{1 + \sigma}{E}\,T_{kj} \quad (i \neq j \neq k \neq i), \quad (10)$$

where the quantities

$$E = \frac{B(3A + 2B)}{A + B}, \qquad \sigma = \frac{A}{2(A + B)} \quad (11)$$

are elastic constants known respectively as *Young's modulus* and *Poisson's ratio*. The physical significance of E and σ may be ascertained by supposing a long rod to be under tension in the x_1 direction, the line of its axis, only, so that $T_{22} = T_{33} = 0$. From (10), with $k = 1$, we obtain $e_{11} = (T_{11}/E)$, so that E is the stress per unit strain, both parallel to a given direction in the situation described. With $k = 2$, then $k = 3$, in (10) we obtain $e_{22} = e_{33} = -(\sigma/E)T_{11}$; that is, the strain of contraction in any direction transverse to the lone direction of tensile stress is σ times the strain of extension in the direction of the stress.

Since A and B are positive, it follows from the second of (11) that $0 < \sigma < \frac{1}{2}$ for all substances.

(e) It follows from the definition (5) of the strain-tensor elements e_{jk} that these quantities cannot be completely arbitrary as functions of position within a body if they are to be continuously twice differentiable as required—*i.e.*, if the components u_1, u_2, u_3 are to be continuously three-times differentiable. Because the order of mixed partial differentiation is immaterial, it follows[1] from (5) that

$$\frac{\partial^2 e_{kk}}{\partial x_i \, \partial x_j} = \frac{\partial}{\partial x_k} \left(-\frac{\partial e_{ij}}{\partial x_k} + \frac{\partial e_{jk}}{\partial x_i} + \frac{\partial e_{ki}}{\partial x_j} \right) \qquad (i \neq j \neq k \neq i) \qquad (12)$$

and

$$2 \frac{\partial^2 e_{jk}}{\partial x_j \, \partial x_k} = \frac{\partial^2 e_{jj}}{\partial x_k^2} + \frac{\partial^2 e_{kk}}{\partial x_j^2} \qquad (j \neq k). \qquad (13)$$

The sets of equations (12) and (13) are known as "equations of compatibility." It may be shown[2] that they are sufficient, as well as necessary, conditions for the existence of functions u_1, u_2, u_3 suitable for describing the displacements of the points of an elastic solid in a state of strain.

10-2. General Equations of Motion and Equilibrium

(a) In order to arrive at the equations of motion—and of equilibrium as a special case—of an elastic solid, we make use of the extended Hamilton's principle enunciated in 6-7. Playing the role of the generalized-force components are the body- and surface-force distributions, defined in 10-1(a), which act upon the solid as influences of external agencies. The generalized coordinates are the components of displacement u_1, u_2, u_3. We proceed to obtain expressions for the elastic potential and kinetic energies of a solid in a given state of deformation.

As stated in 10-1(c), only terms involving the elements of strain e_{jk} in the expressions (7) for the relative displacement of two close neighboring points of the body give rise to stresses in the body. For this reason any function representing the potential energy of deformation must depend

[1] See exercise 3 at end of chapter.
[2] Sokolnikoff, pp. 24–28.

only on the elements e_{jk} of the strain tensor.   In considering the existence of such a function one must take into account the fact that the deformation of an elastic solid is accompanied by the development of heat energy within the solid.   There are, however, two limiting cases in which this fact presents no difficulty and in which the existence of the strain potential-energy function may be established: (i) a state of vibration so rapid that the time of a single cycle is too small for heat to flow out of the body (adiabatic) and (ii) a deformation which occurs so slowly that the temperature of the body remains uniform and equal to the temperature of the surroundings (isothermal).   The achievement of any equilibrium state, for example, falls within the latter category.   We assume in all that follows that either of the two situations obtains, so that we may define the strain potential energy per unit volume[1]

$$W = W(e_{11}, e_{22}, e_{33}, e_{12}, e_{23}, e_{31}, e_{21}, e_{32}, e_{13}). \tag{14}$$

(A specific form for the function W is derived in 10-3(b) below.)   Thus the total potential energy of deformation is given by

$$V = \iiint\limits_R W \, dx_1 \, dx_2 \, dx_3.$$

If ρ is the density (mass per unit volume) of the body, the kinetic energy of a volume element $dx_1 \, dx_2 \, dx_3$ is $\frac{1}{2}\rho(\dot{u}_1^2 + \dot{u}_2^2 + \dot{u}_3^2)dx_1 \, dx_2 \, dx_3$, so that the total elastic kinetic energy of the solid is given by

$$T - \tfrac{1}{2} \iiint\limits_R \rho \sum_{k=1}^{3} \dot{u}_k^2 \, dx_1 \, dx_2 \, dx_3.$$

Since the components of the body- and surface-force distributions are assumed to depend only upon the variables x_1, x_2, x_3, t and not upon the displacements u_1, u_2, u_3, we may employ the form (60) of 6-7 for the integral to be extremized according to the extended Hamilton's principle.
Since we deal with a continuous distribution of mass, the sum $\sum_{k=1}^{s} G_k q_k$ in (60) of 6-7 must be replaced by the sum of the volume and surface integrals

$$\iiint\limits_R \sum_{k=1}^{3} F_k u_k \, dx_1 \, dx_2 \, dx_3 + \iint\limits_B \sum_{k=1}^{3} T_k u_k \, dS.$$

[1] Although $e_{21} = e_{12}$, $e_{32} = e_{23}$, $e_{13} = e_{31}$, inclusion of the final trio of arguments of the function W is useful below.

Thus the integral which is extremized, according to the extended Hamilton's principle, by the functions u_1, u_2, u_3 describing the actual motion of an elastic solid body is

$$I = \int_{t_1}^{t_2} \left\{ \iiint_R \left[\tfrac{1}{2}\rho \sum_{k=1}^{3} \dot{u}_k^2 - W + \sum_{k=1}^{3} F_k u_k \right] dx_1\, dx_2\, dx_3 \right.$$

$$\left. + \iint_B \sum_{k=1}^{3} T_k u_k\, dS \right\} dt. \quad (15)$$

The extremization is carried out, according to 6-7, with respect to sufficiently regular functions u_1, u_2, u_3 which describe the actual elastic configurations at $t = t_1$ and $t = t_2$. Moreover, there may be portions B' of the boundary surface B at which the eligible functions u_1, u_2, u_3 may be required to possess prescribed values. For an elasticity problem is generally posed as a boundary-value problem wherein at each point of the boundary surface either the three components of the surface-force distribution or the three components of the displacement are given. At those portions B'' of B at which the components of the surface-force distribution are given, no restriction is made upon the surface values of the functions eligible for the extremization; in fact, one result of the extremization process is the derivation of boundary conditions which must be satisfied on B''.

(b) To effect the extremization of (15) we replace each u_k in the integrand of I by the one-parameter family of comparison functions $U_k = u_k + \epsilon\eta_k$ $(k = 1,2,3)$ and so form the integral $I(\epsilon)$. Here the $u_k = u_k(x_1,x_2,x_3,t)$ are assumed to be the actual extremizing functions, while the $\eta_k = \eta_k(x_1,x_2,x_3,t)$ are arbitrary to within continuous differentiability and restrictions based on the following considerations: Since, according to the requirements of the extended Hamilton's principle, u_1, u_2, u_3 are prescribed at $t = t_1$ and $t = t_2$, we must require $\eta_1 = \eta_2 = \eta_3 = 0$ at $t = t_1$ and $t = t_2$. Moreover, the three η_k must vanish over those portions B' of the boundary surface B upon which the displacement components are prescribed. (The values of the η_k may be chosen arbitrarily over those portions B'' of B at which the components of the surface-force distribution are prescribed.)

Further, we introduce the notation

$$E_{jk} = \frac{1}{2}\left(\frac{\partial U_k}{\partial x_j} + \frac{\partial U_j}{\partial x_k}\right) = \frac{1}{2}\left(\frac{\partial u_k}{\partial x_j} + \frac{\partial u_j}{\partial x_k}\right) + \epsilon\frac{1}{2}\left(\frac{\partial \eta_k}{\partial x_j} + \frac{\partial \eta_j}{\partial x_k}\right),$$

by the definition of U_1, U_2, U_3 above. We therefore have the result

$$\frac{\partial E_{jk}}{\partial \epsilon} = \frac{1}{2}\left(\frac{\partial \eta_k}{\partial x_j} + \frac{\partial \eta_j}{\partial x_k}\right). \quad (16)$$

Thus when we replace (u_1, u_2, u_3) by (U_1, U_2, U_3) in the function W of (14) —i.e., when we replace each e_{jk} by the corresponding E_{jk}—we obtain

$$\frac{\partial W}{\partial \epsilon} = \frac{1}{2} \sum_{j=1}^{3} \sum_{k=1}^{3} \frac{\partial W}{\partial E_{jk}} \left(\frac{\partial \eta_k}{\partial x_j} + \frac{\partial \eta_j}{\partial x_k} \right) = \sum_{j=1}^{3} \sum_{k=1}^{3} \frac{\partial W}{\partial E_{jk}} \frac{\partial \eta_k}{\partial x_j}, \qquad (17)$$

since† $E_{jk} = E_{kj}$.

The integral $I(\epsilon)$, formed as described above, is clearly an extremum when $\epsilon = 0$, so that we have

$$I'(0) = 0. \qquad (18)$$

With the definitions

$$f = \frac{1}{2} \rho \sum_{k=1}^{3} \dot{u}_k^2 + \sum_{k=1}^{3} F_k u_k, \qquad y = \sum_{k=1}^{3} T_k u_k, \qquad (19)$$

introduced for the sake of brevity, we proceed to form the derivative $I'(\epsilon)$, then set $\epsilon = 0$ (which means: replace U_k by u_k, E_{jk} by e_{jk}). Since $(\partial U_k / \partial \epsilon) = \eta_k$ and $(\partial \dot{U}_k / \partial \epsilon) = \dot{\eta}_k$, we have, with the aid of (17),

$$I'(0) = \int_{t_1}^{t_2} \left\{ \iiint_R \sum_{k=1}^{3} \left[\frac{\partial f}{\partial u_k} \eta_k + \frac{\partial f}{\partial \dot{u}_k} \dot{\eta}_k - \sum_{j=1}^{3} \frac{\partial W}{\partial e_{jk}} \frac{\partial \eta_k}{\partial x_j} \right] dx_1\, dx_2\, dx_3 \right.$$

$$\left. + \iint_B \sum_{k=1}^{3} \frac{\partial g}{\partial u_k} \eta_k\, dS \right\} dt = 0, \qquad (20)$$

because of (18).

Integrating by parts with respect to t, we obtain, since each $\eta_k = 0$ at $t = t_1$ and $t = t_2$,

$$\int_{t_1}^{t_2} \iiint_R \frac{\partial f}{\partial \dot{u}_k} \dot{\eta}_k\, dx_1\, dx_2\, dx_3\, dt = - \int_{t_1}^{t_2} \iiint_R \eta_k \frac{\partial}{\partial t} \left(\frac{\partial f}{\partial \dot{u}_k} \right) dx_1\, dx_2\, dx_3\, dt. \qquad (21)$$

According to Green's theorem (29) of 2-14(b) we have, further,

$$- \iiint_R \sum_{j=1}^{3} \frac{\partial W}{\partial e_{jk}} \frac{\partial \eta_k}{\partial x_j} dx_1\, dx_2\, dx_3 = \iiint_R \eta_k \sum_{j=1}^{3} \frac{\partial}{\partial x_j} \left(\frac{\partial W}{\partial e_{jk}} \right) dx_1\, dx_2\, dx_3$$

$$- \iint_B \eta_k \sum_{j=1}^{3} \frac{\partial W}{\partial e_{jk}} \cos (n, j) dS, \qquad (22)$$

† See exercise 4 at end of chapter.

where cos (n,j) is the cosine of the angle between the x_j direction and the outward normal to B, as a function of position on B.

With the results (21) and (22) equation (20) becomes

$$\int_{1}^{t_2}\left\{\iiint_R \sum_{k=1}^{3} \eta_k\left[\frac{\partial f}{\partial u_k} - \frac{\partial}{\partial t}\left(\frac{\partial f}{\partial \dot{u}_k}\right) + \sum_{j=1}^{3}\frac{\partial}{\partial x_j}\left(\frac{\partial W}{\partial e_{jk}}\right)\right] dx_1\,dx_2\,dx_3\right.$$

$$\left. + \iint_{B''} \sum_{k=1}^{3} \eta_k\left[\frac{\partial g}{\partial u_k} - \sum_{j=1}^{3}\frac{\partial W}{\partial e_{jk}}\cos(n,j)\right] dS\right\} dt = 0,$$

where the surface integral over B is replaced by the corresponding integral over B'', because $\eta_1 = \eta_2 = \eta_3 = 0$ on B', the remainder of B. Since the η_k are arbitrary in R and on B'', it follows from an obvious generalization of the basic lemma of 3-1(c) that[1]

$$\frac{\partial f}{\partial u_k} - \frac{\partial}{\partial t}\left(\frac{\partial f}{\partial \dot{u}_k}\right) + \sum_{j=1}^{3}\frac{\partial}{\partial x_j}\left(\frac{\partial W}{\partial e_{jk}}\right) = 0 \qquad \text{in } R \qquad (k = 1,2,3) \qquad (23)$$

and

$$\frac{\partial g}{\partial u_k} - \sum_{j=1}^{3}\frac{\partial W}{\partial e_{jk}}\cos(n,j) = 0 \qquad \text{on } B'' \qquad (k = 1,2,3). \qquad (24)$$

Since t_1 and t_2 are arbitrary, these results hold for all t. With the definitions (19) of f and g, equations (23) and (24) read respectively

$$F_k + \sum_{j=1}^{3}\frac{\partial}{\partial x_j}\left(\frac{\partial W}{\partial e_{jk}}\right) = \rho\ddot{u}_k \qquad \text{in } R \qquad (k = 1,2,3) \qquad (25)$$

and

$$T_k = \sum_{j=1}^{3}\frac{\partial W}{\partial e_{jk}}\cos(n,j) \qquad \text{on } B'' \qquad (k = 1,2,3). \qquad (26)$$

With the aid of the boundary condition we show in (d) below that

$$\frac{\partial W}{\partial e_{jk}} = T_{jk} \qquad (j,k = 1,2,3, \text{ independently}), \qquad (27)$$

the jk element of the stress tensor defined in 10-1(b) above. With the validity of (27) therefore assumed at this point the equations of elastic

---

[1] The argument is essentially an extension of that which follows directly after equation (54) in 3-8(a) or that which follows equation (72) in 9-5(b).

motion (25) read

$$F_k + \frac{\partial T_{1k}}{\partial x_1} + \frac{\partial T_{2k}}{\partial x_2} + \frac{\partial T_{3k}}{\partial x_3} = \rho \frac{\partial^2 u_k}{\partial t^2} \quad \text{in } R \quad (k = 1,2,3), \quad (28)$$

while the boundary conditions (26) become

$$T_k = T_{1k} \cos (n,1) + T_{2k} \cos (n,2) + T_{3k} \cos (n,3) \quad \text{on } B'', \quad (29)$$

for $(k = 1,2,3)$. On the remainder B' of B it is assumed that u_1, u_2, u_3 are prescribed. It may happen that B' (or B'') coincides with B; that is, u_1, u_2, u_3 may be prescribed everywhere (or nowhere) on B, in which event (29) holds nowhere (or everywhere) on B.

(c) The equations of equilibrium, of which we make some use in succeeding sections, may be derived from the equations of motion (28) by setting the acceleration components $(\partial^2 u_k/\partial t^2)$ equal to zero for $k = 1, 2, 3$:

$$F_k + \frac{\partial T_{1k}}{\partial x_1} + \frac{\partial T_{2k}}{\partial x_2} + \frac{\partial T_{3k}}{\partial x_3} = 0 \quad \text{in } R \quad (k = 1,2,3). \quad (30)$$

The boundary conditions (29), in conjunction with the discussion which follows (29), remain valid for the equilibrium case.

The solution of the equilibrium equations (30) subject to given boundary conditions is uniquely determined, provided the equations of compatibility (12) and (13) of 10-1(e) are also satisfied by the e_{jk} related to the T_{jk} through the Hooke's law relations (10) of 10-1(d).

In the work of the present chapter no use is made of the equations of motion (28) as they stand; we employ, instead, a special method for handling the dynamical problems (vibrating rod, vibrating plate) which come under our consideration. In both cases the special method is developed with the aid of results obtained in the study of problems described by the equations of equilibrium (30). Fuller discussion of the method, in its general aspects, is found in 10-3(a) below.

(d) To derive the relation (27) of (b) above we consider the arbitrary elastic solid R^* whose boundary surface B^* is everywhere interior to the boundary B of a given solid R, of which R^* is clearly an interior portion. Since R^* is completely surrounded by, and is everywhere contiguous with, portions of R, we cannot prescribe the displacement components on B^*; instead, it follows from (26) of (b) that

$$T_k = \sum_{j=1}^{3} \frac{\partial W}{\partial e_{jk}} \cos (n,j) \quad (k = 1,2,3) \quad (31)$$

everywhere on B^*, where T_1, T_2, T_3 are the components of surface-force density exerted upon R^* by the contiguous portions of R.

From the definition of the stress vector in 10-1(b) it follows that T_k at any point of B^* is identical with the x_k component T_{nk} of the stress vector computed with respect to an element of area of B^* at the point; the positive normal direction n is taken as outward from R^*. For T_k is by definition the x_k component of the force per unit area applied to R^* across B^*; but the force applied to R^* across B^* is exerted by the contiguous portion of R on the positive (outer) side of B^*. We therefore have from the definition given in 10-1(b) that the x_k component of the applied force per unit area is precisely T_{nk}; that is, $T_k = T_{nk}$, and thus, according to (31),

$$T_{nk} = \sum_{j=1}^{3} \frac{\partial W}{\partial e_{jk}} \cos{(n,j)} \qquad \text{on } B^* \qquad (k = 1,2,3). \tag{32}$$

Since R^* may be formed in any desired manner within the body R, we may choose B^* so that it passes through any point of R (exclusive of the boundary B) with arbitrary orientation at the point. In particular we pass B^* through an arbitrary point P so that its (outward) normal direction coincides with the positive x_p direction. In this case we have $\cos{(n,p)} = 1$ and $\cos{(n,j)} = 0$ if $j \neq p$; thus (32) reads

$$T_{pk} = \frac{\partial W}{\partial e_{pk}} \qquad \text{at } P \qquad (k = 1,2,3). \tag{33}$$

Since we may successively choose $p = 1,2,3$, and since both members of (33) are defined independently of the auxiliary surface B^*, the relation (27) of (b) is established for all *interior* points of R. The continuity of the quantities involved furnishes the validity of (27) on the boundary surface B, as well.

10-3. General Aspects of the Approach to Certain Dynamical Problems

(a) In a first study of the transverse vibrations of a thin bar or of a thin plate we bypass the general dynamical equations (28) of 10-2(b). The reason for doing this lies in the nature of the approximations we can afford to make in such vibrational problems. The general equations (28) describe every minute detail of displacement as functions of time and position within a vibrating body, thereby providing (if we are able to solve the equations!) a description far more detailed than is generally required for the bar or plate. We can well afford, for example, to ignore the distortion of the bar cross sections during vibrational motion if the cross-sectional dimensions are negligible compared with the bar's length and—which is of no small significance—especially when refusal to consider such distortion leads to equations which are reasonably tractable and which describe the *essential* features of the vibration with a high degree of accuracy. The transverse vibrations of a thin plate, for example, are generally accompanied by elastic waves which travel in the plane of the plate but which, if the plate is sufficiently thin, may be ignored; the *essential* feature of the motion resides in the successive shapes into which the plane of the plate is distorted during the motion.

In order to solve the dynamical problem involving a specific type of motion of a given elastic body we first solve an equilibrium problem which corresponds to the dynamical problem in the following sense: The equilibrium strain configuration at any point must be representative of the *essential* features of the *instantaneous* strain configuration at any point of the body during vibration. Actually, solution of the equilibrium problem needs to be carried only far enough to provide a calculation of the strain potential-energy density as a function of the position and displacement variables in terms of which the essential features of the dynamical motion are to be described. Once an expression for the total potential energy is available, together with the corresponding expression for the kinetic energy, Hamilton's principle may be applied in order to derive the pertinent equations of motion and boundary conditions.

The specific manner in which simplifying approximations (which ignore all but the essential features of the motion under study) are introduced is illustrated below in our studies of the vibrating bar and plate. We merely state here the underlying principle by means of which the most important approximations are effected: We make the very reasonable (and successful!) assumption that *the strain potential-energy density at any point depends only on the essential features of the instantaneous configuration of strain at the point and not upon the specific agencies which induce the strain.* The usefulness of this assumption is greatest, clearly, in those cases in which the features of the strain configuration which are essential to the problem at hand are easily distinguished from the unessential features; the latter are thus readily ignored. (The validity of our assumption, admittedly, is extremely weak in the close neighborhood of points of application of a straining agency, but this fact is unimportant if certain linear dimensions of the vibrating body are large compared with the distances over which straining agencies are applied. The limitation does not concern us in our study of long, thin rods and thin plates.)[1]

(b) From Hooke's law (10-1(d)) and the result

$$\frac{\partial W}{\partial e_{jk}} = T_{jk} \qquad (j,k = 1,2,3, \text{ independently}) \qquad (34)$$

[1] Our assumption is very closely linked with the celebrated principle of Saint-Venant. See, for example, Sokolnikoff, pp. 95, 99.

derived in 10-2(d) we may establish the general form of the strain-energy function W. Since the stress elements T_{jk} are *linear homogeneous* functions of the strain elements e_{jk}, according to (9) of 10-1(d), it follows from (34) that W contains no terms in the e_{jk} of higher order than quadratic and no terms which are linear in the e_{jk}. Moreover, in requiring that W vanish in the unstrained state, we set the arbitrary additive constant equal to zero. Thus we conclude that W is a *homogeneous quadratic* function of the strain elements e_{jk} ($j,k = 1,2,3$, independently), so that we may apply Euler's theorem (2-5) to obtain

$$2W = \sum_{j=1}^{3} \sum_{k=1}^{3} \frac{\partial W}{\partial e_{jk}} e_{jk}. \tag{35}$$

With (34) equation (35) thus becomes the explicit formula

$$W = \tfrac{1}{2} \sum_{j=1}^{3} \sum_{k=1}^{3} T_{jk} e_{jk}. \tag{36}$$

A form of (36) more useful for purposes below is obtained by substituting for the e_{jk} from the Hooke's law equations (10) of 10-1(d):

$$W = \frac{1}{2E} (T_{11}^2 + T_{22}^2 + T_{33}^2) + \frac{1 + \sigma}{E} (T_{12}^2 + T_{23}^2 + T_{31}^2)$$

$$- \frac{\sigma}{E} (T_{11}T_{22} + T_{22}T_{33} + T_{33}T_{11}). \tag{37}$$

10-4. Bending of a Cylindrical Bar by Couples

(*a*) We consider a homogeneous isotropic bar of unstrained cylindrical shape with plane end faces perpendicular to the generators of the cylinder. Any plane section of the bar parallel to the end faces we call a cross section. A cartesian coordinate system is set up so that one end face lies in the x_1x_2 plane while the other is in the plane $x_3 = L > 0$ (see Fig. 10-1). The

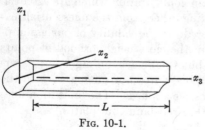

Fig. 10-1.

origin is so located that the x_3 axis passes through the centroid of every cross section; *i.e.*, we have for each cross section

$$\iint_D x_1 \, dx_1 \, dx_2 = \iint_D x_2 \, dx_1 \, dx_2 = 0 \qquad (x_3 = \text{constant}), \tag{38}$$

where the integrals extend over the domain D of the cross section. The orientation of the x_1 and x_2 axes is such that each is parallel to one of the two principal axes of inertia of every cross section; *i.e.*, we have for each cross section[1]

$$\iint_D x_1 x_2 \, dx_1 \, dx_2 = 0 \qquad (x_3 = \text{constant}). \tag{39}$$

(*b*) We proceed to investigate the elastic-displacement configuration of the bar when the system of stresses

$$T_{11} = T_{22} = T_{12} = T_{23} = T_{31} = 0, \qquad T_{33} = Px_2 \tag{40}$$

prevails; P is a given constant, positive or negative. (If $P > 0$, the stress distribution clearly describes tension in the x_3 direction for $x_2 > 0$, pressure in the x_3 direction for $x_2 < 0$; there are no pressures or tensions in the x_1 or x_2 directions and no shearing stresses anywhere. Thus the portion of the bar lying on one side of the plane $x_2 = 0$ is stretched, while the remaining portion is compressed. The stress system obviously arises as the result of the bar's being bent in the $x_2 x_3$ plane.) We completely neglect body forces[2] ($F_1 = F_2 = F_3 = 0$), so that direct substitution into (30) of 10-2(*c*) verifies that the distribution (40) is consistent with elastic equilibrium.

To ascertain the surface-force distribution required to give rise to (40) we note first that on the lateral (cylindrical) surface of the boundary we have $\cos (n,3) = 0$, so that, according to (29) of 10-2(*b*),

$$T'_k = T_{1k} \cos (n,1) + T_{2k} \cos (n,2) = 0 \qquad (k = 1,2,3),$$

because of (40). That is, the lateral surface is completely free of external agencies. Next, on the face $x_3 = L$ we have

$$\cos (n,1) = \cos (n,2) = 0, \qquad \cos (n,3) = 1,$$

so that the general boundary conditions (29) read

$$T_1 = T_{31} = 0, \qquad T_2 = T_{32} = 0, \qquad T_3 = T_{33} = Px_2 \qquad (x_3 = L), \tag{41}$$

because of (40). Similarly, on the face $x_3 = 0$ we have

$$T_1 = -T_{31} = 0, \quad T_2 = -T_{32} = 0, \quad T_3 = -T_{33} = -Px_2 \quad (x_3 = 0); \tag{42}$$

---

[1] The principal axes of inertia of an area are defined as a pair of perpendicular lines in the plane of the area which intersect at the centroid and whose orientation is such that the integral in (39) vanishes; x_1 and x_2 are coordinates measured from the respective lines. The appropriate orientation can always be found, but it is not necessarily unique; any pair of perpendicular diameters, for example, are principal axes of a circle.

[2] The influence of gravity is generally negligible.

the change of sign results from the fact that $\cos(n,3) = -1$ on the face $x_3 = 0$.

The *total* external force on the face $x_3 = L$ is zero; for since $T_1 = T_2 = 0$, the force on $x_3 = L$ is obtained by integrating the component T_3 of surface-force density over this end face. But, with T_3 given by (41), we have

$$\iint_D T_3 \, dx_1 \, dx_2 = P \iint_D x_2 \, dx_1 \, dx_2 = 0 \qquad (x_3 = L),$$

according to (38) as applied to the end cross section. Similarly, with the aid of (42) we find that the total force on the end face $x_3 = 0$ is also zero.

The total effects of the surface-force distributions (41) and (42) are best described in terms of their *bending moments* about the coordinate axes. By definition the three components M_1, M_2, M_3 of the moment of a given surface-force distribution T_1, T_2, T_3 are given by

$$M_1 = \iint_B (T_3 x_2 - T_2 x_3) dS, \qquad M_2 = \iint_B (T_1 x_3 - T_3 x_1) dS,$$

$$M_3 = \iint_B (T_2 x_1 - T_1 x_2) dS, \quad (43)$$

where the integrals extend over the surface B at which the surface-force distribution is applied. Thus the moment components of the distribution (41) on the end face $x_3 = L$ are

$$M_1 = P \iint_D x_2^2 \, dx_1 \, dx_2 = PJ_1 \qquad (x_3 = L), \tag{44}$$

where

$$J_1 = \iint_D x_2^2 \, dx_1 \, dx_2 \tag{45}$$

is by definition the *area moment of inertia* of the face $x_3 = L$ with respect to its principal axis parallel to the x_1 axis;

$$M_2 = -P \iint_D x_2 x_1 \, dx_1 \, dx_2 = 0,$$

because of (39); and $M_3 = 0$.

It therefore follows that the total effect of the distribution (41) upon the bar face $x_3 = L$ is that of a bending moment of magnitude PJ_1 directed along the x_1 axis. Moreover, since the moment of the surface-force distribution on $x_3 = L$ is unchanged[1] by any translation of the

---

[1] The proof is left for end-chapter exercise 6.

x_1 axis parallel to itself, the bending moment is termed a *couple* about the x_1 axis; the magnitude of the couple[1] is PJ_1. Similarly, we find that the surface-force distribution (42) on the end cross section $x_3 = 0$ gives rise to an equal but oppositely directed couple $-PJ_1$ about the x_1 axis.

We note that the quantity J_1, given by (45), is purely geometrical in character and is a constant which is the same for every cross section ($x_3 =$ constant) of the bar.

(c) Using the Hooke's law equations (10) of 10-1(d), we derive from (40) the strain-tensor elements

$$e_{jk} = 0 \ (j \neq k), \qquad e_{11} = e_{22} = -\frac{\sigma P}{E} x_2, \qquad e_{33} = \frac{P}{E} x_2, \qquad (46)$$

which describe the state of strain within the bar under consideration. Through direct substitution into (12) and (13) we verify that the strain elements (46) satisfy the equations of compatibility and are therefore suitable for the description of a physically feasible state of strain.

We now proceed to demonstrate that the bar in which the strain distribution is given by (46) is one which has undergone bending in the $x_2 x_3$ plane. Specifically, we show that every line parallel to the length of the bar in its unstrained state—*i.e.*, every line described by the equations $x_1 =$ constant, $x_2 =$ constant—is in the strained state a parabolic arc lying in a plane parallel to the $x_2 x_3$ plane.

First, to prove that any line $x_1 =$ constant, $x_2 =$ constant becomes a plane curve parallel to the $x_2 x_3$ plane, we must show that u_1 is a constant with respect to x_3 for given constant values of x_1 and x_2. That is, the displacement u_1 in the x_1 direction must be the same for every point of the line in question. Second, if we prove that u_2 is, for given constant values of x_1 and x_2, a quadratic function of x_3, we thereby show that any line $x_1 =$ constant, $x_2 =$ constant becomes a parabolic arc in a plane parallel to the $x_2 x_3$ plane; *i.e.*, we have merely to show that $(\partial^2 u_2/\partial x_3^2)$ is independent of x_3.

With the definition (5) of 10-1(c) of the strain elements in terms of the displacement derivatives the relations (46) read

$$\frac{\partial u_1}{\partial x_2} + \frac{\partial u_2}{\partial x_1} = \frac{\partial u_2}{\partial x_3} + \frac{\partial u_3}{\partial x_2} = \frac{\partial u_3}{\partial x_1} + \frac{\partial u_1}{\partial x_3} = 0, \qquad (47)$$

$$\frac{\partial u_1}{\partial x_1} = \frac{\partial u_2}{\partial x_2} = -\frac{\sigma P}{E} x_2, \qquad \frac{\partial u_3}{\partial x_3} = \frac{P}{E} x_2. \qquad (48)$$

[1] The reference here to the concepts of moment and couple is actually unessential to the continuity of this chapter. The reference merely aims to supply a physical picture of the basis for the stress distribution (40) for the reader who is already acquainted with the concepts.

From (47), (48), and the reversibility of partial differentiation it directly follows that

$$\frac{\partial}{\partial x_1}\left(\frac{\partial u_1}{\partial x_3}\right) = \frac{\partial}{\partial x_3}\left(\frac{\partial u_1}{\partial x_1}\right) = \frac{\partial}{\partial x_3}\left(-\frac{\sigma P}{E}x_2\right) = 0, \tag{49}$$

$$\frac{\partial}{\partial x_3}\left(\frac{\partial u_1}{\partial x_3}\right) = -\frac{\partial}{\partial x_3}\left(\frac{\partial u_3}{\partial x_1}\right) = -\frac{\partial}{\partial x_1}\left(\frac{\partial u_3}{\partial x_3}\right) = -\frac{\partial}{\partial x_1}\left(\frac{P}{E}x_2\right) = 0, \tag{50}$$

$$\frac{\partial}{\partial x_2}\left(\frac{\partial u_1}{\partial x_3}\right) = \frac{\partial}{\partial x_3}\left(\frac{\partial u_1}{\partial x_2}\right) = -\frac{\partial}{\partial x_3}\left(\frac{\partial u_2}{\partial x_1}\right) = -\frac{\partial}{\partial x_1}\left(\frac{\partial u_2}{\partial x_3}\right)$$

$$= \frac{\partial}{\partial x_1}\left(\frac{\partial u_3}{\partial x_2}\right) = \frac{\partial}{\partial x_2}\left(\frac{\partial u_3}{\partial x_1}\right) = -\frac{\partial}{\partial x_2}\left(\frac{\partial u_1}{\partial x_3}\right),$$

whence

$$\frac{\partial}{\partial x_2}\left(\frac{\partial u_1}{\partial x_3}\right) = 0. \tag{51}$$

With the results (49), (50), and (51) we conclude that $(\partial u_1/\partial x_3)$ is a constant independent of x_1, x_2, x_3. By arranging the orientation[1] of the strained bar in such fashion that $(\partial u_1/\partial x_3) = 0$ at a single point, we therefore have that

$$\frac{\partial u_1}{\partial x_3} = 0 \qquad \text{in } R.$$

Or, along any line $x_1 = $ constant, $x_2 = $ constant, we have $u_1 = $ constant; this proves the assertion of the preceding paragraph that any line parallel to the length of the bar in the unstrained state becomes a curve lying in a plane parallel to the x_2x_3 plane as the result of strain.

That this plane curve is a parabola follows directly from (47) and (48), for

$$\frac{\partial^2 u_2}{\partial x_3^2} = \frac{\partial}{\partial x_3}\left(\frac{\partial u_2}{\partial x_3}\right) = -\frac{\partial}{\partial x_3}\left(\frac{\partial u_3}{\partial x_2}\right) = -\frac{\partial}{\partial x_2}\left(\frac{\partial u_3}{\partial x_3}\right) = -\frac{P}{E}. \tag{52}$$

We thus have that u_2 is a quadratic function of x_3.

(d) In accordance with the method outlined in 10-3(a) we proceed to compute an expression for the strain potential energy of the bent bar which depends only on the strain configuration and not upon the agency which gives rise to the strain. Substituting (40) of (b) above into (37) of 10-3(b), we obtain for the strain potential energy per unit volume

$$W = \frac{P^2}{2E}x_2^2. \tag{53}$$

---

[1] The general solution of the six equations (47), (48) contains six arbitrary constants which may be evaluated by specifying the position and orientation of the strained bar *as a whole*. See exercise 7 at end of chapter.

To obtain a quantity which is of direct use in 10-5(b) below we integrate (53) over an arbitrary cross section D (x_3 = constant) of the bar to obtain the potential energy *per unit length* of the bar

$$W_L = \iint\limits_D W \, dx_1 \, dx_2 = \frac{P^2}{2E} \iint\limits_D x_2^2 \, dx_1 \, dx_2 = \frac{P^2 J_1}{2E}, \tag{54}$$

where J_1, defined by (45) of (b) above, is the area moment of inertia of the cross section with respect to its principal axis parallel to the x_1 axis.

Finally, to eliminate the dependence of W_L on the constant P in favor of a quantity which describes the local bending configuration of the bar, we substitute from (52) into (54) to obtain

$$W_L = \frac{1}{2} EJ_1 \left(\frac{\partial^2 u_2}{\partial x_3^2} \right)^2. \tag{55}$$

We employ the result (55) in 10-5 below in the study of transverse vibrations of a bar; discussion of the validity of its use is reserved for that section. (The product EJ_1 is called the *flexural rigidity* of the bar with respect to bending in the $x_2 x_3$ plane.)

A fuller discussion of the bending of a bar can be found in the literature.[1] Further development of the foregoing results is left for the end-chapter exercises.

10-5. Transverse Vibrations of a Bar

To derive the equations of motion and boundary conditions for the transverse vibrations of a bar we appeal, as in the case of the vibrating string and the vibrating membrane, to Hamilton's principle (6-2) as applied to a system involving a continuous distribution of mass. It is our first task, then, to obtain expressions for the kinetic and potential energies of the bar.

(a) We consider a cylindrical bar, or rod, free from (net) longitudinal pressure or tension, the linear dimensions of whose cross section are small compared with its length. As in 10-4, we ignore the influence of body forces. The only external influences to which the bar is subjected are constraints which may be applied to one end, both ends, or neither. We consider chiefly two types of constraint: (i) The "hinge," whereby the effect is merely to hold in fixed position the end of the bar to which it is applied; the orientation of the bar at this end is not influenced by this type of constraint. (ii) The "clamp," whereby the effect is not only to hold the end in fixed position but is also to fix the *orientation* of the bar

[1] See, for example, Sokolnikoff, Chap. 4.

at the end to which it is applied. We suppose that the vibration is parallel to a principal plane of the bar—*i.e.*, to a plane which contains one set of parallel principal axes of every cross section of the bar.[1]

In considering the transverse, or flexural, vibrations of the rod, we ignore the possible distortion of the cross sections and suppose that each element of volume contained between two closely neighboring cross sections moves as a rigid entity. The validity of this simplification rests upon the assumption of small cross-section dimensions made in the preceding paragraph. If the cross section is small, the contribution to the bar's potential and kinetic energies owing to its distortion may be neglected. It is this neglect which enables us to use the formula (55) of 10-4(*d*) for W_L, the potential energy per unit length of the bar: The potential energy is assumed to depend only on the configuration of bending in the plane of vibration.

In accordance with the assumption that each cross-sectional element of volume moves as a rigid entity, we may employ a single variable to describe the shape of the bar as a function of the longitudinal distance from one end and of the time variable *t*. For this we employ $u = u(x,t)$ to denote the transverse displacement of a point of the central line[2] relative to its equilibrium position; here *x* is the distance of the point from the end designated by $x = 0$; the other end of the bar is at $x = L$.

(*b*) If we denote by γ the constant mass per unit length of the rod, the translational kinetic energy of the volume element of thickness dx at *x* is $\frac{1}{2}\dot{u}^2\gamma \, dx$, so that the total kinetic energy is

$$T = \tfrac{1}{2}\gamma \int_0^L \dot{u}^2 \, dx. \tag{56}$$

In employing the expression (55) of 10-4(*d*)—which, in the notation of the present section, reads

$$W_L = \frac{1}{2} EJ_1 \left(\frac{\partial^2 u}{\partial x^2}\right)^2$$

—for the strain potential energy per unit length of the bar, we have for the total potential energy[3]

$$V = \tfrac{1}{2}EJ_1 \int_0^L u_{xx}^2 \, dx. \tag{57}$$

With (56) and (57) the Hamilton's integral (7) of 6-2(*a*) becomes

$$I = \tfrac{1}{2} \int_{t_1}^{t_2} \int_0^L (\gamma \dot{u}^2 - EJ_1 u_{xx}^2) dx \, dt. \tag{58}$$

[1] See footnote 1, p. 213, for the definition of principal axes of a cross section.

[2] The central line is the locus of cross-section centroids.

[3] As in preceding chapters, we employ subscripts to indicate partial differentiation.

According to Hamilton's principle (6-2) the extremum of I with respect to functions $u(x,t)$ which describe the actual bar configurations at $t = t_1$ and $t = t_2$ is supplied by the particular $u(x,t)$ which describes the bar configuration for all t.

Possible end-point conditions which must be satisfied by the functions eligible for the extremization of (58) depend upon the physical constraints which may be placed upon the ends of the bar. We consider the following possibilities:

(i) Free end: If either end of the bar is free, no constraint is made upon the displacement u or the slope u_x of the bar at that end. Accordingly, both u and u_x, evaluated at a free end, are completely arbitrary if u is an eligible function.

(ii) "Hinged" end: Here the constraint is such as to prescribe the value of u, whereas u_x is arbitrary for the eligible functions u.

(iii) "Clamped" end: Here displacement and orientation (slope) are both prescribed, so that the eligible functions u must be selected from among those which have particular given values of both u and u_x at the end in question.

Any one of the conditions (i) to (iii) may prevail at either end of a given bar, independently of which of the three applies at the other end.

For the process of extremizing (58) we form the integral $I(\epsilon)$ by replacing u in the integrand of (58) by the one-parameter family of comparison functions $U = u(x,t) + \epsilon\eta(x,t)$, where $u(x,t)$ is assumed to be the actual extremizing function and $\eta(x,t)$ is arbitrary, with the exception that $\eta(x,t_1) = \eta(x,t_2) = 0$—a requirement of Hamilton's principle. Further, η may be subject to end-point conditions, depending on which of the conditions (i) to (iii) listed above happens to be imposed. Briefly, we have at a

(i') Free end: η arbitrary

(ii') Hinged end: $\eta = 0$, η_x arbitrary

(iii') Clamped end: $\eta = 0$, $\eta_x = 0$.

Clearly, $I(\epsilon)$ is an extremum for $\epsilon = 0$, so that

$$I'(0) = 0. \tag{59}$$

Noting that $(\partial \dot{U}/\partial\epsilon) = \dot{\eta}$ and $(\partial U_{xx}/\partial\epsilon) = \eta_{xx}$, we form the integral $I'(\epsilon)$, then set $\epsilon = 0$ by replacing $\dot{U}$ by $\dot{u}$, U_{xx} by u_{xx}. Thus, with (59), we obtain

$$I'(0) = \int_{t_1}^{t_2}\int_0^L \left(\frac{\partial f}{\partial \dot{u}}\,\dot{\eta} + \frac{\partial f}{\partial u_{xx}}\,\eta_{xx} \right) dx\,dt = 0, \tag{60}$$

where, for abbreviation, we write

$$f = \tfrac{1}{2}(\gamma \dot{u}^2 - EJ_1 u_{xx}^2). \tag{61}$$

Since $\eta = 0$ at $t = t_1$ and $t = t_2$, integration by parts gives

$$\int_{t_1}^{t_2} \int_0^L \frac{\partial f}{\partial \dot{u}} \, \dot{\eta} \, dx \, dt = -\int_{t_1}^{t_2} \int_0^L \eta \frac{\partial}{\partial t} \left(\frac{\partial f}{\partial \dot{u}}\right) dx \, dt. \tag{62}$$

Also, we have through twice-performed integration by parts

$$\int_0^L \frac{\partial f}{\partial u_{xx}} \eta_{xx} \, dx = \left[\frac{\partial f}{\partial u_{xx}} \eta_x\right]_0^L - \int_0^L \eta_x \frac{\partial}{\partial x} \left(\frac{\partial f}{\partial u_{xx}}\right) dx$$

$$= \left[\frac{\partial f}{\partial u_{xx}} \eta_x - \eta \frac{\partial}{\partial x} \left(\frac{\partial f}{\partial u_{xx}}\right)\right]_0^L + \int_0^L \eta \frac{\partial^2}{\partial x^2} \left(\frac{\partial f}{\partial u_{xx}}\right) dx. \tag{63}$$

With (62) and (63) equation (60) becomes

$$\int_{t_1}^{t_2} \left\{ \left[\frac{\partial f}{\partial u_{xx}} \eta_x - \eta \frac{\partial}{\partial x} \left(\frac{\partial f}{\partial u_{xx}}\right)\right]_0^L \right.$$

$$\left. - \int_0^L \eta \left[\frac{\partial}{\partial t} \left(\frac{\partial f}{\partial \dot{u}}\right) - \frac{\partial^2}{\partial x^2} \left(\frac{\partial f}{\partial u_{xx}}\right)\right] dx \right\} dt = 0, \tag{64}$$

for arbitrary choice of the function $\eta(x,t)$ consistent with restrictions placed above. Since (64) must hold for those η which, together with η_x, vanish at $x = 0$ and $x = L$, it follows from a simple extension[1] of the basic lemma of 3-1(c) that

$$\frac{\partial}{\partial t} \left(\frac{\partial f}{\partial \dot{u}}\right) - \frac{\partial^2}{\partial x^2} \left(\frac{\partial f}{\partial u_{xx}}\right) = 0 \qquad (0 \leqq x \leqq L). \tag{65}$$

Moreover, if η and η_x are arbitrary at both $x = 0$ and $x = L$, it follows[2] that the coefficients of $\eta(0,t)$, $\eta(L,t)$, $\eta_x(0,t)$, $\eta_x(L,t)$ all vanish separately. By writing, accordingly,

$$\frac{\partial f}{\partial u_{xx}} \eta_x = 0, \qquad \eta \frac{\partial}{\partial x} \left(\frac{\partial f}{\partial u_{xx}}\right) = 0 \qquad (x = 0,L), \tag{66}$$

we take into account the possibility that either or both of η, η_x may be required to vanish at either or both of $x = 0$, $x = L$—in which event the corresponding coefficient in (64) need not vanish.

Substituting (61) into (65), we obtain the differential equation of motion of the vibrating bar under consideration:

$$\gamma \frac{\partial^2 u}{\partial t^2} + EJ_1 \frac{\partial^4 u}{\partial x^4} = 0 \qquad (0 \leqq x \leqq L). \tag{67}$$

---

[1] See exercise 10 at the end of this chapter.

[2] The proof is left for the reader. See, for example, the argument leading to the result (76) of 3-10(a); the argument must be slightly modified and extended for the present case.

The end-point conditions (66) read

$$\frac{\partial^2 u}{\partial x^2} \eta_x = 0, \qquad \frac{\partial^3 u}{\partial x^3} \eta = 0 \qquad (x = 0, L). \tag{68}$$

(c) The most important examples of constraint of the end of a vibrating bar consist of hinging which renders the end-point displacement equal to zero, and clamping which renders both the end-point displacement and the end-point slope equal to zero; we confine our attention to these types of constraint. With the aid of (68) we list the boundary conditions which apply to these constraints as well as to the case of a free end of the bar:

(i) Free end: Since both η and η_x are arbitrary at an unconstrained end, (68) implies that

$$\frac{\partial^2 u}{\partial x^2} = 0, \qquad \frac{\partial^3 u}{\partial x^3} = 0 \qquad \text{(free end)}. \tag{69}$$

(ii) Hinged end: Since η_x is arbitrary, (68) implies that its coefficient vanishes. Since, also, the displacement of the end is maintained at zero, we have

$$\frac{\partial^2 u}{\partial x^2} = 0, \qquad u = 0 \qquad \text{(hinged end)}. \tag{70}$$

(iii) Clamped end: Here, both displacement and slope are maintained at zero, so that

$$\frac{\partial u}{\partial x} = 0, \qquad u = 0 \qquad \text{(clamped end)}. \tag{71}$$

10-6. The Eigenvalue-Eigenfunction Problem for the Vibrating Bar

(a) We begin the attack upon the vibrating-bar equation (67), subject to any of the boundary conditions (69), (70), (71) applied independently at $x = 0$ and $x = L$, in the manner in which we handle the vibrating-membrane equation in 9-4(a); that is, we seek solutions of the form

$$u = \phi(x)q(t), \tag{72}$$

where $\phi(x)$ satisfies one from each of the two groups of end-point conditions which follow:

(i) $\phi''(0) = \phi'''(0) = 0$ (free)   (i') $\phi''(L) = \phi'''(L) = 0$;

(ii) $\phi(0) = \phi''(0) = 0$ (hinged)   (ii') $\phi(L) = \phi''(L) = 0$; (73)

(iii) $\phi(0) = \phi'(0) = 0$ (clamped)   (iii') $\phi(L) = \phi'(L) = 0$.

(For example, if the end $x = 0$ is clamped while $x = L$ is free, $\phi(x)$ satisfies both (iii) and (i').) For convenience we suppose that ϕ is normalized so that

$$\int_0^L \phi^2 \, dx = 1. \tag{74}$$

Substituting (72) into (67), we obtain, on dividing through by $\gamma \phi q$,

$$\frac{EJ_1}{\gamma} \frac{\phi''''}{\phi} = -\frac{\ddot{q}}{q}. \tag{75}$$

Since the left-hand member of (75) is independent of t and the right-hand is independent of x, the two members must be equal to a constant, which we denote by λ. Thus (75) implies the two ordinary differential equations

$$EJ_1\phi'''' - \gamma\lambda\phi = 0 \tag{76}$$

and

$$\ddot{q} + \lambda q = 0. \tag{77}$$

In (b) below it is shown that $\lambda > 0$ (with the excepted possibility of $\lambda = 0$ if the bar is free at both ends or free at one end and hinged at the other), so that the general solution of the time-dependent equation (77) is

$$q = A \cos \sqrt{\lambda} \, t + B \sin \sqrt{\lambda} \, t, \tag{78}$$

where A and B are arbitrary constants.

(b) The determination of the permissible values of λ—and thus, according to (78), the list of natural vibration frequencies of the bar—is an eigenvalue-eigenfunction problem of the type encountered in the three chapters preceding. That is, any value of λ for which there exists a function ϕ which satisfies (76) and (74), together with the appropriate set of end-point conditions from among (73), is an eigenvalue of λ; the solution ϕ is the corresponding eigenfunction.

Explicit solution of the eigenvalue-eigenfunction problem for the bar is left for the end-chapter exercises, but we prove here that there can be no negative eigenvalues of λ—a fact used in arriving at (78):

With the assumption that ϕ satisfies (74) and (76), we multiply the latter by ϕ and integrate from $x = 0$ to $x = L$ to obtain

$$\lambda = \frac{EJ_1}{\gamma} \int_0^L \phi\phi'''' \, dx. \tag{79}$$

After two successive integrations by parts (79) becomes

$$\lambda = \frac{EJ_1}{\gamma} \left\{ [\phi\phi''' - \phi'\phi'']_0^L + \int_0^L (\phi'')^2 \, dx \right\}. \tag{80}$$

If, further, ϕ satisfies any set of end-point conditions from among (73)—

one from among each of the two groups—the integrated part of (80) is seen to vanish; it therefore follows that $\lambda \geqq 0$, since E, J_1, γ are all positive.[1]

(c) The sequence of eigenfunctions ϕ_1, ϕ_2, . . . , ϕ_k, . . . for the vibrating bar form an orthogonal set in $0 \leqq x \leqq L$ with respect to a constant weight function. That is, if ϕ_j and ϕ_k are two different eigenfunctions of the problem, we have

$$\int_0^L \phi_j \phi_k \, dx = 0 \qquad (j \neq k). \tag{81}$$

To prove (81) we multiply the equations satisfied by the two eigenfunctions—namely, according to (76),

$$EJ_1\phi_j'''' = \gamma\lambda_j\phi_j, \qquad EJ_1\phi_k'''' = \gamma\lambda_k\phi_k \tag{82}$$

—by ϕ_k and ϕ_j, respectively. Subtracting the results and integrating from $x = 0$ to $x = L$, we obtain

$$\gamma(\lambda_j - \lambda_k) \int_0^L \phi_j \, \phi_k \, dx = EJ_1 \int_0^L (\phi_j''''\phi_k - \phi_j \phi_k'''')dx$$
$$= EJ_1 [\phi_j'''\phi_k - \phi_j\phi_k''' + \phi_j'\phi_k'' - \phi_j''\phi_k']_0^L, \tag{83}$$

as we find on twice integrating by parts each term of the integral on the right. But if ϕ_j and ϕ_k satisfy the same set of end-point conditions from among (73)—as they must, since they are assumed to be eigenfunctions of the same problem—it is clear that the final member of (83) is zero. Thus, since $\lambda_j \neq \lambda_k$ if $j \neq k$—a fact proved in exercise 14 at the end of this chapter—the orthogonality (81) follows directly.

A second proof of the orthogonality is based upon the fact that the eigenvalue-eigenfunction problem for the vibrating bar is equivalent to an isoperimetric problem. Namely, an extremum of the integral

$$\int_0^L (\phi'')^2 \, dx \tag{84}$$

with respect to functions ϕ which satisfy the normalization condition

$$\int_0^L \phi^2 \, dx = 1$$

is effected by a function ϕ which satisfies the differential equation (76).

To verify this fact we use the result of exercise 9, Chap. 4; namely, if we introduce $(\gamma/EJ_1)\lambda$ as undetermined Lagrange multiplier, the extremization process leads us directly to (76). Further, if at a given end point ($x = 0$ or $x = L$) no *a priori* restrictions are placed upon the functions ϕ eligible for the extremization of (84), the extremizing functions must

---

[1] Proof that $\lambda = 0$ can hold only in the cases of a bar with both ends free or one end free and the other hinged is left for exercise 13(c) at the end of this chapter.

satisfy $\phi'' = \phi''' = 0$ at that end; *i.e.*, the free-end conditions (i) or (i′) of (73) must be satisfied. Similarly, we see that the appropriate end-point conditions for a hinged or clamped end must likewise be satisfied by the extremizing functions. For example, at a hinged end, the functions eligible for the extremization must satisfy $\phi = 0$, while ϕ' is arbitrary; according to exercise 9, Chap. 4, we obtain the additional condition $\phi'' = 0$ at the hinged end for the extremizing functions—in accord with (ii) or (ii′) of (73).

For the orthogonality proof we use the result (73) of exercise 9, Chap. 4. Namely, we have

$$\int_0^L \left(\phi'' \eta'' - \frac{\gamma\lambda}{EJ_1} \phi\eta \right) dx = 0, \tag{85}$$

where ϕ is any extremizing function with λ the corresponding eigenvalue and η is arbitrary, except that it must satisfy any *a priori* end-point restrictions which may be placed upon the eligible functions ϕ. In (85) we may therefore write $\phi = \phi_j$, $\lambda = \lambda_j$, $\eta = \phi_k$ and then rewrite the same result with the indices j and k interchanged ($j \neq k$). Subtracting the two results thus obtained, we get

$$\frac{\gamma}{EJ_1} (\lambda_j - \lambda_k) \int_0^L \phi_j \phi_k \, dx = 0,$$

whence the orthogonality[1] (81) follows, inasmuch as $\lambda_j \neq \lambda_k$.

10-7. Bending of a Rectangular Plate by Couples

The problem of the transverse vibration of a thin plate is most easily approached through consideration of the bending of a rectangular plate by couples applied at its edge surfaces. Just as the problem of the bar bent by couples (10-5) leads to an expression for the strain potential-energy function applicable to the theory of the vibrating thin rod (10-6), so also does the study of the rectangular plate bent by couples lead to a suitable potential-energy function for the vibrating thin plate. In both cases the connection between the static problem and the corresponding vibration problem is developed on the basis of the general principle enunciated in 10-3(a).

(a) We consider a rectangular plate of uniform thickness $2h$ situated, in its unstrained state, with its middle plane in $x_3 = 0$; thus the faces of the unstrained plate lie in the planes $x_3 = \pm h$, respectively. The bound-

---

[1] The main advantage of the second proof is that no explicit mention of the boundary (end-point) conditions is required—an advantage of great significance in the demonstration (10-9(b) below) that the vibrating-plate eigenfunctions form an orthogonal set.

ary-edge surfaces lie in the planes $x_1 = 0$, $x_1 = L_1$, $x_2 = 0$, $x_2 = L_2$, respectively (see Fig. 10-2). We neglect the influence of body forces.

With proper regard for the altered orientation of coordinate axes, reference to (40) of 10-4(b) reveals that a state of stress in which

$$T_{22} = P_2 x_3 \qquad (P_2 = \text{constant}), \tag{86}$$

with $T_{11} = T_{33} = T_{12} = T_{23} = T_{31} = 0$, describes the bending of the plate as if it were a bar extending in the x_2 direction. As described in 10-4(c) every line $x_1 = $ constant,

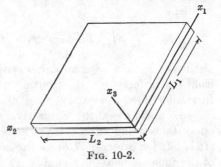

FIG. 10-2.

$x_3 = $ constant (in the unstrained state) is strained into parabolic shape in a plane parallel to the $x_2 x_3$ plane. According to 10-4(b) the bending results from a pair of equal but oppositely directed couples about the x_1 axis.

Further, we superimpose upon the bending of the plate described in the preceding paragraph an additional bending which arises from a pair of equal but oppositely directed couples about the x_2 axis. That is, we withdraw the condition $T_{11} = 0$ and replace it with

$$T_{11} = P_1 x_3 \qquad (P_1 = \text{constant}). \tag{87}$$

We proceed to investigate the condition of the strained plate under the system of stresses given by (86), (87), and

$$T_{33} = T_{12} = T_{23} = T_{31} = 0. \tag{88}$$

Substitution of (86), (87), (88) into (30) of 10-2(c)—with F_1, F_2, F_3 set equal to zero, since we ignore the influence of body forces—verifies directly that the given stress distribution is consistent with static equilibrium.

(b) To ascertain the surface-force distribution which gives rise to the stress distribution (86), (87), (88) we employ (29) of 10-2(b), from which, together with (88), it follows that the plate is completely free of surface forces on the faces $x_3 = \pm h$. Further, we have from (86), (87), (88), and (29) that

$$
\begin{aligned}
T_1 = \quad & T_{11} = \quad P_1 x_3, \quad & T_2 = T_3 = 0 \quad & \text{on } x_1 = L_1, \\
T_2 = \quad & T_{22} = \quad P_2 x_3, \quad & T_1 = T_3 = 0 \quad & \text{on } x_2 = L_2, \\
T_1 = \; & -T_{11} = -P_1 x_3, \quad & T_2 = T_3 = 0 \quad & \text{on } x_1 = 0, \\
T_2 = \; & -T_{22} = -P_2 x_3, \quad & T_1 = T_3 = 0 \quad & \text{on } x_2 = 0,
\end{aligned}
\tag{89}
$$

since $\cos(n,1) = 1$, $\cos(n,2) = \cos(n,3) = 0$ on $x_1 = L_1$, etc.

The total force on the face $x_1 = L_1$ is given by

$$\int_{-h}^{h} \int_{0}^{L_2} T_1 \, dx_2 \, dx_3 = P_1 \int_{-h}^{h} \int_{0}^{L_2} x_3 \, dx_2 \, dx_3 = 0,$$

according to the first line of (89). Similar computations reveal that the total force on each of the three remaining edge faces is likewise zero. From the definitions given in (43) of 10-4(b) we find the effect of the surface-force distribution on $x_1 = L_1$ (first line of (89)) to be a bending moment

$$M_2 = \int_{-h}^{h} \int_{0}^{L_2} T_1 x_3 \, dx_2 \, dx_3 = P_1 \int_{-h}^{h} \int_{0}^{L_2} x_3^2 \, dx_2 \, dx_3 = \tfrac{2}{3} P_1 L_2 h^3$$

about the x_2 axis, with an equal but oppositely directed moment arising from the surface-force distribution on the face $x_1 = 0$ (third line of (89)). In similar fashion we may show[1] that the surface-force distributions on $x_2 = L_2$ and $x_2 = 0$ are respectively describable in terms of equal but opposite bending moments of magnitude $\tfrac{2}{3} P_2 L_1 h^3$ about the x_1 axis. (It is left for exercise 17(b) at the end of this chapter to show that each of the four moments is a couple, as defined in 10-4(b).)

(c) Using the Hooke's law equations (10) of 10-1(d), we derive from (86), (87), (88) the strain-tensor elements

$$e_{jk} = 0 \ (j \neq k), \qquad e_{11} = \frac{1}{E} (P_1 - \sigma P_2) x_3,$$

$$e_{22} = \frac{1}{E} (P_2 - \sigma P_1) x_3, \qquad e_{33} = -\frac{\sigma}{E} (P_1 + P_2) x_3. \tag{90}$$

Substitution of (90) into (12) and (13) of 10-1(e) verifies that these strain elements satisfy the compatibility equations and are therefore suitable for the description of a physically feasible state of strain.

In the manner of 10-4(c)—with the details left for end-chapter exercise 18—we can show with the aid of (90) and the definitions (5) of 10-1(c) that every unstrained line $x_1 =$ constant, $x_3 =$ constant is bent into parabolic shape (in a plane $x_1 =$ constant) for which

$$\frac{\partial^2 u_3}{\partial x_2^2} = -\frac{1}{E} (P_2 - \sigma P_1). \tag{91}$$

Also, the lines $x_2 =$ constant, $x_3 =$ constant become parabolas (in planes $x_2 =$ constant) for which

$$\frac{\partial^2 u_3}{\partial x_1^2} = -\frac{1}{E} (P_1 - \sigma P_2). \tag{92}$$

---

[1] See exercise 17(a) at end of chapter.

Further, we obtain the result

$$\frac{\partial^2 u_3}{\partial x_1 \, \partial x_2} = 0,$$ (93)

from which we ascertain, by definition, that the planes of principal curvature of any plane x_3 = constant are respectively parallel to the $x_1 x_3$ and $x_2 x_3$ planes.   This fact comes in for special consideration in the derivation in (d) below of the strain potential-energy function applicable to the study of the vibrating thin plate.

(d) In accordance with the method outlined in 10-3(a) we proceed to compute an expression for the strain potential energy of the bent plate which depends only on the strain configuration and not upon the agencies which give rise to the strain.   Since we wish to apply the result to the vibration of thin plates of arbitrary shape, we must obtain an expression for the strain potential energy which is independent of the orientation of the x_1 and x_2 axes.   Even in the case of a rectangular vibrating plate we cannot be sure that the planes of principal curvature will at every point be respectively parallel to the $x_1 x_3$ and $x_2 x_3$ planes—a fact upon which the results (91), (92), (93) of (c) above depend.

Substituting (86), (87), and (88) into (37) of 10-3(b), we obtain for the strain potential energy per unit volume

$$W = \frac{1}{2E} \, (P_1^2 + P_2^2 - 2\sigma P_1 P_2) x_3^2.$$

Integrating over the thickness of the plate—from $x_3 = -h$ to $x_3 = h$, that is—we obtain the strain potential energy per unit area of plate surface

$$W_A = \int_{-h}^{h} W \, dx_3 = \frac{h^3}{3E} \, (P_1^2 + P_2^2 - 2\sigma P_1 P_2).$$ (94)

Solving (91) and (92) of (c) above for P_1 and P_2, we obtain

$$W_A = \frac{h^3 E}{3(1 - \sigma^2)} \left[\left(\frac{\partial^2 u_3}{\partial x_1^2} + \frac{\partial^2 u_3}{\partial x_2^2} \right)^2 - 2(1 - \sigma) \left(\frac{\partial^2 u_3}{\partial x_1^2} \right) \left(\frac{\partial^2 u_3}{\partial x_2^2} \right) \right],$$ (95)

through substitution of the results into (94).

In order to free the result (95) from the specificity inherent in the circumstance (93) of (c) above we replace the coordinates x_1, x_2 by a pair of cartesian coordinates x, y related to x_1 and x_2 through the equations

$$x_1 = x \cos Q - y \sin Q, \qquad x_2 = x \sin Q + y \cos Q.$$ (96)

Q, considered arbitrary, is the angle through which the x_1 and x_2 axes must be rotated about the x_3 axis to be brought into coincidence with the new x and y axes, respectively. Using the results of exercise $2(a,b)$, Chap. 9, we have

$$\frac{\partial^2 u_3}{\partial x_1^2} + \frac{\partial^2 u_3}{\partial x_2^2} = \frac{\partial^2 u_3}{\partial x^2} + \frac{\partial^2 u_3}{\partial y^2}$$

and

$$\frac{\partial^2 u_3}{\partial x_1^2} \frac{\partial^2 u_3}{\partial x_2^2} - \left(\frac{\partial^2 u_3}{\partial x_1 \, \partial x_2}\right)^2 = \frac{\partial^2 u_3}{\partial x^2} \frac{\partial^2 u_3}{\partial y^2} - \left(\frac{\partial^2 u_3}{\partial x \, \partial y}\right)^2,$$

where both results are independent of the angle Q. Thus, because of (93), we may rewrite (95) as

$$W_A = \frac{1}{2} D_0 \left\{ \left(\frac{\partial^2 u}{\partial x^2} + \frac{\partial^2 u}{\partial y^2}\right)^2 - 2(1 - \sigma) \left[\frac{\partial^2 u}{\partial x^2} \frac{\partial^2 u}{\partial y^2} - \left(\frac{\partial^2 u}{\partial x \, \partial y}\right)^2\right] \right\} \qquad (97)$$

—with u written for u_3 and D_0 for $[2h^3E/3(1 - \sigma^2)]$—and so obtain an expression for the strain potential energy per unit plate area which is independent of the orientation of the coordinate axes lying in the middle plane of the plate. The constant D_0 is called the *flexural rigidity* of the plate.

We employ the result (97) in the section which follows directly.

10-8. Transverse Vibrations of a Thin Plate

(a) In considering the transverse vibrations of a thin plate, it is convenient—as well as consistent with all of the simplifying assumptions generally made in a first approach to the phenomenon—to suppose the entire mass of the plate to be concentrated in the plane midway between the parallel plane faces of the plate. We suppose that in its equilibrium position the plate covers the domain D with boundary curve C in the xy plane. The deviation from equilibrium during vibration is described by the function $u = u(x,y,t)$, where u is the transverse displacement of a point located (in equilibrium) at (x,y). Thus, if the constant mass per unit area of the plate is μ, the total kinetic energy of the plate is given by

$$T = \tfrac{1}{2}\mu \iint\limits_{D} \dot{u}^2 \, dx \, dy. \qquad (98)$$

To obtain an expression for the potential energy of the plate, we employ the assumption introduced in 10-3(a) that the strain potential-energy density at a point depends only on the strain configuration at the point. With this assumption we may thus employ the expression (97) of 10-7(d) for the potential energy per unit plate area, so that the total potential energy is given by

$$V = \tfrac{1}{2}D_0 \iint\limits_{D} [(\nabla^2 u)^2 - 2(1 - \sigma)(u_{xx}u_{yy} - u_{xy}^2)]dx \, dy, \qquad (99)$$

where $\nabla^2 u$ is written for $(u_{xx} + u_{yy})$ and the subscripts indicate partial differentiation.

With (98) and (99) we have for the hamiltonian integral (7) of 6-2(a)

$$I = \tfrac{1}{2} \int_{t_1}^{t_2} \iint_D \left\{ \mu \dot{u}^2 - D_0[(\nabla^2 u)^2 - 2(1 - \sigma)(u_{xx} u_{yy} - u_{xy}^2)] \right\} dx\, dy\, dt; \quad (100)$$

according to Hamilton's principle (6-2(a)), (100) is extremized with respect to those functions $u(x,y,t)$ which describe the actual plate configuration at $t = t_1$ and $t = t_2$ by the particular function $u(x,y,t)$ which describes the actual configuration for all t. Owing to possible physical constraints placed upon the plate at its boundary edge C the functions u eligible for the extremization of I may be required to satisfy certain conditions on C; explicit consideration of such conditions is carried out in (e) below.

(b) To effect the extremization of (100) we form the integral $I(\epsilon)$ by replacing u in the integrand of (100) by the one-parameter family of comparison functions

$$U = u(x,y,t) + \epsilon \eta(x,y,t), \quad (101)$$

where $u(x,y,t)$ is the actual extremizing function and $\eta(x,y,t)$ is arbitrary to within twice continuous differentiability and the requirement (according to Hamilton's principle) that $\eta(x,y,t_1) = \eta(x,y,t_2) = 0$. Further, both η and/or its normal derivative $(\partial \eta / \partial n) = \eta_n$ (taken with respect to the outward normal to C) may be subject to boundary conditions consistent with restrictions on C imposed upon the functions eligible for the extremization of I; these are discussed in (e) below. It is clear that $I(\epsilon)$ is an extremum for $\epsilon = 0$, so that

$$I'(0) = 0. \quad (102)$$

Writing

$$f(\dot{u}, u_{xx}, u_{yy}, u_{xy}) = \tfrac{1}{2}\{\mu \dot{u}^2 - D_0[(\nabla^2 u)^2 - 2(1 - \sigma)(u_{xx} u_{yy} - u_{xy}^2)]\}, \quad (103)$$

according to (100), and using (101) to compute

$$\frac{\partial U_{xx}}{\partial \epsilon} = \eta_{xx}, \qquad \frac{\partial U_{yy}}{\partial \epsilon} = \eta_{yy}, \qquad \frac{\partial U_{xy}}{\partial \epsilon} = \eta_{xy}, \qquad \frac{\partial \dot{U}}{\partial \epsilon} = \dot{\eta},$$

we form $I'(\epsilon)$ and then set $\epsilon = 0$ (replacing U by u, according to (101)) to obtain

$$I'(0) = \int_{t_1}^{t_2} \iint_D \left(\frac{\partial f}{\partial \dot{u}} \dot{\eta} + \frac{\partial f}{\partial u_{xx}} \eta_{xx} + \frac{\partial f}{\partial u_{yy}} \eta_{yy} + \frac{\partial f}{\partial u_{xy}} \eta_{xy} \right) dx\, dy\, dt = 0, \quad (104)$$

according to (102).

Through integration by parts over t we get

$$\int_{t_1}^{t_2} \iint_D \frac{\partial f}{\partial \dot{u}} \dot{\eta} \, dx \, dy \, dt = - \int_{t_1}^{t_2} \iint_D \eta \frac{\partial}{\partial t} \left(\frac{\partial f}{\partial \dot{u}} \right) dx \, dy \, dt,$$

since $\eta = 0$ for $t = t_1$ and $t = t_2$. To transform the final three terms of (104) we employ the forms of Green's theorem given in (25), (26), and (27) of 2-13(e), respectively:

$$\iint_D \frac{\partial f}{\partial u_{xx}} \eta_{xx} \, dx \, dy = \iint_D \eta \frac{\partial^2}{\partial x^2} \left(\frac{\partial f}{\partial u_{xx}} \right) dx \, dy + \int_C \left[\eta_x \frac{\partial f}{\partial u_{xx}} - \eta \frac{\partial}{\partial x} \left(\frac{\partial f}{\partial u_{xx}} \right) \right] dy,$$

$$\iint_D \frac{\partial f}{\partial u_{yy}} \eta_{yy} \, dx \, dy = \iint_D \eta \frac{\partial^2}{\partial y^2} \left(\frac{\partial f}{\partial u_{yy}} \right) dx \, dy - \int_C \left[\eta_y \frac{\partial f}{\partial u_{yy}} - \eta \frac{\partial}{\partial y} \left(\frac{\partial f}{\partial u_{yy}} \right) \right] dx,$$

$$\iint_D \frac{\partial f}{\partial u_{xy}} \eta_{xy} \, dx \, dy = \iint_D \eta \frac{\partial^2}{\partial x \, \partial y} \left(\frac{\partial f}{\partial u_{xy}} \right) dx \, dy + \frac{1}{2} \int_C \frac{\partial f}{\partial u_{xy}} (\eta_y \, dy - \eta_x \, dx)$$

$$+ \frac{1}{2} \int_C \eta \left[\frac{\partial}{\partial x} \left(\frac{\partial f}{\partial u_{xy}} \right) dx - \frac{\partial}{\partial y} \left(\frac{\partial f}{\partial u_{xy}} \right) dy \right]. \quad (105)$$

Further, we have direct use for the result

$$\frac{\partial^2}{\partial x^2} \left(\frac{\partial f}{\partial u_{xx}} \right) + \frac{\partial^2}{\partial y^2} \left(\frac{\partial f}{\partial u_{yy}} \right) + \frac{\partial^2}{\partial x \, \partial y} \left(\frac{\partial f}{\partial u_{xy}} \right)$$

$$= -D_0 \left[\frac{\partial^4 u}{\partial x^4} + 2 \frac{\partial^4 u}{\partial x^2 \, \partial y^2} + \frac{\partial^4 u}{\partial y^4} \right] = -D_0 \nabla^2 (\nabla^2 u) = -D_0 \nabla^4 u,$$

as we find[1] on direct computation from (103). From (103) we also obtain

$$\frac{\partial}{\partial t} \left(\frac{\partial f}{\partial \dot{u}} \right) = \mu \ddot{u}.$$

With the results of the preceding paragraph we may rewrite (104) as

$$\int_{t_1}^{t_2} \Bigg\{ \iint_D \eta(-\mu \ddot{u} - D_0 \nabla^4 u) dx \, dy$$

$$+ \int_C \left\{ \eta \left[\frac{\partial}{\partial y} \left(\frac{\partial f}{\partial u_{yy}} \right) + \frac{1}{2} \frac{\partial}{\partial x} \left(\frac{\partial f}{\partial u_{xy}} \right) \right] - \eta_y \frac{\partial f}{\partial u_{yy}} - \frac{1}{2} \eta_x \frac{\partial f}{\partial u_{xy}} \right\} dx$$

$$+ \int_C \left\{ \eta_x \frac{\partial f}{\partial u_{xx}} + \frac{1}{2} \eta_y \frac{\partial f}{\partial u_{xy}} - \eta \left[\frac{\partial}{\partial x} \left(\frac{\partial f}{\partial u_{xx}} \right) + \frac{1}{2} \frac{\partial}{\partial y} \left(\frac{\partial f}{\partial u_{xy}} \right) \right] \right\} dy \Bigg\} dt = 0;$$

$$(106)$$

the explicit expressions for the integrands of the line integrals along C are obtained from (103) in (d) below. Since (106) must hold for arbi-

---

[1] The expression $\nabla^4 u$ is merely an abbreviation for

$$\nabla^2 (\nabla^2 u) = \frac{\partial^2}{\partial x^2} (u_{xx} + u_{yy}) + \frac{\partial^2}{\partial y^2} (u_{xx} + u_{yy}).$$

trary η, it must, in particular, hold for those η for which $\eta = \eta_x = \eta_y = 0$ on C. For such η, (106) reduces to the integral over D (in turn integrated from $t = t_1$ to $t = t_2$); an obvious extension of the basic lemma of 3-1(c) yields the result

$$\mu \frac{\partial^2 u}{\partial t^2} + D_0 \nabla^4 u = 0 \qquad \text{in } D \qquad \text{(all } t) \qquad (107)$$

as necessary for the fulfillment of (106). The fourth-order partial differ-

ential equation (107) is thus, according to Hamilton's principle, the equation of motion of a vibrating thin plate.

Derivation of the various sets of boundary conditions for the vibrating plate depends upon the integrals over C which appear in (106). The treatment of these integrals is greatly facilitated by the transformations carried out in (c) following.

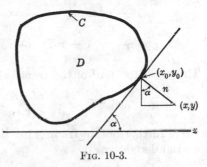

FIG. 10-3.

(c) While the use of the cartesian coordinates x and y may be continued with profit in a problem involving a rectangular plate, it is essential for more general purposes to introduce as coordinate variables the arc length s of the boundary curve C of D and the distance n measured from C along the normal to C. Given a point (x,y), we determine its (n,s) coordinates by drawing the shortest normal to C through (x,y); the intersection of this normal with C determines the s coordinate of the point, while the n coordinate is the distance from (x,y) to C along the normal (positive if (x,y) is exterior to D, negative if interior). (See Fig. 10-3.) Clearly, it is necessary that through each point (x,y) there be a uniquely determined shortest normal to C, which is the case if (x,y) is not separated from C by the evolute of C (locus of centers of curvature). Since our purpose here is merely to transform the line integrals along C of (106), this condition of nonseparation may be considered fulfilled inasmuch as we employ the (n,s) system in the evaluation of quantities on C only—that is, for $n = 0$.[1]

If we represent C in the parametric form $x = x_0(s)$, $y = y_0(s)$ and let $\alpha = \alpha(s)$ be the angle made by the tangent to C with the positive x direction,[2] we derive from Fig. 10-3 the transformation equations

---

[1] At a corner of C the normal direction is undefined. We assume that C consists of a finite number of smooth arcs and therefore possesses at most a finite number of corners.

[2] We measure α counterclockwise from the positive x axis to the *positive* direction (direction of increasing s) of the tangent. With this definition α is surely a continuous function of s ($0 \leq \alpha < 2\pi$), except, possibly, at a finite number of corners.

$$x = x_0(s) + n \sin \alpha(s),$$
$$y = y_0(s) - n \cos \alpha(s). \tag{108}$$

With the use of the elementary relationships

$$\frac{dx_0}{ds} = \cos \alpha, \qquad \frac{dy_0}{ds} = \sin \alpha, \qquad \frac{d\alpha}{ds} = K(s), \tag{109}$$

where K is the curvature of C, we obtain from (108) the partial derivatives

$$\frac{\partial x}{\partial n} = \sin \alpha, \qquad \frac{\partial y}{\partial n} = -\cos \alpha,$$

$$\frac{\partial x}{\partial s} = (1 + nK) \cos \alpha, \qquad \frac{\partial y}{\partial s} = (1 + nK) \sin \alpha. \tag{110}$$

Thus the jacobian[1] of the transformation (108) is

$$\frac{\partial(x,y)}{\partial(n,s)} = 1 + nK. \tag{111}$$

We further obtain from (110) the equations of transformation of first partial derivatives

$$u_n = u_x \sin \alpha - u_y \cos \alpha, \qquad u_s = (1 + nK)(u_x \cos \alpha + u_y \sin \alpha).$$

Solving these equations for u_x and u_y, we get

$$u_x = u_n \sin \alpha + u_s \frac{\cos \alpha}{1 + nK}, \qquad u_y = -u_n \cos \alpha + u_s \frac{\sin \alpha}{1 + nK}. \tag{112}$$

(We have occasion below to employ the relations (112) with u replaced by the function η; the replacement is valid because (112) hold for *any* differentiable function u.)

To transform *second* partial derivatives to the (n,s) coordinate system we employ the result of 9-2(b), with the identification $n = r_1$, $s = r_2$ and the suppression of the third independent variable r_3. With the use of (112) we successively substitute

(i) $f = \dfrac{1}{2} u_x^2 = \dfrac{1}{2} \left\{ u_n \sin \alpha + u_s \dfrac{\cos \alpha}{1 + nK} \right\}^2,$

(ii) $f = \dfrac{1}{2} u_y^2 = \dfrac{1}{2} \left\{ -u_n \cos \alpha + u_s \dfrac{\sin \alpha}{1 + nK} \right\}^2,$

(iii) $f = \dfrac{1}{2} u_x u_y = \dfrac{1}{2} \left\{ u_n \sin \alpha + u_s \dfrac{\cos \alpha}{1 + nK} \right\} \left\{ -u_n \cos \alpha + u_s \dfrac{\sin \alpha}{1 + nK} \right\}$

into (22) of 9-2(b), with the jacobian—denoted by D in 9-2(b)—given by $(1 + nK)$, according to (111). The results of these substitutions are

[1] See 2-8(f). Also see exercise 20 at the end of the present chapter for a discussion of the required nonvanishing property of this jacobian.

respectively

(i)　$u_{xx} = u_{nn} \sin^2 \alpha + u_{ss} \dfrac{\cos^2 \alpha}{(1 + nK)^2} + 2u_{sn} \dfrac{\sin \alpha \cos \alpha}{1 + nK}$

$$+ Ku_n \dfrac{\cos^2 \alpha}{1 + nK} - u_s \left[\dfrac{2K \sin \alpha \cos \alpha}{(1 + nK)^2} + \dfrac{nK' \cos^2 \alpha}{(1 + nK)^3} \right],$$

(ii)　$u_{yy} = u_{nn} \cos^2 \alpha + u_{ss} \dfrac{\sin^2 \alpha}{(1 + nK)^2} - 2u_{sn} \dfrac{\sin \alpha \cos \alpha}{1 + nK}$

$$+ Ku_n \dfrac{\sin^2 \alpha}{1 + nK} + u_s \left[\dfrac{2K \sin \alpha \cos \alpha}{(1 + nK)^2} - \dfrac{nK' \sin^2 \alpha}{(1 + nK)^3} \right],$$

(iii)　$u_{xy} = -u_{nn} \sin \alpha \cos \alpha + u_{ss} \dfrac{\sin \alpha \cos \alpha}{(1 + nK)^2} + u_{sn} \dfrac{\sin^2 \alpha - \cos^2 \alpha}{1 + nK}$

$$+ Ku_n \dfrac{\sin \alpha \cos \alpha}{1 + nK} + u_s \left[\dfrac{K(\cos^2 \alpha - \sin^2 \alpha)}{(1 + nK)^2} - \dfrac{nK' \sin \alpha \cos \alpha}{(1 + nK)^3} \right],$$

(113)

where $K' = (dK/ds) = (d^2\alpha/ds^2)$. In all that follows, our use of the results (113) is restricted to the boundary curve C—that is, for the special case $n = 0$.

(d) In applying the results of (c) to the transformation of the line integrals of (106), we first note that along C we have $n = 0$, $dn = 0$, so that, according to the second line of (110),

$$dx = ds \cos \alpha, \qquad dy = ds \sin \alpha \qquad \text{(along } C\text{).} \qquad (114)$$

Further, we have on direct (although somewhat lengthy) computation from (103), with the aid of (114), that

$$\left[\frac{\partial}{\partial y} \left(\frac{\partial f}{\partial u_{yy}} \right) + \frac{1}{2} \frac{\partial}{\partial x} \left(\frac{\partial f}{\partial u_{xy}} \right) \right] dx - \left[\frac{\partial}{\partial x} \left(\frac{\partial f}{\partial u_{xx}} \right) + \frac{1}{2} \frac{\partial}{\partial y} \left(\frac{\partial f}{\partial u_{xy}} \right) \right] dy$$

$$= D_0 \left[\sin \alpha \frac{\partial}{\partial x} (u_{xx} + u_{yy}) - \cos \alpha \frac{\partial}{\partial y} (u_{xx} + u_{yy}) \right] ds$$

$$= D_0 \left[\frac{\partial}{\partial n} (\nabla^2 u) \right] ds; \qquad (115)$$

the final form of (115) springs from the fact that

$$\frac{\partial}{\partial n} = \frac{\partial x}{\partial n} \frac{\partial}{\partial x} + \frac{\partial y}{\partial n} \frac{\partial}{\partial y} = \sin \alpha \frac{\partial}{\partial x} - \cos \alpha \frac{\partial}{\partial y},$$

according to the first line of (110).

Setting $n = 0$ and replacing u by η in (112), we obtain

$$\eta_x = \eta_n \sin \alpha + \eta_s \cos \alpha, \qquad \eta_y = -\eta_n \cos \alpha + \eta_s \sin \alpha \qquad \text{(along } C\text{).} \qquad (116)$$

Another lengthy computation, with the aid of (116), (114), and (103), brings us to the result

$$- \left(\eta_y \frac{\partial f}{\partial u_{yy}} + \frac{1}{2} \eta_x \frac{\partial f}{\partial u_{xy}} \right) dx + \left(\eta_x \frac{\partial f}{\partial u_{xx}} + \frac{1}{2} \eta_y \frac{\partial f}{\partial u_{xy}} \right) dy$$

$$= -D_0\{\eta_n[\nabla^2 u - (1 - \sigma)(u_{xx} \cos^2 \alpha + 2u_{xy} \sin \alpha \cos \alpha + u_{yy} \sin^2 \alpha)]$$

$$+ (1 - \sigma)\eta_s[u_{xy}(\sin^2 \alpha - \cos^2 \alpha) + (u_{xx} - u_{yy}) \sin \alpha \cos \alpha]\} ds$$

$$= -D_0\{\eta_n[\nabla^2 u - (1 - \sigma)(u_{ss} + Ku_n)] + (1 - \sigma)\eta_s(u_{sn} - Ku_s)\} ds, \quad (117)$$

where the final form is achieved through the transformations (113) with n set equal to zero.

Still another aid to the transformation of the line integrals of (106) is the integration by parts

$$\int_C \eta_s(u_{sn} - Ku_s)ds = - \int_C \eta \frac{\partial}{\partial s} (u_{sn} - Ku_s)ds. \quad (118)$$

The integrated term vanishes because C is a *closed* curve, and it is assumed that each factor of $[\eta(u_{sn} - Ku_s)]$ is a single-valued continuous function of position along C.[1] (At a corner—point of discontinuous normal direction—of C the required continuity of $(u_{sn} - Ku_s)$ involves special consideration; in the case of the rectangular plate with boundary edge free (10-10(e) below) such consideration leads to a new boundary condition.)[2]

With the results (115), (117), and (118), in conjunction with (107), we may write (106) of (b) above as

$$\int_{t_1}^{t_2} \int_C \left\{ \eta \left[\frac{\partial}{\partial n} (\nabla^2 u) + (1 - \sigma) \frac{\partial}{\partial s} (u_{sn} - Ku_s) \right] + \eta_n[(1 - \sigma)(u_{ss} + Ku_n) - \nabla^2 u] \right\} ds\, dt = 0. \quad (119)$$

The interpretation of this result becomes meaningful only on the basis of discussion of the various types of physical constraints which may be imposed at the boundary edge of the plate; this discussion follows directly:

(e) The most important types of physical constraint which may be imposed along C are (i) clamping, (ii) simple support, (iii) complete freedom.

(i) At a clamped edge not only is C constrained to remain in the equilibrium ($u = 0$) plane, but, in addition, the normal derivative u_n of the transverse displacement is held to zero at the boundary. Thus the functions u eligible for the extremization of (100) of (a) above must satisfy

---

[1] Although the subscript s of η_s indicates *partial* (n held constant) differentiation with respect to s, the fact that $n = 0$ along C reduces partial differentiation with respect to s to ordinary differentiation in this case.

[2] See also exercise 21 at the end of this chapter for a proof that $(u_{sn} - Ku_s)$ cannot be discontinuous along C even if the restriction to continuity is not made *a priori* in the free-edge problem.

$u = u_n = 0$ along C.   It therefore follows that

$$\eta = \eta_n = 0 \text{ along } C \text{ (clamped edge)}. \tag{120}$$

(ii) At a simply supported edge the only constraint is the holding of C fixed in the equilibrium plane; no condition is imposed upon the normal derivative of the transverse displacement.   The functions u eligible for the extremization of (100) must therefore satisfy $u = 0$ (with u_n arbitrary) along C; that is,

$$\eta = 0, \ \eta_n \text{ arbitrary along } C \text{ (simply supported edge)}. \tag{121}$$

(iii) At a free edge there are no physical constraints, so that the eligible functions u are required to satisfy no special conditions.   Thus we have both of

$$\eta, \ \eta_n \text{ arbitrary along } C \text{ (free edge)}. \tag{122}$$

Because of (120) equation (119) reduces to a triviality in the case of the clamped-edge plate (i).   The boundary conditions, therefore, are those which are imposed from the outset—namely

$$u = 0, \ u_n = 0 \text{ along } C \text{ (clamped edge)}. \tag{123}$$

With (121) taken into account (119) directly implies that the coefficient of η_n vanishes[1] along C.   Together with the condition imposed at the outset this result gives us for the simply supported plate (ii)

$$u - 0, \qquad (1 - \sigma)(u_{ss} + Ku_n) - \nabla^2 u = 0 \qquad \text{along } C. \tag{124}$$

The second condition may be simplified as follows: From the first two identities of (113), with $n = 0$, it follows that

$$\nabla^2 u = u_{xx} + u_{yy} = u_{nn} + u_{ss} + Ku_n \qquad \text{along } C. \tag{125}$$

But since $u = 0$ along C, it follows that u_s, and therefore u_{ss}, must vanish on C.   In view of (125), (124) thus reads

$$u = 0, \qquad K\sigma u_n + u_{nn} = 0 \qquad \text{along } C \text{ (simply supported edge)}. \tag{126}$$

Because of (122) the relation (119) implies the separate vanishing of the respective coefficients of η and η_n in the case of the free boundary edge (iii):

$$\left. \begin{array}{l} \dfrac{\partial}{\partial n}(\nabla^2 u) + (1 - \sigma)\dfrac{\partial}{\partial s}(u_{sn} - Ku_s) = 0 \\[2mm] (1 - \sigma)(u_{ss} + Ku_n) - \nabla^2 u = 0 \end{array} \right\} \text{along } C \text{ (free edge)}. \tag{127}$$

We may, of course, deal with a plate whose boundary conditions are mixed—whereby, that is, one of the three sets of conditions above may

[1] See 3-1(c).

apply to certain portions of C while one or both of the other two sets may apply to the remaining portions.

10-9. The Eigenvalue-Eigenfunction Problem for the Vibrating Plate

(a) As in the case of the vibrating-membrane problem $(9\text{-}4(a))$, the first step in the solution of the vibrating-plate equation (107) of $10\text{-}8(b)$, in conjunction with any one of the sets of homogeneous boundary conditions (123), (126), or (127), is to seek a solution of the form

$$u = \phi(x,y)q(t). \tag{128}$$

Substituting (128) into (107), we obtain, on division by $D_0\phi q$,

$$\frac{\nabla^4\phi}{\phi} = -\frac{\mu}{D_0}\frac{\ddot{q}}{q}. \tag{129}$$

Since the left-hand member of (129) is independent of t and the right-hand member depends upon t alone, it follows that the two members are equal to a constant, which—in view of the proof in (b) below that this constant cannot be negative—we denote by β^4. We therefore conclude that (129) implies two separate differential equations,

$$\nabla^4\phi - \beta^4\phi = 0 \tag{130}$$

and

$$\ddot{q} + \omega^2 q = 0, \tag{131}$$

where we write

$$\omega^2 = \frac{\beta^4 D_0}{\mu}. \tag{132}$$

The general solution of the time-dependent equation (131) is

$$q = A \cos \omega t + B \sin \omega t, \tag{133}$$

where A and B are arbitrary constants.

Since the sets of boundary conditions (123), (126), and (127) of $10\text{-}8(e)$ are all homogeneous, substitution of (128) involves cancellation of the time-dependent factor $q(t)$ and thus directly yields the three sets of boundary conditions for $\phi(x,y)$:

(i) $\phi = 0, \qquad \dfrac{\partial\phi}{\partial n} = 0$ along C (clamped edge),

(ii) $\phi = 0, \qquad K\sigma\dfrac{\partial\phi}{\partial n} + \dfrac{\partial^2\phi}{\partial n^2} = 0$ along C (simply supported edge),

(iii) $\dfrac{\partial}{\partial n}(\nabla^2\phi) + (1-\sigma)\dfrac{\partial}{\partial s}\left(\dfrac{\partial^2\phi}{\partial s\,\partial n} - K\dfrac{\partial\phi}{\partial s}\right) = 0$ $\left.\begin{array}{l} \\ \\ \end{array}\right\}$ along C

$\qquad\qquad (1-\sigma)\left(\dfrac{\partial^2\phi}{\partial s^2} + K\dfrac{\partial\phi}{\partial n}\right) - \nabla^2\phi = 0$ (free edge).

$$\tag{134}$$

Inasmuch as the equations which must be satisfied by ϕ—namely, (130) and one set from among (134)—are homogeneous, we can without loss of generality impose the convenient normalization condition

$$\iint\limits_{D} \phi^2\, dx\, dy = 1. \tag{135}$$

The problem of solving the fourth-order partial differential equation (130) in conjunction with any one (or combination) of the sets of boundary conditions (134) constitutes an eigenvalue-eigenfunction problem of the type encountered in connection with the vibrating string, membrane, and rod. Any value of β for which there exists a function ϕ, normalized according to (135), which satisfies (130) and the single required set of boundary conditions from among (134) is an eigenvalue of β; ϕ is the corresponding eigenfunction. There may, in some cases, exist several linearly independent eigenfunctions corresponding to a single eigenvalue of β; that is, the eigenvalues of the vibrating plate may exhibit degeneracy.

According to (128), (133), and (132), it is clear that each eigenfunction ϕ describes a mode of single-frequency vibration which the given plate (under given boundary conditions) is capable of executing. The frequency $(\omega/2\pi)$ of each mode is related to the corresponding eigenvalue of β through (132). The degeneracy of a given eigenvalue implies the existence of more than one independent mode of vibration associated with the given frequency.

(b) We may characterize the eigenvalue-eigenfunction problem for the vibrating plate as an isoperimetric problem:

The extrema of

$$I = \iint\limits_{D} [(\Delta^2\phi)^2 - 2(1 - \sigma)(\phi_{xx}\phi_{yy} - \phi_{xy}^2)]dx\, dy$$

with respect to functions ϕ which are four-times differentiable in D, which satisfy the normalization condition

$$\iint\limits_{D} \phi^2\, dx\, dy = 1, \tag{136}$$

and which, in the case of the
  (i) Clamped plate, satisfy $\phi = (\partial\phi/\partial n) = 0$ on C,
  (ii) Simply supported plate, satisfy $\phi = 0$ on C,
are supplied by the eigenfunctions of (130) taken in conjunction with the appropriate set of boundary conditions from among (134).

The proof of the above characterization is left for exercise 24 at the end of this chapter.   At an early stage of this proof we achieve the result

$$\iint_D [(\nabla^2\phi)(\Delta^2\eta) - (1 - \sigma)(\phi_{xx}\eta_{yy} + \phi_{yy}\eta_{xx} - 2\phi_{xy}\eta_{xy}) - \lambda\phi\eta]dx \, dy = 0,$$

(137)

where ϕ is an extremizing function and η is arbitrary to within twice differentiability and, in the case of the

(i) Clamped plate, $\eta = (\partial\eta/\partial n) = 0$ on C,
(ii) Simply supported plate, $\eta = 0$ on C;

the constant λ, originally introduced as a Lagrange undetermined multiplier, is at a later step in the proof shown to be identical with the parameter β^4 which appears in (130).   That is,

$$\lambda = \beta^4.$$

(138)

Because of the arbitrary character of the function η we may set $\eta = \phi$ in (137) to obtain, after solving for λ,

$$\lambda = \iint_D [(\nabla^2\phi)^2 - 2(1 - \sigma)(\phi_{xx}\phi_{yy} - \phi_{xy}^2)]dx \, dy,$$

(139)

with the aid of (136).   The integrand of (139) may be rewritten as

$$\phi_{xx}^2 + 2\sigma\phi_{xx}\phi_{yy} + \phi_{yy}^2 + 2(1 - \sigma)\phi_{xy}^2$$
$$= (\phi_{xx} + \sigma\phi_{yy})^2 + (1 - \sigma^2)\phi_{yy}^2 + 2(1 - \sigma)\phi_{xy}^2 \geqq 0,$$

since† $0 < \sigma < \frac{1}{2}$.   It therefore follows from (139) that $\lambda \geqq 0$ and, through (138), that the substitution of β^4 for the common value of the two members of (129) in (*a*) above is justified.

We may also use the result (137) to establish the orthogonality of the plate eigenfunctions: If $\phi^{(j)}$ and $\phi^{(k)}$ are eigenfunctions associated respectively with a pair of *distinct* eigenvalues of β^4—which, in accordance with (138), we denote by λ_j and λ_k—we replace ϕ by $\phi^{(j)}$ and therefore λ by λ_j in (137), along with the substitution $\phi^{(k)}$ for the arbitrary‡ η.   We effect a second substitution of the same type into (137) by reversing the indices j and k in the initial substitution.   Subtracting the two special cases of (137) so obtained, we achieve the result

† See 10-1(*d*), final paragraph.
‡ In those cases (clamped or simply supported plate) in which η is required to satisfy special conditions ($\eta = \eta_n = 0$, or $\eta = 0$) on C, the eigenfunctions $\phi^{(k)}$ and $\phi^{(j)}$ must satisfy these same conditions, so that the substitution $\eta = \phi^{(k)}$ or $\eta = \phi^{(j)}$ into (137) is justified.

$$(\lambda_j - \lambda_k) \iint\limits_{D} \phi^{(j)} \phi^{(k)} \, dx \, dy = 0.$$

Since $\lambda_j \neq \lambda_k$, the integral must vanish, and the orthogonality follows.

In the case of degeneracy—the existence of more than one linearly independent eigenfunction corresponding to a single eigenvalue of β— the orthogonalization process delineated in 9-6(c) is applicable. With the result of the preceding paragraph, therefore, we conclude

$$\iint\limits_{D} \phi^{(j)} \phi^{(k)} \, dx \, dy = 0 \qquad (j \neq k),$$

for the plate eigenfunctions $\phi^{(1)}$, $\phi^{(2)}$, . . . , $\phi^{(k)}$, . . . associated with any one of the three types of boundary conditions considered here.

(c) The isoperimetric characterization of the vibrating-plate eigenvalue-eigenfunction problem may be sharpened into a minimum characterization which reads as follows:

We arrange the totality of eigenvalues of β^4 associated with the plate problem, for any one of the three types of boundary situations considered above (clamped, simply supported, or free edge), in the ascending order $\beta_1^4 \leqq \beta_2^4 \leqq \ldots \leqq \beta_k^4 \leqq \ldots$; each degenerate eigenvalue appears consecutively in the list a number of times equal to the number of independent eigenfunctions associated with it. *The kth eigenvalue β_k^4 is the minimum of the integral*

$$I = \iint\limits_{D} [(\nabla^2 \phi)^2 - 2(1 - \sigma)(\phi_{xx}\phi_{yy} - \phi_{xy}^2)] dx \, dy \qquad (140)$$

with respect to those functions ϕ which satisfy the normalization condition

$$\iint\limits_{D} \phi^2 \, dx \, dy = 1 \qquad (141)$$

and the $(k - 1)$ orthogonality relations

$$\iint\limits_{D} \phi^{(m)} \phi \, dx \, dy = 0 \qquad (m = 1,2, \ldots, k - 1),$$

where $\phi^{(m)}$ $(m = 1,2,3, \ldots)$ is the eigenfunction which satisfies

$$\nabla^4 \phi^{(m)} - \beta_m^4 \phi^{(m)} = 0$$

and the appropriate set of boundary conditions from among (134). Further, the functions ϕ eligible for the minimization of I must be, together with their first partial derivatives ϕ_x and ϕ_y, continuous everywhere in D;

the higher order partial derivatives may exhibit finite discontinuities at an isolated number of points or across a finite number of smooth arcs in D.

For the plate with boundary edge clamped the *additional* restrictions $\phi = (\partial\phi/\partial n) = 0$ on C must be imposed on the eligible functions; for the plate whose edge is simply supported the eligible functions are restricted to those which satisfy $\phi = 0$ on C. *No* special boundary restrictions are imposed upon the eligible functions ϕ if the edge of the plate is completely free.

The minimum β_k^4 of I under the stated restrictions is achieved when $\phi = \phi^{(k)}$.

A proof of the above characterization of the plate eigenvalues runs along the lines of the corresponding proof of the minimum characterization of the vibrating-membrane eigenvalues which appears in 9-9(b). It thus depends upon the validity of an expansion theorem for the plate eigenfunctions analogous to the expansion theorem for the membrane eigenfunctions stated (without proof) in 9-6(d). Explicit statement of the required expansion theorem is found in exercise 25 at the end of this chapter; proof of the minimum characterization of the plate eigenvalue-eigenfunction problem is reserved for the same exercise.

(d) Finally, it is possible to characterize the eigenvalue-eigenfunction problem for the vibrating plate in terms of a maximum-minimum principle which corresponds to the maximum-minimum characterization of the membrane eigenvalues demonstrated in 9-11(a). Explicit statement of the principle for the plate, together with the proof, is reserved for exercise 26 at the end of this chapter.

10-10. The Rectangular Plate. Ritz Method of Approximation

As compared with the success in solving the vibrating-membrane problem in several cases, there are relatively few examples of the eigenvalue-eigenfunction problem for the vibrating plate which have been solved rigorously. The problem of the circular plate, considered in end-chapter exercise 29, is the one case in which a complete solution has been achieved for each of the three types of boundary situations (clamped, simply supported, free) introduced in 10-8 above. For the rectangular shape, however, only the problem of the simply supported plate has been completely solved. Partly responsible for the lack of solutions for the free- and clamped-rectangular-plate problems is the easily verified fact that the partial differential equation (130) of 10-9(a)—the equation satisfied by the plate eigenfunctions—is not separable[1] in rectangular coordinates.

In the absence of a method for obtaining a precise analytical solution,

---

[1] The meaning of separability in this sense is given in 9-8(a), second paragraph.

W. Ritz[1] was the first to employ the minimum property enunciated in 10-9(c) as an aid to approximating the eigenvalues and eigenfunctions of the free-edge–rectangular-plate problem.   (It is for this reason that any method of eigenvalue-eigenfunction approximation based upon the direct minimization of integrals—such as those employed in 7-6, 9-13, 11-5, as well as below in the present section—is generally termed a "Ritz method.")[2]   Because of the ready accessibility of Ritz's monumental work —but especially because of the almost overwhelming amount of detailed computation involved—we limit ourselves to merely a few remarks concerning the problem of the rectangular plate with boundary edge free; these are found in (e) below.   The main portion of this section is devoted to the Ritz method as applied to the square plate with boundary edge clamped.

(a) By setting $u = \eta = \phi$ and $f = (\phi_{xx}\phi_{yy} - \phi_{xy}^2)$ in the trio of Green's-theorem results (105) of 10-8(b), we transform the integral (140) of 10-9(c) as follows:

$$I = \iint_D (\nabla^2\phi)^2 \, dx \, dy$$

$$- (1 - \sigma) \int_C [(\phi_x\phi_{yy} - \phi_{xy}\phi_y)dy + (\phi_x\phi_{xy} - \phi_y\phi_{xx})dx]. \quad (142)$$

Further, we employ the relations (112) and (113) of 10-8(c)—with the function u replaced by ϕ and with $n = 0$ (along C)—together with (114) of 10-8(d) to bring (142) into the form

$$I = \iint_D (\nabla^2\phi)^2 \, dx \, dy - (1 - \sigma) \int_C [\phi_n(\phi_{ss} + K\phi_n) + \phi_s(K\phi_s - \phi_{sn})]ds$$

$$= \iint_D (\nabla^2\phi)^2 \, dx \, dy - (1 - \sigma) \int_C \left[\phi_n(\phi_{ss} + K\phi_n) - \phi\frac{\partial}{\partial s}(K\phi_s - \phi_{sn}) \right] ds;$$

$$(143)$$

the final form is reached through an integration by parts.[3]

In the case of a clamped plate the functions ϕ eligible for the minimization of (140) of 10-9(c), which is the original form of (143), must satisfy $\phi = \phi_n = 0$ on C.   Thus, in this case, the line integral of (143) vanishes, so that the integral whose successive minima, in the sense of 10-9(c), are the eigenvalues of the clamped-plate problem is simply

---

[1] *Annalen der Physik*, Bd. 28, p. 737, 1909; or *Gesamelte Werke Walther Ritz*, p. 265, Paris, 1911.

[2] Such a method is frequently called a "Rayleigh-Ritz method."

[3] See discussion directly following (118) of 10-8(d), with accompanying footnote.

$$I = \iint_D (\nabla^2 \phi)^2 \, dx \, dy. \tag{144}$$

(b) The method we employ for the approximation of the successive eigenvalues of a given clamped-plate problem is completely analogous to the method developed in 9-13 for approximating the eigenvalues of a given membrane problem. Thus we replace the class of functions ϕ eligible for the minimization of (144) by the subclass of functions ψ which exhibit the form

$$\psi = c_1\Phi_1 + c_2\Phi_2 + \cdots + c_s\Phi_s, \tag{145}$$

where $\Phi_1(x,y)$, $\Phi_2(x,y)$, $\ldots$, $\Phi_s(x,y)$ are s given functions, continuously differentiable in D; $c_1, c_2, \ldots, c_s$ are arbitrary constants consistent with the required normalization condition (141) of 10-9(c) with ϕ replaced by ψ. The functions $\Phi_1, \Phi_2, \ldots, \Phi_s$ satisfy the clamped-plate requirement of vanishing, together with the normal derivative of each, on the boundary curve C.

We denote by $\psi_1, \psi_2, \ldots, \psi_s$ the first s approximate eigenfunctions[1] sought, and the corresponding approximate eigenvalues (of the parameter β^4) by $\Lambda_1, \Lambda_2, \ldots, \Lambda_s$. In accordance with (145) we write

$$\psi_m = \sum_{j=1}^{s} c_j^{(m)}\Phi_j \qquad (m = 1,2, \ldots ,s), \tag{146}$$

so that the problem of finding each minimizing ψ_m is equivalent to that of determining the set of values $c_1^{(m)}$, $c_2^{(m)}$, $\ldots$, $c_s^{(m)}$ for the coefficients $c_1, c_2, \ldots, c_s$, respectively, in (145), for each m. Since the functions ψ eligible for the kth minimization of (144) must be orthogonal in D to the first $(k-1)$ approximate eigenfunctions $\psi_1, \psi_2, \ldots, \psi_{k-1}$, we have, because of (145) and (146),

$$\iint_D \psi\psi_m \, dx \, dy = \sum_{i=1}^{s} \sum_{j=1}^{s} c_i c_j^{(m)}\sigma_{ij} = 0 \qquad (m = 1,2, \ldots ,k-1), \tag{147}$$

where we define

$$\sigma_{ji} = \sigma_{ij} = \iint_D \Phi_i\Phi_j \, dx \, dy. \tag{148}$$

Substitution of (145) for ϕ in the normalization condition (141) gives, further, the requirement

$$\sum_{i=1}^{s} \sum_{j=1}^{s} c_i c_j \sigma_{ij} = 1. \tag{149}$$

[1] See 7-6 (b).

Finally, if we define

$$\Gamma_{ji} = \Gamma_{ij} = \iint_D \nabla^2 \Phi_j \nabla^2 \Phi_i \, dx \, dy, \tag{150}$$

substitution of (145) into (144) gives

$$I = \sum_{i=1}^{s} \sum_{j=1}^{s} c_i c_j \Gamma_{ij} \tag{151}$$

for the quantity whose successive minima we seek.

Comparison of the foregoing paragraph with 9-13(a) reveals that the problem of minimizing (151) with respect to the set of $c_1, c_2, \ldots, c_s$ which satisfy the subsidiary conditions (147) and (149) is identical with the minimization of (196) of the earlier section under the restrictions (192) and (194) of that section. For this reason we may repeat, in essence, the paragraph of 9-13(a) which follows the equations referred to:

The first s approximate eigenvalues $\Lambda_1, \Lambda_2, \ldots, \Lambda_s$ of the clamped-plate problem are given by the s roots of the equation in Λ

$$\begin{vmatrix} -\Gamma_{11} + \Lambda\sigma_{11} & -\Gamma_{12} + \Lambda\sigma_{12} & \cdots & -\Gamma_{1s} + \Lambda\sigma_{1s} \\ -\Gamma_{21} + \Lambda\sigma_{21} & -\Gamma_{22} + \Lambda\sigma_{22} & \cdots & -\Gamma_{2s} + \Lambda\sigma_{2s} \\ \cdots & \cdots & \cdots & \cdots \\ -\Gamma_{s1} + \Lambda\sigma_{s1} & -\Gamma_{s2} + \Lambda\sigma_{s2} & \cdots & -\Gamma_{ss} + \Lambda\sigma_{ss} \end{vmatrix} = 0. \tag{152}$$

The coefficients $c_1^{(k)}, c_2^{(k)}, \ldots, c_s^{(k)}$, which—when k ranges over the values $1, 2, \ldots, s$—supply, through (146), the corresponding approximate eigenfunctions $\psi_1, \psi_2, \ldots, \psi_s$, are obtained by solving the system of s linear homogeneous equations

$$\sum_{j=1}^{s} (\Lambda_k \sigma_{ij} - \Gamma_{ij}) c_j^{(k)} = 0 \qquad (i = 1, 2, \ldots, s)$$

in conjunction with the normalization requirement

$$\sum_{i=1}^{s} \sum_{j=1}^{s} c_i^{(k)} c_j^{(k)} \sigma_{ij} = 1,$$

for each k. The constants σ_{ij} are computed by means of their definition (148); the Γ_{ij} are computed from (150).

As in the case of the approximate membrane eigenvalues obtained in 9-13(a), each approximate plate eigenvalue is an approximation *from above*; that is, $\beta_k^4 \leq \Lambda_k$, for all k. This fact is a direct consequence of the maximum-minimum characterization of the plate eigenvalues stated, but not proved,[1] in 10-9(d). The larger we choose the value of s, clearly, the greater is the accuracy, in general, of each approximate eigenvalue.

(c) We apply the method outlined in (b) above in order to approximate the eigenvalues (of the parameter β^4) associated with the clamped square plate of side a. For the sake of simplicity of the computation involved we confine our attention to the degree of approximation achieved by the value s = 3. (In order to keep in focus the wider generality of the method, however, we do not specify the value of s until the very point at which it is quite necessary to particularize to the value s = 3.)

An almost obvious choice[2] for the set of functions $\Phi_1(x,y)$, $\Phi_2(x,y)$, . . . , $\Phi_s(x,y)$—in linear terms of which we seek to express the approximate eigenfunctions ψ_1, ψ_2, . . . , ψ_s—is the set of products of eigenfunctions of the clamped-vibrating-bar problem; namely, if $w_q(x)$ is the qth of the orthonormal eigenfunctions for the clamped bar of length a along the x direction—so that $w_r(y)$ is the rth such eigenfunction for the clamped bar of length a along the y direction—we employ the products $w_q(x)w_r(y)$ in the following fashion:

$$\Phi_1 = w_1(x)w_1(y), \qquad \Phi_2 = w_1(x)w_2(y), \qquad \Phi_3 = w_2(x)w_1(y),$$

and, in general,

$$\Phi_j = w_q(x)w_r(y) \qquad [j = \tfrac{1}{2}(q + r - 2)(q + r - 1) + q], \qquad (153)$$

where

$$j = 1, 2, \ldots, s; \qquad r = 1, 2, \ldots, \tfrac{1}{2}(\sqrt{8s + 1} - 1);$$
$$q = 1, 2, \ldots, \tfrac{1}{2}(\sqrt{8s + 1} + 1) - r.$$

(Thus we restrict the choice of s to values for which $(8s + 1)$ is the square of an integer.) Since $w_q(x)$ and $w'_q(x)$ both vanish for x = 0 and x = a, and since $w_r(y)$ and $w'_r(y)$ both vanish[3] for y = 0 and y = a, it follows that the products of the form (153) satisfy the required conditions of vanishing, together with their normal derivatives, on the boundary[4] of the square plate of side a.

If we write p_q^4 for $(\gamma/EJ_1)\lambda_q$ and w_q for ϕ_q, it is clear from (82) of 10-6(c) that the orthonormal clamped-bar eigenfunctions w_q satisfy the differential equation

$$\frac{d^4w_q}{d\xi^4} = p_q^4 w_q \qquad (q = 1,2,3, \ldots), \qquad (\xi = x \text{ or } y), \qquad (154)$$

---

[1] See, however, end-chapter exercise 26.

[2] See end-chapter exercise 32 for still another choice.

[3] These properties of $w_q(x)$ and $w_r(y)$ spring from the fact that they are eigenfunctions of the clamped-bar problem (see (73), line (iii), of 10-6(a)).

[4] See the final sentence of the opening paragraph of (b) above.

together with the end-point conditions

$$w_q(0) = w_q(a) = 0, \qquad w_q'(0) = w_q'(a) = 0. \tag{155}$$

The explicit form of $w_q(\xi)$ is given in (162) below.

In accordance with the parenthetic portion of (153) we write

$$i = \tfrac{1}{2}(q' + r' - 2)(q' + r' - 1) + q', \tag{156}$$

where q' and r' run through the same sets of values as do q and r, respectively. Thus, with (153), we obtain from (148)

$$\sigma_{ji} = \sigma_{ij} = \int_0^a w_q(x)w_{q'}(x)dx \int_0^a w_r(y)w_{r'}(y)dy = \delta_{qq'}\delta_{rr'},$$

because of the orthonormality of the bar eigenfunctions. It further follows, since $i = j$ if and only if both $q = q'$ and $r = r'$, that

$$\sigma_{ji} = \sigma_{ij} = \delta_{ij}. \tag{157}$$

Since it follows from (153) that

$$\nabla^2\Phi_i = w_q''(x)w_r(y) + w_q(x)w_r''(y),$$

we have from (150), with the aid of (156),

$$\begin{aligned}
\Gamma_{ji} = \Gamma_{ij} &= \int_0^a \int_0^a [w_q''(x)w_r(y)+w_q(x)w_r''(y)][w_{q'}''(x)w_r(y)+w_{q'}(x)w_{r'}''(y)]dx\,dy \\
&= \delta_{rr'}\int_0^a w_q''(x)w_{q'}''(x)dx + \delta_{qq'}\int_0^a w_r''(y)w_{r'}''(y)dy \\
&\quad + \int_0^a w_q''(x)w_{q'}(x)dx \int_0^a w_r(y)w_{r'}''(y)dy \\
&\quad + \int_0^a w_q(x)w_{q'}''(x)dx \int_0^a w_r''(y)w_{r'}(y)dy \\
&= \delta_{rr'}H_{qq'} + \delta_{qq'}H_{rr'} + L_{q'q}L_{rr'} + L_{qq'}L_{r'r}, \tag{158}
\end{aligned}$$

where we define

$$H_{mm'} = \int_0^a w_m''(\xi)w_{m'}''(\xi)d\xi, \qquad L_{mm'} = \int_0^a w_m(\xi)w_{m'}''(\xi)d\xi; \tag{159}$$

the Kronecker deltas appearing in the second line of (158) arise from the orthonormality of the functions $w_r(\xi)$. If we twice integrate by parts each of the integrals appearing in (159) we find, on using (155), that

$$H_{mm'} = \int_0^a \frac{d^4w_m}{d\xi^4} w_{m'}(\xi)d\xi, \qquad L_{mm'} = \int_0^a w_m''(\xi)w_{m'}(\xi)d\xi; \tag{160}$$

that is, $L_{mm'} = L_{m'm}$. Also, with the aid of (154), we obtain

$$H_{mm'} = p_m^4 \int_0^a w_m(\xi)w_{m'}(\xi)d\xi = p_m^4\delta_{mm'}.$$

With the last two results (158) reads

$$\Gamma_{ji} = \Gamma_{ij} = \delta_{rr'}\delta_{qq'}(p_r^4 + p_q^4) + 2L_{qq'}L_{rr'}$$
$$= \delta_{ij}(p_q^4 + p_r^4) + 2L_{qq'}L_{rr'}, \tag{161}$$

since $i = j$ if and only if both $q = q'$ and $r = r'$. It thus remains to evaluate the quantities $L_{mm'}$ defined in (159); for this we require an explicit expression for $w_q(\xi)$.

The orthonormal clamped-bar eigenfunctions, which satisfy (154) and (155), are given explicitly by[1]

$$w_q(\xi) = \frac{1}{\sqrt{a}} \left\{ \left[\frac{\sin p_q\left(\xi - \dfrac{a}{2}\right)}{\sin \frac{1}{2}p_q a} - \frac{\sinh p_q\left(\xi - \dfrac{a}{2}\right)}{\sinh \frac{1}{2}p_q a} \right] \cos^2 \frac{1}{2} q\pi \right.$$
$$\left. + \left[\frac{\cos p_q\left(\xi - \dfrac{a}{2}\right)}{\cos \frac{1}{2}p_q a} - \frac{\cosh p_q\left(\xi - \dfrac{a}{2}\right)}{\cosh \frac{1}{2}p_q a} \right] \sin^2 \frac{1}{2} q\pi \right\}, \tag{162}$$

for $q = 1, 2, 3, \ldots$; (ap_q) is the qth positive root of the transcendental equation

$$\tan^2 (\tfrac{1}{2}ap) = \tanh^2 (\tfrac{1}{2}ap) \tag{163}$$

or of the equivalent equation $\cosh (ap) = \sec (ap)$.

For the purposes of the computation carried out below we require that the quantities $L_{mm'} = L_{m'm}$ be evaluated only for $m, m' = 1,2$, independently. These cases are fully covered by the following results:

$$L_{mm'} = 0 \qquad \text{if } (m + m') \text{ odd}, \tag{164}$$

$$L_{mm} = \begin{cases} \left(\dfrac{2}{a}\right) p_m \cot \dfrac{1}{2} ap_m - p_m^2 \cot^2 \dfrac{1}{2} ap_m & (m \text{ even}), \\[2ex] -\left(\dfrac{2}{a}\right) p_m \tan \dfrac{1}{2} ap_m - p_m^2 \tan^2 \dfrac{1}{2} ap_m & (m \text{ odd}), \end{cases} \tag{165}$$

the computations of which are left for exercise 16 at the end of this chapter. (Both (164) and (165) may be obtained directly from (162) and (160) as follows: (164) results from the fact that $w_m''(\xi)w_{m'}(\xi)$ is an *odd* function with respect to $\xi = \tfrac{1}{2}a$ if $(m + m')$ is odd. Direct integration yields (165) on use of the fact, which follows from (163), that $\tan \tfrac{1}{2}ap_m = (-1)^m \tanh \tfrac{1}{2}ap_m$.)

(d) At this point we specify the value $s = 3$; that is, i and j take on, independently, the values 1, 2, 3. By means of the parenthetic part of (153) and (156) we establish the following tabulation of correspondences:

---

[1] See end-chapter exercise 15. This result may, of course, be verified by direct substitution into (154) and (155).

$j = 1$: $r = q = 1$; $j = 2$: $r = 2$, $q = 1$; $j = 3$: $r = 1$, $q = 2$;
$i = 1$: $r' = q' = 1$; $i = 2$: $r' = 2$, $q' = 1$; $i = 3$: $r' = 1$, $q' = 2$.

From (161), with the aid of (164), we thus obtain

$$\Gamma_{11} = 2p_1^4 + 2L_{11}^2, \qquad \Gamma_{22} = (p_1^4 + p_2^4) + 2L_{11}L_{22} = \Gamma_{33},$$
$$\Gamma_{12} = \Gamma_{21} = \Gamma_{13} = \Gamma_{31} = 2L_{11}L_{12} = 0, \qquad \Gamma_{23} = \Gamma_{32} = 2L_{12}^2 = 0.$$

With these results taken in conjunction with (157) of (c) above, the determinantal equation (152) of (b) above (with $s = 3$) assumes the particularly simple form

$$(\Lambda - 2p_1^4 - 2L_{11}^2)(\Lambda - p_1^4 - p_2^4 - 2L_{11}L_{22})^2 = 0. \tag{166}$$

The numerical values of (ap_1) and (ap_2) are given by[1]

$$ap_1 = 4.7300408, \qquad ap_2 = 7.8532046,$$

to seven decimal places. Use of these figures gives (with somewhat less accuracy), according to (165),

$$a^2L_{11} = -12.31, \qquad a^2L_{22} = 46.05.$$

Thus (166) becomes

$$\left(\Lambda - \frac{1304}{a^4}\right)\left(\Lambda - \frac{5438}{a^4}\right)^2 = 0,$$

so that we achieve the following approximations to the three lowest eigenvalues (of the parameter β^4) of the clamped square plate:

$$\Lambda_1 = \frac{1304}{a^4}, \qquad \Lambda_2 = \Lambda_3 = \frac{5438}{a^4}. \tag{167}$$

Without further analysis we cannot, of course, estimate the degree of accuracy of the approximations (167). It is beyond the scope of our study, unfortunately, to consider a method which has been developed[2] for approximating *from below* the eigenvalues for the clamped square plate. This method gives the results

$$\Lambda_1' = \frac{1295}{a^4}, \qquad \Lambda_2' = \Lambda_3' = \frac{4910}{a^4} \tag{168}$$

---

[1] See Rayleigh, Vol. I, pp. 277, 278.

[2] A. Weinstein, *Mémorial des sciences mathématiques*, Vol. 88, "Études des spectres des équations aux dérivées partielles de la théorie des plaques élastiques," Gauthier-Villars, Paris, 1937. The figures quoted in (168) are derived from Weinstein's *Mémorial* volume, in which he has given the results of computations for $a = \pi$ (pp. 54, 56).

as approximations from below to the first three eigenvalues of β^4. From (167) and (168) we therefore conclude

$$\frac{1295}{a^4} \leqq \beta_1^4 \leqq \frac{1304}{a^4}, \qquad \frac{4910}{a^4} \leqq \beta_{2,3}^4 \leqq \frac{5438}{a^4}. \tag{169}$$

According to (132) of 10-9(a) we have $\omega_k^2 = (D_0/\mu)\beta_k^4$ ($k = 1,2,3, \ldots$), where ($\omega_k/2\pi$) is the frequency of the plate's kth natural mode of vibration, μ is the mass per unit area, and D_0 is the flexural rigidity (defined just following (97) of 10-7(d)). Thus we have for an approximation to the fundamental frequency of the clamped square plate

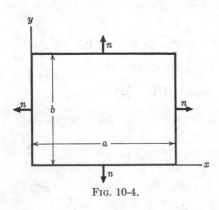

$$\frac{\omega_1}{2\pi} \cong \frac{5.75}{a^2} \sqrt{\frac{D_0}{\mu}},$$

Fig. 10-4.

according to (167); according to the result (169) this approximation has an accuracy of better than 0.4 per cent. Approximate computations of the second and third natural frequencies may be similarly computed from (167).

(e) In accord with the opening remarks of the present section (page 241) consideration of the rectangular vibrating plate with boundary edge free is limited here to an enunciation of the boundary conditions which are satisfied by the eigenfunctions of the problem. We note, first, that the curvature K, introduced in (109) of 10-8(c), is zero along the four edges of the plate. Further, if the rectangular domain D is given by $0 \leqq x \leqq a$, $0 \leqq y \leqq b$, we have the following set of relations between the coordinates x, y, and the n, s variables introduced in 10-8(c): If we measure the arc length s from the origin—that is, if $s = 0$ at $x = 0$, $y = 0$—it is clear from Fig. 10-4 that

$$\begin{aligned}
&\text{along } y = 0, &&s = x, &&n = -y; \\
&\text{along } x = a, &&s = a + y, &&n = x - a; \\
&\text{along } y = b, &&s = 2a + b - x, &&n = y - b; \\
&\text{along } x = 0, &&s = 2a + 2b - y, &&n = -x.
\end{aligned} \tag{170}$$

With the aid of (170), and $K = 0$, we apply the general boundary conditions (134,iii) of 10-9(a) to the rectangular plate:

$$\frac{\partial}{\partial y}\left(\nabla^2\phi\right) + (1-\sigma)\frac{\partial}{\partial x}\left(\frac{\partial^2\phi}{\partial x\,\partial y}\right) = 0 \atop (1-\sigma)\frac{\partial^2\phi}{\partial x^2} - \nabla^2\phi = 0 \Bigg\} \quad \text{along } y = 0,\, y = b;$$

$$\frac{\partial}{\partial x}\left(\nabla^2\phi\right) + (1-\sigma)\frac{\partial}{\partial y}\left(\frac{\partial^2\phi}{\partial x\,\partial y}\right) = 0 \atop (1-\sigma)\frac{\partial^2\phi}{\partial y^2} - \nabla^2\phi = 0 \Bigg\} \quad \text{along } x = 0,\, x = a.$$

Since $\nabla^2\phi = \phi_{xx} + \phi_{yy}$, the second of the two pairs of bracketed equations may be replaced by $\phi_{yy} + \sigma\phi_{xx} = 0$ and $\phi_{xx} + \sigma\phi_{yy} = 0$, respectively.

An additional boundary condition arises from the required continuity of the mixed partial derivative ϕ_{sn}, pointed out in 10-8(d) above.[1] From (170) we perceive that ϕ_{sn} assumes the successive forms $-\phi_{xy}$, ϕ_{yx}, $-\phi_{xy}$, ϕ_{yx} on the four sides of the rectangle as we traverse C counterclockwise from the origin. Since the order of partial differentiation is immaterial, we therefore conclude $\phi_{xy} = -\phi_{xy}$, or

$$\phi_{xy} = 0, \qquad \text{at the four corners.[2]}$$

EXERCISES

1. Let (x_1, x_2, x_3) and (x_1', x_2', x_3') be two neighboring points of an unstrained solid, in the sense of 10-1(c); after strain the points are located respectively at $x_i + u_i$ and $x_i' + u_i'$ $(i = 1,2,3)$. Use (7) to show that the change of the distance between the points which results from the strain is independent of the quantities ω_1, ω_2, ω_3—provided we neglect squares and products such as ω_1^2, $\omega_1\omega_3$, $e_{12}\omega_2$, etc. (Thus ω_1, ω_2, ω_3 are called *components of rotation.*) HINT: The change involved is

$$\delta = \sqrt{\sum_{i=1}^{0}\left[(x_i + u_i) - (x_i' + u_i')\right]^2} - \sqrt{\sum_{i=1}^{3}(x_i - x_i')^2},$$

but we may here approximate $[(x_i + u_i) - (x_i' + u_i')]^2$ by

$$(x_i - x_i')^2 + 2(x_i - x_i')(u_i - u_i').$$

From (7) it follows that $\displaystyle\sum_{i=1}^{3}(x_i - x_i')(u_i - u_i')$ is independent of ω_1, ω_2, ω_3.

2. Use (7) to prove that a diagonal element e_{kk} of the strain tensor represents an extension per unit length in the x_k direction. Show also that an off-diagonal element

[1] The quantity $(u_{sn} - Ku_n)$ is originally required to be continuous along C. Here, however, $K = 0$, and u_{sn} differs from ϕ_{sn} merely through a factor which depends upon the time t alone, as in (128) of 10-9(a).

[2] A brief account of the controversial history of the boundary conditions applicable to the plate with free edge may be found in Rayleigh, Vol. I, pp. 369–371. It is interesting to note that not even Rayleigh's derivation and statement (Vol. I, pp. 352–357) of these conditions are completely correct.

e_{jk} $(j \neq k)$ represents a shear in the x_jx_k plane, whereby lines parallel respectively to the x_j and x_k axes in the unstrained state are each rotated through an angle e_{jk} in opposite senses, so that in the strained state the angles between the lines are $(\frac{1}{2}\pi \pm 2e_{jk})$. HINT: In each proof, set equal to zero ω_1, ω_2, ω_3 and all strain elements, except for the one on which attention is focused.

3. (a) Employ the definitions (5) to verify the equations of compatibility (12) and (13).

(b) Show that (12) and (13) represent exactly six independent relations for *all* possible choices of i,j,k, so that there are six equations of compatibility in all. (By "*all* possible choices" is meant the inclusion of combinations of i, j, k which violate the parenthetic inequalities of (12) and (13).)

4. Prove the equality of the second and third members of (17). HINT: Consider the second member as the sum of two double sums and reverse the indices of summation in the second of these; use the fact that $(\partial W/\partial E_{jk}) = (\partial W/\partial E_{kj})$.

5. Obtain an independent proof of the elementary relation (2) on the basis of the results (32) and (33) derived—without the use of (2)—in 10-2(d).

6. (a) Show that the component of bending moment (arising from a given surface-force distribution) about the line $x_2 = x_2'$, $x_3 = x_3'$ (arbitrary line parallel to x_1 axis, if x_2' and x_3' are arbitrary constants) is given, according to the definition (43), by

$$M_1' = \iint_B [T_3(x_2 - x_2') - T_2(x_3 - x_3')]dS.$$

HINT: Translate the x_1 axis so as to coincide with $x_2 = x_2'$, $x_3 = x_3'$.

(b) Use the result of part (a) to prove the assertion made in 10-4(b) to the effect that the surface-force distribution (41) on $x_3 = L$ gives rise to a bending moment PJ_1 taken about *any* axis parallel to the original x_1 axis; J_1 is defined by (45). HINT: Use (38) of 10-4(a).

(c) Similarly prove that the distribution (42) on $x_3 = 0$ gives rise to an equal, but oppositely directed, moment along *any* such axis. (It is thus shown that the bar of 10-4 is bent by equal and opposite *couples* applied at the two end faces, as stated in the text.)

7. (a) From (49), (50), and (51) it follows that

$$\frac{\partial u_1}{\partial x_3} = C_2. \tag{171}$$

(In this exercise C_1, C_2, C_3, b_1, b_2, b_3 are used to denote arbitrary constants.) In the manner of achieving (49), (50), (51) use (47) and (48) to derive that

$$\frac{\partial}{\partial x_1}\left(\frac{\partial u_1}{\partial x_2}\right) = -\frac{\sigma P}{E}, \qquad \frac{\partial}{\partial x_2}\left(\frac{\partial u_1}{\partial x_2}\right) = \frac{\partial}{\partial x_3}\left(\frac{\partial u_1}{\partial x_2}\right) = 0$$

and therefore that

$$\frac{\partial u_1}{\partial x_2} = -\frac{\sigma P}{E} x_1 - C_3. \tag{172}$$

Derive, with the aid of (171), (172), and (48), that

$$u_1 = -\frac{\sigma P}{E} x_1x_2 + C_2x_3 - C_3x_2 + b_1. \tag{173}$$

(b) Use (47) and (48) to derive

$$\frac{\partial u_2}{\partial x_3} = -\frac{P}{E} x_3 - C_1. \tag{174}$$

From (47) and (172) show also that

$$\frac{\partial u_2}{\partial x_1} = \frac{\sigma P}{E} x_1 + C_3.$$ (175)

With the aid of (174), (175), and (48) derive the result

$$u_2 = \frac{\sigma P}{2E} (x_1^2 - x_2^2) - \frac{P}{2E} x_3^2 + C_3 x_1 - C_1 x_3 + b_2.$$ (176)

(c) From (171), (174), and (47) derive

$$\frac{\partial u_3}{\partial x_1} = - C_2, \qquad \frac{\partial u_3}{\partial x_2} = \frac{P}{E} x_3 + C_1,$$

and therefore, with the aid of (48), that

$$u_3 = \frac{P}{E} x_2 x_3 + C_1 x_2 - C_2 x_1 + b_3.$$ (177)

(d) Substitute the solutions (173), (176), (177) back into the differential equations (47), (48) and so verify that the constants C_1, C_2, C_3, b_1, b_2, b_3 are mutually independent.

8. (a) Prove that the cross sections $x_3 =$ constant of the bar of 10-4 remain plane in the strained state described in that section. HINT: Use either (47), (48), or the results of exercise 7 to prove that $(\partial^2 u_3/\partial x_1^2) = (\partial^2 u_3/\partial x_2^2) = 0$.

(b) For the same bar prove that any line $x_2 =$ constant, $x_3 =$ constant is strained into a parabolic shape whose curvature is oppositely directed from that of the strained shape of a line $x_1 =$ constant, $x_2 =$ constant. HINT: Use either (47), (48), or the results of exercise 7 to prove that $(\partial^2 u_2/\partial x_1^2) = (\sigma P/E)$; complete by using part (a) and comparing with (52).

9. Derive, for the bar of 10-4, the relation

$$\frac{\partial^2 u_2}{\partial x_3^2} = - \frac{M_1}{E J_1},$$

where M_1 is given by (44) and J_1 is defined in (45). (This result, known as the Bernoulli-Euler equation, is the usual starting point of the "engineering theory" of the bending of thin rods: M_1 is interpreted as the total bending moment exerted by the portion of the bar to the right of (x_3 larger) a given cross section upon the portion of the bar which lies to the left of (x_3 smaller) the cross section. The axis with respect to which M_1 is computed (see exercise 6(a)) passes through the centroid of the cross section and is parallel to the x_1 axis. Thus—unlike the case of the bar of 10-4—M_1 may be a function of x_3.)

10. State and prove the extension to the basic lemma of 3-1 required for the derivation of (65) and (66) of 10-5(b).

11. Derive the differential equation of motion for the vibrating rod of 10-5 if, in addition to (56), we take into account the kinetic-energy term

$$\tfrac{1}{2}\rho J_1 \int_0^L \dot{u}_x^2 \, dx$$

which arises from the rotational motion of the various cross-sectional volume elements; here ρ is the mass per unit volume of the rod. ANSWER:

$$E J_1 \frac{\partial^4 u}{\partial x^4} + \gamma \frac{\partial^2 u}{\partial t^2} - \rho J_1 \frac{\partial^4 u}{\partial t^2 \, \partial x^2} = 0.$$

12. Derive the equation of motion for the vibrating rod (neglecting the rotational kinetic energy introduced in exercise 11) for the case in which both (EJ_1) and γ are functions of x. (Assume the validity of (58).) Answer:

$$\frac{\partial^2}{\partial x^2}\left(EJ_1\frac{\partial^2 u}{\partial x^2}\right) + \gamma\frac{\partial^2 u}{\partial t^2} = 0.$$

13. (a) Prove that $\phi = M(x) + N(x)$ is the solution of (76), with p^4 written for $(\gamma\lambda/EJ_1)$, where

$$\frac{d^2 M}{dx^2} - p^2 M = 0, \qquad \frac{d^2 N}{dx^2} + p^2 N = 0; \tag{178}$$

and, further (if $p \neq 0$),

$$M = \frac{1}{2}\left(\phi + \frac{1}{p^2}\frac{d^2\phi}{dx^2}\right), \qquad N = \frac{1}{2}\left(\phi - \frac{1}{p^2}\frac{d^2\phi}{dx^2}\right). \tag{179}$$

Thus derive the general solution of (76):

$$\phi = A\cosh px + B\sinh px + C\cos px + D\sin px, \tag{180}$$

where A, B, C, D are arbitrary constants.

(b) Prove that, if $M(0) = M(L) = 0$, then M must vanish identically in $0 \leq x \leq L$. Thus use (179) and the second of (178) to prove that the eigenfunctions of the hinged vibrating bar are identical with those of the uniform vibrating string whose ends are fixed (see exercise 9(a), Chap. 7). Hint: For the hinged bar, $M(0) = M(L) = 0$, so that $\phi = N$.

Show, however, that the ratios of successive vibration frequencies of the hinged bar are different from the corresponding ratios for the string of equal length.

(c) Solve (76) with $\lambda = 0$ and show that no such solution is an eigenfunction except in each of the two cases

(i) $\phi''(0) = \phi'''(0) = \phi''(L) = \phi'''(L) = 0$ (both ends free),
(ii) $\phi''(0) = \phi'''(0) = \phi(L) = \phi''(L) = 0$ (one end free, the other hinged).

Show that case (i) violates the general rule (exercise 14 below) of no more than one linearly independent eigenfunction per bar eigenvalue. Use the Schmidt process of 9-6(c) to obtain a pair of orthonormal eigenfunctions corresponding to the eigenvalue $\lambda = 0$ for the bar free at both ends. Answer (not unique):

$$\phi_1 = L^{-\frac{1}{2}}, \quad \phi_2 = 2\sqrt{3}\,L^{-\frac{3}{2}}(\tfrac{1}{2}L - x).$$

The eigenvalue $\lambda = 0$ is of no interest in the study of vibration.

14. (a) Fill in the details of the proof of the following theorem:
If p is an eigenvalue of the differential equation

$$\frac{d^4\phi}{dx^4} = p^4\phi \tag{181}$$

with the end-point conditions

$$\phi(0) = \phi'(0) = \phi(L) = \phi'(L) = 0, \tag{182}$$

there exists only one linearly independent eigenfunction $\phi(x)$ satisfying both (181) and (182).

Suppose that there are two linearly independent functions $\phi_1(x)$ and $\phi_2(x)$ which satisfy (181) and (182). Consider also

$$\psi_1(x) = \sinh px, \qquad \psi_2(x) = \sin px,$$

which satisfy the differential equation (181) but not all of (182). We notice

$$\psi_1(0) = \psi_2(0) = 0. \tag{183}$$

We have that there exist constants A_1, A_2, B_1, B_2, not all zero, such that

$$A_1\phi_1(x) + A_2\phi_2(x) + B_1\psi_1(x) + B_2\psi_2(x) = 0, \tag{184}$$

identically; for the wronskian (see 2-8(e))

$$w = \begin{vmatrix} \phi_1 & \phi_2 & \psi_1 & \psi_2 \\ \phi_1' & \phi_2' & \psi_1' & \psi_2' \\ \phi_1'' & \phi_2'' & \psi_1'' & \psi_2'' \\ \phi_1''' & \phi_2''' & \psi_1''' & \psi_2''' \end{vmatrix}$$

may be shown to vanish identically. (Form the derivative (dw/dx) by using the rule (2-8(d)) for differentiating a determinant, and use the fact that each of ϕ_1, ϕ_2, ψ_1, ψ_2 satisfies (181): $(dw/dx) = 0$, so that $w = $ constant. But $w = 0$ at $x = 0$ because of (183) and (182) as satisfied by ϕ_1 and ϕ_2.)

Since (184) holds identically, we have, because ϕ_1 and ϕ_2 satisfy (182),

$$B_1\psi_1'(0) + B_2\psi_2'(0) = 0, \qquad B_1\psi_1'(L) + B_2\psi_2'(L) = 0.$$

But this implies, it is easily shown, either $\cos pL = \cosh pL$—an impossibility for $p \neq 0$ (why?)—or

$$B_1 = B_2 = 0.$$

Thus, from (184), there exist A_1 and A_2, not both zero, such that

$$A_1\phi_1(x) + A_2\phi_2(x) = 0,$$

identically; that is, ϕ_1 and ϕ_2 are *not* linearly independent.

(b) Prove the theorem corresponding to that of part (a) for the eigenfunctions of (181) with each of the following sets of end point conditions:

(i) $\quad \phi''(0) = \phi'''(0) = \phi''(L) = \phi'''(L) = 0 \qquad (p \neq 0)$
(ii) $\quad \phi(0) \quad = \phi''(0) \quad = \phi(L) \quad = \phi''(L) = 0$
(iii) $\quad \phi''(0) = \phi'''(0) = \phi(L) \quad = \phi''(L) = 0$
(iv) $\quad \phi(0) \quad = \phi''(0) \quad = \phi(L) \quad = \phi'(L) = 0$
(v) $\quad \phi''(0) = \phi'''(0) = \phi(L) \quad = \phi'(L) = 0$

HINT: For (ii) use the result of exercise 13(b) together with the theorem proved in exercise 3, Chapter 7. For (iv) use $\psi_1 = \sinh p(L - x)$, $\psi_2 = \sin p(L - x)$.

15. (a) Determine explicitly, to within a multiplicative factor, the eigenfunctions for the vibrating bar clamped at $x = 0$ and $x = L$. HINT: Use (180) of exercise 13 to obtain the solution of (181) with the end-point conditions (182).

Show that

$$\phi = A(\cosh px - \cos px) + B(\sinh px - \sin px),$$

where

$$-\frac{A}{B} = \frac{\sinh pL - \sin pL}{\cosh pL - \cos pL} = \frac{\cosh pL - \cos pL}{\sinh pL + \sin pL}, \tag{185}$$

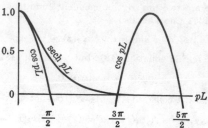

Each value of pL at which the two curves
intersect is a solution of (186).

FIG. 10-5.

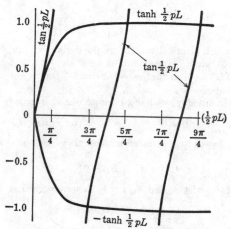

Each value of $\frac{1}{2}pL$ at which an intersection
occurs is a solution of (187).

FIG. 10-6.

whence

$$\cosh pL = \sec pL, \qquad \text{or} \qquad \operatorname{sech} pL = \cos pL. \tag{186}$$

Show that (186) is equivalent to

$$\tanh^2 (\tfrac{1}{2}pL) = \tan^2 (\tfrac{1}{2}pL). \tag{187}$$

(b) From the graphical solution of (186), Fig. 10-5, ascertain the following relation involving (p_qL), the qth positive solution of (186):

$$p_qL = (2q + 1)\tfrac{1}{2}\pi - (-1)^q\alpha_q \qquad (q = 1,2,3, \ldots), \tag{188}$$

where

$$0 < \alpha_q < \tfrac{1}{2}\pi, \qquad \alpha_1 > \alpha_2 > \alpha_3 > \ldots, \qquad \lim_{q \to \infty} \alpha_q = 0.$$

Thus show, with the aid of (187) and its graphical solution, Fig. 10-6, that

$$\tanh (\tfrac{1}{2}p_qL) = (-1)^q \tan (\tfrac{1}{2}p_qL). \tag{189}$$

(c) Show, with the aid of (186), that (185) reads, with $p = p_q$,

$$\frac{A}{B} = \frac{\cos pL - \dfrac{|\sin pL|}{\sin pL}}{\sin pL}$$

$$= \begin{cases} -\tan \frac{1}{2}pL & (q = 2,4,6, \ldots, \text{ whence } \sin pL > 0), \\ \cot \frac{1}{2}pL & (q = 1,3,5, \ldots, \text{ whence } \sin pL < 0), \end{cases}$$

as we find with the aid of (188). Thus show, with $p = p_q$,

$$\phi_q = C_q \begin{cases} (\cosh px - \cos px) \sin \dfrac{pL}{2} - (\sinh px - \sin px) \cos \dfrac{pL}{2} \\ (\cosh px - \cos px) \cos \dfrac{pL}{2} + (\sinh px - \sin px) \sin \dfrac{pL}{2} \end{cases}, \quad (190)$$

where C_q is constant: The upper form is taken if q is even, the lower if q is odd. With the aid of (189) bring (190) into the form

$$\phi_q = C_q' \begin{cases} \dfrac{\sin p(x - \frac{1}{2}L)}{\sin \frac{1}{2}pL} - \dfrac{\sinh p(x - \frac{1}{2}L)}{\sinh \frac{1}{2}pL} & (q \text{ even}) \\ \dfrac{\cos p(x - \frac{1}{2}L)}{\cos \frac{1}{2}pL} - \dfrac{\cosh p(x - \frac{1}{2}L)}{\cosh \frac{1}{2}pL} & (q \text{ odd}) \end{cases}, \quad (191)$$

with C_q' constant and $p = p_q$.
(d) Show that

$$C_q' = \frac{1}{\sqrt{L}} \quad (192)$$

supplies the normalization condition

$$\int_0^L \phi_q^2 \, dx = 1.$$

(e) Obtain (191), together with (189), *ab initio* by considering the bar which is clamped at $x = -\frac{1}{2}L$ and $x = \frac{1}{2}L$. HINT: Use (180), with $\phi = \phi' = 0$ at $x = -\frac{1}{2}L$ and $x = \frac{1}{2}L$; then shift the origin of x.

16. Evaluate the integral

$$L_{qq'} = \int_0^L \phi_q'' \phi_{q'} \, dx,$$

where ϕ_q is given by (191) and C_q' by (192). HINT: Write $\phi_q = u_q - v_q$, where $u_q'' = -p_q^2 u_q$, $v_q'' = p_q^2 v_q$, and $u_q(0) = v_q(0)$, $u_q(L) = v_q(L)$, $u_q'(0) = v_q'(0)$, $u_q'(L) = v_q'(L)$ because of (182). Use the differential equations to show that

$$L_{qq'} = -\frac{p_q^2}{L} \int_0^L (u_q + v_q)(u_{q'} - v_{q'}) \, dx$$

and, for example,

$$\int_0^L u_q u_{q'} \, dx = -\int_0^L v_q v_{q'} \, dx = \frac{[u_{q'} u_q' - u_{q'}' u_q]_0^L}{p_{q'}^2 - p_q^2}, \text{ etc.} \quad (q \neq q').$$

ANSWER:

$$L_{qq'} = L_{q'q} = \begin{cases} 0 & (q + q' \text{ odd}), \\[2mm] \dfrac{8p_q^2 p_{q'}^2 (p_{q'} \cot \frac{1}{2} p_{q'} L - p_q \cot \frac{1}{2} p_q L)}{L(p_{q'}^4 - p_q^4)} & (q, q' \text{ even}), \\[2mm] \dfrac{8p_q^2 p_{q'}^2 (p_q \tan \frac{1}{2} p_q L - p_{q'} \tan \frac{1}{2} p_{q'} L)}{L(p_{q'}^4 - p_q^4)} & (q, q' \text{ odd}), \end{cases}$$

$$L_{qq} = \begin{cases} \left(\dfrac{2}{L}\right) p_q \cot \dfrac{1}{2} p_q L - p_q^2 \cot^2 \dfrac{1}{2} p_q L & (q \text{ even}), \\[2mm] -\left(\dfrac{2}{L}\right) p_q \tan \dfrac{1}{2} p_q L - p_q^2 \tan^2 \dfrac{1}{2} p_q L & (q \text{ odd}). \end{cases}$$

17. (a) Show that the surface-force distributions (89) on the surfaces $x_2 = L_2$, $x_2 = 0$ of the plate of 10-7 give rise to equal and opposite bending moments of magnitude $\frac{2}{3} P_2 L_1 h^3$ about the x_1 axis. HINT: Use the definitions (43) of 10-4(b).

(b) Show that the bending moments of part (a), as well as those which arise from the distributions (89) on $x_1 = L_1$, $x_1 = 0$, are indeed couples. HINT: See exercise 6 above.

18. Use (90) in conjunction with the definitions (5) of 10-1(c) to derive the results (91), (92), and (93) of 10-7(c). HINT: Proceed in the manner of attaining (52) of 10-4(c) from (47) and (48).

19. State and prove the extension of the basic lemma of 3-1 required for the derivations of (107) of 10-8(b) and the boundary conditions (126) and (127) from (119) of 10-8(d).

20. Prove that the jacobian $(1 + nK)$ given in (111) of 10-8(c) is positive if the point (n,s) is not separated from C by the evolute (locus of centers of curvature) of C. (It is required to investigate the geometric factors which determine the sign of K, defined in (109).) Show that $(1 + nK) = 0$ if (n,s) lies on the evolute.

21. (a) Suppose that the quantity $(u_{sn} - Ku_s)$ is allowed a jump discontinuity of magnitude δ at a single point of C. Show, by making the appropriate change in (118) of 10-8(d), that a term $\pm(1 - \sigma)\eta_0\delta$ is introduced into the time integral (but exterior to the integral along C) of (119); that is, (119) reads

$$\int_{t_1}^{t_2} \left[\pm(1 - \sigma)\eta_0\delta + \int_C (\text{integrand unaltered})ds \right] dt = 0, \tag{193}$$

where η_0 is an arbitrary constant in the free-edge problem and zero in the clamped and simply supported cases.

Show that (193) implies, in the free-edge case, that $\delta = 0$.

(b) Extend the argument of part (a) to cover the case in which $(u_{sn} - Ku_s)$ is allowed an arbitrary finite number of jump discontinuities along C; that is, show that $(u_{sn} - Ku_s)$ must be continuous along C in the free-edge case.

The validity of (118) is thus proved without a *priori* assumption of the continuity of $(u_{sn} - Ku_s)$, for in the two cases not covered—clamped and simply supported edges—we have $\eta = 0$ (and therefore $\eta_0 = 0$) along C. (Tacit in the above proof is the assumption that $(u_{sn} - Ku_s)$ approaches a finite limit as any point of C is approached along C from a given direction—that no discontinuities other than *jump* discontinuities are allowed, that is. Such other discontinuities are ruled out from the start, however, by the physical nature of the problem.)

22. Write down the boundary conditions (134) as they read in the following special cases:

(a) Circular plate $(r \leqq a)$ clamped. ANSWER: $\phi = 0$, $(\partial\phi/\partial r) = 0$ at $r = a$.

(b) Rectangular plate $(0 \leqq x \leqq a;\ 0 \leqq y \leqq b)$ simply supported. ANSWER: $\phi = 0$, $(\partial^2\phi/\partial x^2) = 0$ on $x = 0$ and $x = a$; $\phi = 0$, $(\partial^2\phi/\partial y^2) = 0$ on $y = 0$ and $y = b$. Note also that $(\partial^2\phi/\partial y^2) = 0$ on $x = 0$, $x = a$; and $(\partial^2\phi/\partial x^2) = 0$ on $y = 0$, $y = b$. Why?

(c) Semicircular plate $(r \leqq a;\ 0 \leqq \theta \leqq \pi)$ simply supported. ANSWER: Since $K = (1/a)$ on $r = a$,

$$\phi = 0, \qquad \frac{\sigma}{a}\frac{\partial\phi}{\partial r} + \frac{\partial^2\phi}{\partial r^2} = 0 \text{ on } r = a\ (0 \leqq \theta \leqq \pi),$$

$$\phi = 0, \qquad \frac{1}{r^2}\frac{\partial^2\phi}{\partial\theta^2} = 0 \text{ on } \theta = 0,\ \theta = \pi.$$

(d) Circular plate $(r \leqq a)$ free. ANSWER:

$$\left.\begin{aligned}\frac{\partial}{\partial r}(\nabla^2\phi) + (1 - \sigma)\frac{1}{a^2}\frac{\partial}{\partial\theta}\left(\frac{\partial^2\phi}{\partial\theta\,\partial r} - \frac{1}{a}\frac{\partial\phi}{\partial\theta}\right) = 0 \\ (1 - \sigma)\left(\frac{1}{a^2}\frac{\partial^2\phi}{\partial\theta^2} + \frac{1}{a}\frac{\partial\phi}{\partial r}\right) - \nabla^2\phi = 0\end{aligned}\right\} \text{on } r = a,$$

where

$$\nabla^2\phi = \frac{1}{r}\frac{\partial}{\partial r}\left(r\frac{\partial\phi}{\partial r}\right) + \frac{1}{r^2}\frac{\partial^2\phi}{\partial\theta^2} \qquad [9\text{-}2(e)].$$

23. (Compare exercise 13.) (a) Prove that $\phi = M(x,y) + N(x,y)$ is the solution of (130), where

$$\nabla^2 M - \beta^2 M = 0, \qquad \nabla^2 N + \beta^2 N = 0;$$

and, further, if $\beta \neq 0$, that

$$M = \frac{1}{2}\left(\phi + \frac{1}{\beta^2}\nabla^2\phi\right), \qquad N = \frac{1}{2}\left(\phi - \frac{1}{\beta^2}\nabla^2\phi\right).$$

(b) Prove that, if $M = 0$ everywhere on the boundary C, it follows that $M = 0$ identically in D. HINT: Use Green's formula (23) of 2-13.

(c) Show that, if we deal with a simply supported rectangular plate, we have $M = 0$ on the boundary and, therefore, $\phi = N$ in D. HINT: See exercise 22(b).

Thus show that the eigenfunctions for the rectangular simply supported plate are identical with those associated with the uniform vibrating membrane of the same size and shape (see (108) of 9-8, with $\sigma = \sigma_0$ and $\beta^2 = (\lambda_0/\tau)$).

(d) Compare the relationship between the various natural vibration frequencies of the hinged rectangular plate and those of the corresponding rectangular membrane. HINT: Use (132) of 10-9(a), (57) of 9-4(a), and (118) of 9-8(b).

24. Use the "$\epsilon\eta$ process"—essentially the same as that which is employed in 10-8(b), but with the integral over t suppressed—to prove the isoperimetric character of the plate eigenvalue-eigenfunction problem, as enunciated in 10-9(b). In particular derive (137) of that section.

25. (a) We assume the validity of the following expansion theorem: Let $\phi^{(1)}$, $\phi^{(2)}$, . . . , $\phi^{(m)}$, . . . be the totality of the orthonormal eigenfunctions associated with a given vibrating-plate problem. Let arbitrary $g\ (x,\ y)$ have continuous first partial derivatives everywhere in the plate domain D; D may be split into a finite number of subdomains with smooth boundaries such that g_{xx}, g_{xy}, and g_{yy} are continuous in each subdomain. Then if we write

$$g(x,y) = \sum_{m=1}^{\infty} c_m\phi^{(m)}(x,y) \qquad \left(c_m = \iint_D \phi^{(m)}g\,dx\,dy\right), \tag{194}$$

the series converges uniformly to $g(x,y)$ in every subdomain of D in which $g(x,y)$ is continuous. Further, in every subdomain of D in which a given partial derivative (first or second order) of g is continuous we may form that derivative by term-by-term differentiation of the series (194); the differentiated series converges uniformly in the subdomain to the corresponding derivative of g.

(b) Use the orthonormality of the eigenfunctions $\phi^{(m)}$ to derive the parenthetic formula for c_m in (194). (Assume, of course, the validity of the series expansion in (194).)

(c) Use the expansion theorem of part (a) above to establish the following expression for the general solution of the equation of motion (107) of 10-8(b) for the vibrating plate (if $\beta^4 = 0$ is *not* an eigenvalue):

$$u(x,y,t) = \sum_{m=1}^{\infty} \left(A_m \cos \sqrt{\frac{D_0}{\mu}} \beta_m^2 t + B_m \sin \sqrt{\frac{D_0}{\mu}} \beta_m^2 t \right) \phi^{(m)}(x,y), \qquad (195)$$

where $\phi^{(m)}$ is the mth eigenfunction of the corresponding eigenvalue-eigenfunction problem. HINT: Use the method of 9-7(a) by substituting

$$u(x,y,t) = \sum_{m=1}^{\infty} c_m(t) \phi^{(m)}(x,y) \qquad (196)$$

into the hamiltonian integral (100) of 10-8(a); proceed to extremize with respect to the functions $c_m(t)$. A tremendous simplification is effected by first using the transformation

$$\iint_D (u_{xx}u_{yy} - u_{xy}^2)dx\,dy = \frac{1}{2} \int_C \left[u_n(u_{ss} + Ku_s) - u\frac{\partial}{\partial s}(Ku_s - u_{sn}) \right]ds,$$

which is derived in the manner in which (143) of 10-10(a) is achieved. It is readily seen that the part of the hamiltonian integral embodied in the line integral around C vanishes on substitution of (196) when we take into account the boundary conditions satisfied on C by the $\phi^{(m)}(x,y)$.

How must (195) be modified if $\beta^4 = 0$ is an eigenvalue?

(d) Use the expansion theorem of part (a) above to establish the minimum characterization of the vibrating-plate eigenvalues enunciated in 10-9(c). HINT: See the proof of the corresponding theorem for the membrane eigenvalues in 9-9(b). Simplicity is greatly served by employing the form (143) of 10-10(a) for the integral (140) of 10-9(c).

Take especial note of the point in the proof at which it is essential to require continuity everywhere in D of the first partial derivatives of the functions ϕ eligible for each minimization.

26. State and prove a maximum-minimum characterization of the eigenvalue-eigenfunction problem for a vibrating plate. This characterization bears the same relationship to the simple minimum characterization of 10-9(c) as does the maximum-minimum characterization of the membrane eigenvalues (9-11) to their minimum characterization (9-9).

27. (a) In the membrane problem the functions eligible for the minimization of the integral I (given by (123) of 9-8(a)) need not have first partial derivatives which are continuous *everywhere* in D. For the vibrating-plate problem, on the other hand, only functions whose first partial derivatives are continuous *everywhere* in D are eligible for the minimization of I (given by (140) of 10-9(c)). To what physical difference between a plate and a membrane does this fact correspond?

(b) It has never been proved that the eigenfunction corresponding to the lowest eigenvalue of the clamped-plate problem for arbitrary domain D has no nodal lines in D. Any attempt at a proof along the lines of the one given in exercise 21(c), Chap. 9 (for the corresponding theorem concerning the first eigenfunction of the general fixed-edge-membrane problem), breaks down because of the required continuity everywhere in D of ϕ_x and ϕ_y for the functions ϕ eligible for the minimization in the plate problem. Demonstrate the occurrence of this breakdown.

28. By the direct substitution $\phi = X(x)Y(y)$ show that the equation (130) of 10-9(a) for the plate eigenfunctions is not separable (in the sense of the second paragraph, 9-8(a)) in rectangular coordinates.

29. We consider, in this exercise, a circular plate of radius a; we employ the polar coordinates (r, θ) with origin at the center of the circle, so that $r = a$ is the equation of the plate boundary.

(a) With the aid of exercise 23(a) and the identity (43) of 9-2(e) show that the solution of $\nabla^4\phi - \beta^4\phi = 0$ which is *independent* of θ is given by

$$\phi(r) = H^+(r) + H^-(r), \tag{197}$$

where H^+ and H^- satisfy the differential equations

$$\frac{d}{dr}\left(r\frac{dH^\pm}{dr}\right) \pm \beta^2 r H^\pm = 0,$$

where upper signs (or lower signs) of the ambiguities ($\pm$) are taken together.

(b) We impose the requirement that ϕ be bounded for $r \leq a$—in particular, for $r = 0$. Show, with the aid of 8-3(c), that (197) becomes

$$\phi(r) = AJ_0(\beta r) + BJ_0(i\beta r) \qquad (i^2 = -1), \tag{198}$$

where A and B are arbitrary constants and $J_0(z)$ is the zero-order Bessel function of the first kind.

(c) Show that $J_0(i\beta r)$ is a real function since β and r are real. HINT: Use (42) of 8-3(c).

$J_n(iz)$ is generally denoted by $i^n I_n(z)$—where I_n is the so-called *modified* nth order Bessel function of the first kind. Thus we may rewrite (198) as

$$\phi(r) = AJ_0(\beta r) + BI_0(\beta r). \tag{199}$$

Use (182) of 8-7(d) to prove that

$$J_0'(z) = -J_1(z), \qquad I_0'(z) = I_1(z), \tag{200}$$

where the prime (') indicates differentiation with respect to the argument of the function involved.

(d) Show that, if (199) is an eigenfunction of the circular-clamped-plate problem, we must have

$$\frac{A}{B} = \frac{I_1(\beta a)}{J_1(\beta a)} = -\frac{I_0(\beta a)}{J_0(\beta a)}. \tag{201}$$

HINT: Use exercise 22(a) together with (200).

The final equation (equality of the second and third members) has for solutions an infinite unbounded set of positive values of βa. These supply the list of eigenvalues for the circularly symmetric (independent of θ) modes of vibration of the clamped circular plate.[1]

---

[1] For numerical results involved in the circular-plate problem, see Philip M. Morse, "Vibration and Sound," 2d ed., p. 210, McGraw-Hill Book Company, Inc., New York, 1948.

(e) Write down the equations which correspond to (201) if (199) is an eigenfunction of (i) the circular-simply-supported-plate problem and (ii) the circular-free-plate problem.

30. We subject a clamped plate to a nonconservative transverse force per unit area given by the expression $F(x,y,t)$. (That is, an element of area $dx\,dy$ experiences the externally applied force $F(x,y,t)dx\,dy$ perpendicular to the xy plane.)

(a) Use the extended Hamilton's principle of 6-7 to show that the equation of motion of the plate so influenced is

$$\mu \frac{\partial^2 u}{\partial t^2} + D_0 \nabla^4 u = F(x,y,t). \tag{202}$$

(b) Extend the method of 9-7 to show that the solution of (202), with $u = (\partial u/\partial n) = 0$ on C, is

$$u = \sum_{m=1}^{\infty} c_m(t)\,\phi^{(m)}(x,y),$$

where

$$\nabla^4 \phi^{(m)} - \beta_m^4 \phi^{(m)} = 0 \text{ in } D, \qquad \phi^{(m)} = \frac{\partial \phi^{(m)}}{\partial n} = 0 \text{ on } C,$$

and

$$\mu \frac{d^2 c_m}{dt^2} + \beta_m^4 D_0 c_m = \iint_D F(x,y,t)\,\phi^{(m)}\,dx\,dy.$$

(Each $\phi^{(m)}$ is normalized in D.)

(c) If the applied transverse-force density is $F(x,y)$—independent of t, that is— show that the equation of equilibrium for the clamped plate is given by

$$D_0 \nabla^4 u = F(x,y) \text{ in } D, \qquad \text{with } u = \frac{\partial u}{\partial n} = 0 \text{ on } C.$$

HINT: Apply to (202) the condition of equilibrium—namely, that u is independent of the time t.

(d) Make the required modifications of statement in parts (a), (b), (c) if a simply supported plate is substituted for the clamped plate.

31. Prove the assertion made in 10-10(b), final paragraph, that $\beta_k^4 \leqq \Lambda_k$. HINT: Use exercise 26 above to develop the same sort of argument as that given in the final paragraph of 9-13(a).

32. (a) Apply the method developed in 10-10(b) for approximating the eigenvalues of the square-clamped-plate problem, with $s = 2$, using

$$\Phi_1 = \left(\frac{x}{a}\right)^2 \left(\frac{y}{a}\right)^2 \left(1 - \frac{x}{a}\right)^2 \left(1 - \frac{y}{a}\right)^2,$$

$$\Phi_2 = \left(\frac{x}{a}\right)^3 \left(\frac{y}{a}\right)^3 \left(1 - \frac{x}{a}\right)^2 \left(1 - \frac{y}{a}\right)^2.$$

ANSWER: $\Lambda_1 = (1296/a^4)$, $\Lambda_2 = (5793/a^4)$.

(b) In place of the functions employed in part (a) use, with $s = 1$,

$$\Phi_1 = \sin^2 \left(\frac{\pi x}{a}\right) \sin^2 \left(\frac{\pi y}{a}\right).$$

ANSWER: $\Lambda_1 = (1385/a^4)$.

CHAPTER 11

QUANTUM MECHANICS

Of the tremendous body of theory known as *quantum* (or *wave*) *mechanics*, we consider in the present chapter a narrow segment impinged upon by the ideas and methods of the calculus of variations as developed in the preceding sections of this work. Roughly speaking, quantum mechanics may be described as the mathematical theory developed in the years following 1925 which has had success in describing accurately the great bulk of extranuclear atomic phenomena. The exceptions to this success—the phenomena not correctly described by the present development of quantum mechanics—although notable in importance, are few in number.

Historically, the role played in the origins of quantum-mechanical theory by the calculus of variations is signal. The Schrödinger differential equation, a cornerstone of the theory, was discovered and first applied by the man whose name it bears as the result of a problem he posed calling for the extremization of an integral with respect to an unknown integrand function. While Schrödinger's proposal of the problem was purely arbitrary in its lack of motivation grounded in physical considerations, it found a *posteriori* justification through its immediate success, with suitable interpretation, in describing the radiation spectrum of the hydrogen atom. Soon after the first discovery of his equation, however, and again with the aid of the calculus of variations, Schrödinger was able to provide insight into the physical basis of the new atomic mechanics and so derive "his" equation anew with some degree of *a priori* physical justification.

In its present form the science of quantum mechanics is based upon a set of simply stated postulates leading to results which include the all-important Schrödinger equation merely as a special case. Nevertheless, the fact that this equation is derivable from a variational problem makes available the calculus of variations as a valuable tool for the approximate solution of many atomic problems. The present chapter treats a few such problems in addition to offering an exposition of the essence of Schrödinger's early work.

In all that follows we avoid consideration of refinements to the elementary theory which take into account relativistic effects and the influence of "spin" (the intrinsic angular momentum of the fundamental particles of matter).

11-1. First Derivation of the Schrödinger Equation for a Single Particle

(a) In his initial paper[1] Schrödinger considers the reduced Hamilton-Jacobi equation (6-5(c)) associated with a single particle of mass m moving under the influence of an arbitrary force field described by the potential energy $V(x,y,z)$; the instantaneous position of the particle is denoted by the cartesian coordinates (x,y,z). According to (38) of 6-5(c), this equation reads

$$\frac{1}{2m}\left[\left(\frac{\partial S^*}{\partial x}\right)^2 + \left(\frac{\partial S^*}{\partial y}\right)^2 + \left(\frac{\partial S^*}{\partial z}\right)^2\right] + V(x,y,z) - E = 0, \qquad (1)$$

where E is the constant total energy of the particle.[2] With the change of dependent variable $S^* = K \log \psi$—with K a constant open for experimental determination[3]—(1) becomes, on multiplication by ψ^2,

$$\frac{K^2}{2m}\left[\left(\frac{\partial \psi}{\partial x}\right)^2 + \left(\frac{\partial \psi}{\partial y}\right)^2 + \left(\frac{\partial \psi}{\partial z}\right)^2\right] + (V - E)\psi^2 = 0. \qquad (2)$$

Ignoring the problem of solving (2), Schrödinger instead considers the volume integral[4] of the left-hand member carried out over all space:

$$I^* = \iiint \left\{\frac{K^2}{2m}\left[\left(\frac{\partial \psi}{\partial x}\right)^2 + \left(\frac{\partial \psi}{\partial y}\right)^2 + \left(\frac{\partial \psi}{\partial z}\right)^2\right] + (V - E)\psi^2\right\} dx\,dy\,dz. \qquad (3)$$

He then poses the question: What differential equation must the function ψ satisfy if I^*, given by (3), is to be an extremum with respect to twice-differentiable functions ψ which vanish at infinity in such fashion that I^* exists? The answer to this question lies in the result of 9-1(b): We substitute the integrand[5]

$$f = \frac{K^2}{2m}(\psi_x^2 + \psi_y^2 + \psi_z^2) + (V - E)\psi^2$$

---

[1] There is available an English translation of the set of Schrödinger's first papers published under the title "Collected Papers on Wave Mechanics," Blackie & Son, Ltd., Glasgow, 1928.

[2] The significance of the dependent variable S^*, of no immediate importance at this point, is given in 6-5.

[3] The reader familiar with quantum theory should soon recognize the identity of K with the well-known $(h/2\pi)$.

[4] In this chapter we uniformly omit explicit indication of the limits of integration whenever a multiple integral is carried out over all space.

[5] As in preceding chapters, we employ subscripts to indicate partial differentiation.

of (3) into the Euler-Lagrange equation (9) of 9-1(b), with w replaced by ψ. We thus obtain

$$-\frac{K^2}{2m}(\psi_{xx} + \psi_{yy} + \psi_{zz}) + (V - E)\psi = 0,$$

or, in abbreviated notation,

$$\frac{K^2}{2m}\nabla^2\psi + (E - V)\psi = 0, \tag{4}$$

as the differential equation—the so-called Schrödinger equation for a single particle—which must be satisfied in order that ψ render (3) an extremum.

(b) For a given potential-energy function V, solutions ψ of the Schrödinger equation (4) which vanish sufficiently rapidly at infinity (for the existence of (3)) exist, in general, for only a privileged discrete set of values of E; that is to say, the solution of (4) under the "boundary" condition that (3) exist is an eigenvalue-eigenfunction problem in which the eigenvalues of E are to be determined. Schrödinger's early assumption—that, namely, the eigenvalues of E in (4) are the physically realizable values of the total energy of a particle under the influence of the potential energy V—is maintained in the theory as it stands today.

On the other hand the physical interpretation of the Schrödinger eigenfunctions ψ—the so-called *wave functions*—was not uniquely assigned in the first days of quantum mechanics; the interpretation which has eventually become accepted universally is elucidated in 11-3(c) below.

The fact that E cannot in general be assigned arbitrarily provides an equivalent, but more useful, extremum problem which leads directly to (4), as Schrödinger points out in an addendum to his first paper: If we extremize the integral

$$I = \iiint \left[\frac{K^2}{2m}(\psi_x^2 + \psi_y^2 + \psi_z^2) + V\psi^2\right] dx\, dy\, dz \tag{5}$$

with respect to functions ψ which satisfy the normalization

$$\iiint \psi^2\, dx\, dy\, dz = 1, \tag{6}$$

we are led, according to 9-1(c), directly to (4), provided we denote by E the undetermined Lagrange multiplier of the problem. Thus the Schrödinger eigenvalue-eigenfunction problem is equivalent to the above isoperimetric problem—a fact which, following a more precise statement in 11-4(d) below, is applied to the approximate solution of certain atomic problems.

(Discussion of the constant K which appears in the Schrödinger equation is reserved for 11-2(d) below.)

11-2. The Wave Character of a Particle. Second Derivation of the Schrödinger Equation

Prior to the discovery of the Schrödinger equation, De Broglie had developed an approach to the theoretical study of the atom which is based upon what he considered a *fictitious* wave character associated with a material particle. The theory was presented as a physically plausible basis for certain inherently arbitrary rules of procedure in the older (1913), narrowly successful, atomic theory of Bohr. Inasmuch as Schrödinger's application of "his" equation to the hydrogen atom (11-3 below) yielded the same (experimentally verified) energy levels as the Bohr theory, he sought to develop a connection between his own work and the wave theory of De Broglie. He found the desired connection with the aid of Fermat's principle (Chap. 5), the principle of least action (6-6), and the form of classical mechanics embodied in the Hamilton-Jacobi differential equation (6-5). The extreme importance of this connection achieved full recognition with the almost simultaneous establishment of the *physical* wave character of electrons through the experiments of Thomson, Davisson and Germer, and others.

(*a*) In order to develop the essence of the connection between the Schrödinger equation and the wave character of material particles we consider briefly a few aspects of the subject of wave phenomena in general. For our present purpose we may define a wave as a "disturbance" $\Psi = \Psi(x,y,z,t)$† which is propagated through space so as to be described by the equation

$$\nabla^2 \Psi = \frac{1}{u^2} \frac{\partial^2 \Psi}{\partial t^2}, \tag{7}$$

where u is a positive constant. (In the case of a plane-polarized light wave, for example, Ψ may represent the associated electric field intensity as a function of position and time. For a sound wave traveling through a gaseous medium, Ψ may represent the longitudinal displacement from equilibrium of the gas particles as a function of position and time, etc.)

An important type of solution of the wave equation (7) is that which can be written in the form

$$\Psi = \psi(x,y,z)e^{-i\omega t}, \tag{8}$$

where

$$e^{-i\omega t} = \cos \omega t - i \sin \omega t \qquad (i^2 = -1) \tag{9}$$

† As usual, t denotes the time variable.

and ω is a positive constant. Substituting (8) into (7), we obtain

$$\Delta^2\psi + \frac{\omega^2}{u^2}\psi = 0 \tag{10}$$

as the equation which must be satisfied by the position-dependent function ψ if (8) is a solution of the wave equation.

(The fact that Ψ is a complex function (of real variables) should not be disturbing. Each of its real and imaginary parts taken separately is a solution of the wave equation[1] (7); either may thus be used to characterize a real physical quantity.)

The equation (10) for ψ we call the time-independent wave equation.

With the aid of (9) we see that the function Ψ given by (8) is a periodic function of time with frequency $(\omega/2\pi)$. Such a solution of the wave equation is generally termed *monochromatic*. A more general solution may be constructed as a linear superposition (either sum or integral) of monochromatic solutions involving more than a single frequency.

To simplify the discussion we temporarily limit consideration to the case in which the disturbance Ψ is a function of only one of the three space variables—x, for the sake of definiteness. In this case we have $\nabla^2\psi = (d^2\psi/dx^2)$, so that (10) becomes

$$\frac{d^2\psi}{dx^2} + \frac{\omega^2}{u^2}\psi = 0.$$

Of this equation we choose the particular solution

$$\psi = Ce^{inx} \qquad \left(p - \frac{\omega}{u}\right),$$

where C is an arbitrary constant, real or complex. With (8) we thus have for a monochromatic solution of (7) which depends on x and t only

$$\Psi = Ce^{i(px-\omega t)} \qquad \left(p = \frac{\omega}{u}\right). \tag{11}$$

As stated above, we may employ either the real or the imaginary part of (11) to represent the physical disturbance which constitutes the wave.

We note the following properties of the disturbance described by (11):

(i) The disturbance is the same at all points lying in any plane $x = $ constant; Ψ is thus said to represent a *plane* wave.

(ii) The amplitude (maximum value with respect to time) is the same at all points of space.

[1] See end-chapter exercise 1.

(iii) The value of the disturbance is the same, at any instant t, on any pair of planes x = constant which are separated by the distance $\lambda = (2\pi/p) = (2\pi u/\omega)$ or any integral multiple thereof. The quantity λ is called the *wavelength* of the disturbance.

(iv) If an observer moves in the x direction with a velocity such that $(px - \omega t)$ remains constant, the disturbance as seen by him at his position is the same at all instants of time; the required velocity—such that $(d/dt)(px - \omega t) = 0$—is clearly $(dx/dt) = (\omega/p) = u$. Since u is a constant, it is thus evident that the entire disturbance pattern is propagated in the positive x direction with velocity u. The quantity $(px - \omega t)$ is called the *phase* of the wave; surfaces of constant phase travel in the positive x direction with velocity u, the so-called *phase velocity*. We note in passing that the direction of the wave motion is normal to the surfaces of constant phase.

(b) In (a) above we consider the phase velocity to be constant, the same at all points of space. To generalize we suppose that $u = u(x)$, a slowly varying positive function of x in the sense that[1]

$$\left|\frac{du}{dx}\right| \ll \frac{u}{\lambda} = \frac{\omega}{2\pi}.$$

With this restriction on the magnitude of (du/dx) it is meaningful to assign an *essentially* constant phase velocity u and a corresponding wavelength $\lambda = (2\pi u/\omega) = (2\pi/p)$ within any region over which the phase (with t = constant) varies by no more than a small integral multiple of 2π; that is, we may speak of a "local" phase velocity and of a "local" wavelength. The frequency $(\omega/2\pi)$ is assumed *strictly* constant.

A second generalization is to replace the constant C by a slowly varying function $C(x)$, with the restriction

$$\left|\frac{dC}{dx}\right| \ll \left|\frac{C}{\lambda}\right|;$$

we may thus also speak of a "local" amplitude. Of especial significance is the fact that (11)—with C, u, and $p = (\omega/u)$ functions of x—is approximately a solution of the wave equation (7) in so far as we may neglect the derivatives (dC/dx) and (du/dx).

For a final generalization we return to the case in which Ψ (and therefore ψ) may depend upon all three space variables x, y, z. For this purpose we consider the monochromatic disturbance

$$\Psi = C(x,y,z)e^{i[\phi(x,y,z)-\omega t]} \qquad (\phi \text{ real}), \qquad (12)$$

---

[1] The symbol $\ll$ is read "is small compared with."

in which C and the first partial derivatives of ϕ are slowly varying functions of position; *i.e.*, the relative variations of C, $(\partial\phi/\partial x)$, $(\partial\phi/\partial y)$, $(\partial\phi/\partial z)$ are all small in any region over which the phase $[\phi(x,y,z) - \omega t]$—with t = constant—varies by no more than a small integral multiple of 2π. We maintain the *strict* constancy of the frequency $(\omega/2\pi)$.

We observe, by direct substitution, that (12) is an approximate solution of the wave equation (7) in so far as we may neglect the first partial derivatives of C and the second partial derivatives of ϕ—provided that[1]

$$u = \frac{\omega}{\sqrt{(\partial\phi/\partial x)^2 + (\partial\phi/\partial y)^2 + (\partial\phi/\partial z)^2}} = \frac{\omega}{|\nabla\phi|}. \tag{13}$$

The direction of wave motion at any point is defined as that of the normal to the surface of constant phase (with t = constant) through the point, in the sense of increasing phase. For the disturbance (12), therefore, the direction of wave motion is clearly in the direction of the gradient[2] of ϕ—namely, $\nabla\psi$. For an observer to travel in the direction of the wave motion so that the phase as seen by him at his position remains constant (along a curved path, in general), his velocity—of which the three components are (dx/dt), (dy/dt), (dz/dt)—must be such that

$$\frac{d}{dt}(\phi - \omega t) = \frac{\partial\phi}{\partial x}\frac{dx}{dt} + \frac{\partial\phi}{\partial y}\frac{dy}{dt} + \frac{\partial\phi}{\partial z}\frac{dz}{dt} - \omega = 0. \tag{14}$$

The first three terms of the middle member of (14) constitute the scalar product[3] of $\nabla\phi$ and the observer's velocity; since this velocity has the same direction as $\nabla\phi$, the scalar product must be equal to its magnitude (ds/dt) multiplied by the magnitude of $\nabla\phi$. That is, (14) gives

$$\frac{ds}{dt} = \frac{\omega}{|\nabla\phi|}. \tag{15}$$

Comparison of (15) with (13) demonstrates the equivalence of u with (ds/dt), the speed with which a point of a given surface of constant phase travels in the direction of the wave motion. We thus conclude that $u = u(x,y,z)$ also plays the role of a local phase velocity in the general case represented by (12).

Under the assumption of an essentially constant $|\nabla\phi|$ made here, we may, as in the preceding case, define a local wavelength by means of the relation $\lambda = (2\pi u/\omega)$ first introduced in (iii) of (a) above. Thus, accord-

[1] See 2-12 for the definition of $\nabla\phi$.
[2] The direction of $\nabla\phi$ at any point (see 2-12(a)) is normal to the surface ϕ = constant through the point, in the sense of increasing ϕ.
[3] See 2-12(b).

ing to (13), we have

$$\lambda = \frac{2\pi}{\sqrt{(\partial\phi/\partial x)^2 + (\partial\phi/\partial y)^2 + (\partial\phi/\partial z)^2}}. \tag{16}$$

It is directly verified that λ, so defined, is the distance between surfaces ϕ = constant for which the difference of phase in (12) is equal to 2π (with t = constant).

The results of the preceding paragraphs are of use in the derivation of the Schrödinger equation which follows.

(c) The discovery and development of a new mechanical theory which would be applicable to the atom was made necessary in the first quarter of the twentieth century by the failure of classical mechanics to provide a description of atomic phenomena consistent with the tremendous body of experimental results which had been compiled. Classical mechanics, completely successful in the description of macroscopic events, broke down in the attempt to apply it to phenomena occurring within the confines of atomic dimensions.

This fact calls to mind the analogous failure of geometrical optics: While it is completely adequate so long as one deals with large-scale optical phenomena, geometrical optics fails to describe the behavior of light in the presence of apertures or obstacles whose linear dimensions are comparable with the wavelengths of light. That is, the phenomenon of the diffraction of light is not at all comprehensible within the framework of geometrical optics. In order to describe and understand diffraction one must appeal to the *wave theory* of light, which, in essence, has its mathematical formulation in the wave equations (7) and (10) of (a) above.

Schrödinger, in one of his early papers, projects the idea of the possible need for a *wave theory of mechanics* to describe the submicroscopic realm of phenomena in which classical mechanics has broken down. In the development of this idea he sets forth the following double analogy: The relationship of geometrical optics to classical mechanics is the same as that of wave optics to the required "wave mechanics." In mathematical terms the analogy is set in the form: Fermat's principle (Chap. 5) bears the same relation to the least-action principle (6-6) as does the time-independent wave equation (10) for light to the required "time-independent wave equation for mechanics."

The development of Schrödinger's double analogy toward the discovery of a time-independent wave equation for mechanics runs essentially as follows:

Fermat's principle, according to 5-2, requires the extremization of the integral

$$I_1 = \int_{s_1}^{s_2} \frac{ds}{u} \tag{17}$$

with respect to paths connecting two given fixed points. Here $u = u(x,y,z)$ is the local phase velocity of the light—the quantity which also appears in the wave equations (7) and (10). According to 6-6(c) the least-action principle, applied to a single particle of mass m, requires the extremization of

$$I_2 = \sqrt{2m} \int_{s_1}^{s_2} \sqrt{E - V}\, ds \tag{18}$$

with respect to paths connecting two given fixed points. Here E is the actual constant total energy, and $V = V(x,y,z)$ is the potential energy of the particle. Thus we may effect the analogy between the optical and mechanical principles by "assigning" to the motion of the particle a local phase velocity which is, according to (17) and (18), inversely proportional to $\sqrt{E - V}$. With Schrödinger, therefore, we define the *particle phase velocity* as

$$u = \frac{A}{\sqrt{E - V}}, \tag{19}$$

where A is a constant whose determination is made directly.

If the phase associated with the particle motion is denoted by $(\phi - \omega t)$, it follows from (13) and (19) that ϕ must satisfy the partial differential equation

$$\left(\frac{\partial \phi}{\partial x}\right)^2 + \left(\frac{\partial \phi}{\partial y}\right)^2 + \left(\frac{\partial \phi}{\partial z}\right)^2 = \frac{\omega^2}{A^2} (E - V). \tag{20}$$

Equation (20), we notice, is identical in form with the reduced Hamilton-Jacobi equation (1) of 11-1(a). This fact makes natural the assumption that

$$\phi = K_1 S^*, \tag{21}$$

from which it follows, according to (1) and (20), that

$$\frac{\omega^2}{A^2} = 2m K_1^2; \tag{22}$$

here K_1 is a constant whose determination is left to experiment.[1]

The Schrödinger double analogy is completed by substituting into the time-independent wave equation (10) the phase velocity given by (19),

[1] The effect of the relation (22) is merely to replace, in the work following, one undetermined constant (A) by another (K_1). The reason for this replacement is subsequent convenience; K_1 turns out to be an easily identified *universal* constant.

together with (22): We obtain

$$\nabla^2\psi + 2mK_1^2(E - V)\psi = 0 \tag{23}$$

as the required time-independent wave equation associated with the motion of a particle of mass m, having total constant energy E and moving in a field of force described by the potential-energy function $V = V(x,y,z)$.

By identifying the constant K_1 with the reciprocal of the constant K introduced in 11-1(a), we indeed observe the identity of the *particle wave equation* (23) with the Schrödinger equation (4), derived independently in 11-1(a). (For reference below we record the equivalence

$$K_1 = \frac{1}{K}.\Bigg) \tag{24}$$

The success of the Schrödinger equation in the description of atomic phenomena thus justifies the optical-mechanical double analogy set forth by Schrödinger as well as the wave concept of matter first conceived by De Broglie and further developed by Schrödinger along the lines indicated in the foregoing paragraphs.

It is to be kept in mind that the above derivation of the Schrödinger equation based upon the analogy of optics with mechanics is purely heuristic, by no means rigorous. In particular it is assumed throughout that the first partial derivatives of the function $\phi(x,y,z)$ are slowly varying functions of position; yet no such assumption underlies the validity of the Schrödinger equation in its application to specific atomic problems. The merit of the Schrödinger equation resides in its description of atomic phenomena consistent with the results of experiment and not upon any particular method of its derivation.

(d) The experiments of Thomson, Davisson and Germer, and others on the diffraction of electrons, executed more or less simultaneously with Schrödinger's early research in quantum mechanics, provide a justification of the line of argument of (c) above even more direct and more striking than the stated success of the Schrödinger equation. The wave character of material particles manifested in these electron-diffraction experiments made necessary the assignment of a numerical wavelength to an electron moving with given speed. From (16) of (b) above, together with (20), (22), and (24) of (c), we have

$$\lambda = \frac{2\pi K}{\sqrt{2m(E - V)}} = \frac{2\pi K}{mv}, \tag{25}$$

where v is the "classical" speed of the particle under consideration.[1]

[1] According to 6-6(c) we have $(E - V) = \frac{1}{2}mv^2$, whence the final form of (25).

The equation (25), derived first by De Broglie on an entirely different basis, is precisely the relationship between wavelength and speed required by the results of the electron-diffraction experiments! The consequent experimental determination of the constant $2\pi K$ shows it to be identical with the Planck constant of action (usually denoted by "h"), a universal constant which made its first appearance (1900) in the Planck theory of black-body radiation[1] and soon after (1905) appeared as a fundamental quantity in the Einstein theory of the photoelectric effect (see (*e*) below). The validity of (25) as applied to atomic and subatomic particles (in addition to electrons) and even to molecules is well established by experiment.

(*e*) A direct extension of the portion of the argument of (*c*) above which employs the reduced Hamilton-Jacobi differential equation leads to the assignment of a specific value to the frequency $(\omega/2\pi)$ associated with the motion of a material particle. Following the success of the assumption (21)—namely, that the space-dependent term ϕ of the phase $(\phi - \omega t)$ is proportional to the solution S^* of the *reduced* Hamilton-Jacobi equation—it appears natural to assume the phase itself to be proportional to the solution S of the full (time-dependent) Hamilton-Jacobi equation (33) of 6-5(*a*), with the same constant of proportionality. According to (36) of 6-5(*b*) we have $S = S^* - Et$, so that the stated assumption reads

$$\phi - \omega t = K_1(S^* - Et),$$

whence, because of (21) and (24),

$$E - K\omega. \tag{26}$$

The relation (26) is identical with the Einstein equation (1905) relating the frequency $(\omega/2\pi)$ of a light wave to the energy E of each associated light corpuscle (photon). Since the frequency, unlike the wavelength λ, associated with the motion of a particle cannot be measured directly, there is no direct experimental verification of (26). The validity of (26) is intimately connected with the validity of quantum mechanics as a whole, on purely theoretical grounds, however.

11-3. The Hydrogen Atom. Physical Interpretation of the Schrödinger Wave Functions

(*a*) In applying the Schrödinger method to a system consisting of a fixed atomic nucleus and a single electron (of which the hydrogen atom is an example), we limit ourselves to the derivation of only those solutions which possess spherical symmetry. That is, we use spherical coordi-

---

[1] See exercise 25(*f*), Chap. 9.

nates (r,θ,ϕ) and seek only those wave functions ψ which are independent of the angles θ and ϕ. An analysis more complete than ours shows that the lowest energy state—the so-called *ground*, or normal, *state*, in which our interest mainly lies—possesses this symmetry, so that we do not miss consideration of it through the restriction

$$\psi = \psi(r). \tag{27}$$

We start here with the extremization problem rather than with the Schrödinger differential equation to which it leads. That is, we seek to extremize the integral (3) of 11-1(*a*)—with the appropriate potential-energy function V inserted—with respect to functions of the form (27). For this purpose we employ the three identities (27) of 9-2(*c*), with $w = \psi$; squaring and adding, we obtain

$$\psi_x^2 + \psi_y^2 + \psi_z^2 = \left(\frac{d\psi}{dr}\right)^2, \tag{28}$$

since, according to (27), $\psi_\theta = \psi_\phi = 0$.

If the magnitude of the fundamental electronic charge measured in electrostatic units is denoted by ϵ, and if the charge on the atomic nucleus under consideration is $Z\epsilon$, the potential energy of an electron moving under the influence of this nucleus is

$$V = -\frac{Z\epsilon^2}{r}, \tag{29}$$

where r is the distance from the nucleus (considered fixed in position) to the electron; both are considered as point charges. With (28) and (29) the integral (3) of 11-1(*a*) becomes

$$I^* = \int_0^\infty \int_0^{2\pi} \int_0^\pi \left[\frac{K^2}{2m} \left(\frac{d\psi}{dr}\right)^2 - \left(\frac{Z\epsilon^2}{r} + E\right) \psi^2 \right] r^2 \sin\theta \, d\theta \, d\phi \, dr. \tag{30}$$

(In accordance with (26) of 9-2(*c*) the jacobian of the transformation from cartesian to spherical coordinates is $r^2 \sin\theta$, so that $dx\, dy\, dz$ in (3) is replaced by $r^2 \sin\theta \, d\theta \, d\phi \, dr$ in (30).) Since ψ is a function of r alone, integration over the angle variables in (30) is carried out directly:

$$I^* = 4\pi \int_0^\infty \left[\frac{K^2}{2m} \left(\frac{d\psi}{dr}\right)^2 - \left(\frac{Z\epsilon^2}{r} + E\right) \psi^2 \right] r^2 \, dr. \tag{31}$$

We seek the functions ψ which extremize I^*.

We introduce into (31) the new independent variable

$$\xi = \frac{r}{a}, \tag{32}$$

where a is a positive constant whose value we determine directly.   Further, we introduce the auxiliary constants

$$\alpha = \frac{2ma^2E}{K^2} \quad \text{and} \quad \beta = \frac{2maZ\epsilon^2}{K^2}, \tag{33}$$

so that, with (32), equation (31) becomes

$$I^* = \frac{2\pi K^2 a}{m} \int_0^\infty [\xi^2\psi'^2 - (\beta\xi + \alpha\xi^2)\psi^2]d\xi, \tag{34}$$

where the prime (') indicates differentiation with respect to ξ.

The possibility of bringing (34) into the form exhibited by (35) of 8-3(b) suggests the substitution

$$\psi = Qe^{-\frac{1}{2}\xi}, \tag{35}$$

whereby (34) becomes

$$I^* = \frac{2\pi K^2 a}{m} \int_0^\infty e^{-\xi} \left\{ \xi^2 Q'^2 - \xi^2 QQ' - \left[\left(\alpha - \frac{1}{4} \right) \xi^2 + \beta\xi \right] Q^2 \right\} d\xi$$

$$= \frac{2\pi K^2 a}{m} \int_0^\infty \left\{ e^{-\xi} \xi^2 Q'^2 - e^{-\xi} \left[\left(\alpha + \frac{1}{4} \right) \xi^2 + (\beta - 1)\xi \right] Q^2 \right.$$

$$\left. - \frac{1}{2} \frac{d}{d\xi} (Q^2\xi^2 e^{-\xi}) \right\} d\xi \tag{36}$$

$$= \frac{2\pi K^2 a}{m} \int_0^\infty e^{-\xi} \xi \left\{ \xi Q'^2 - \left[\left(\alpha + \frac{1}{4} \right) \xi + (\beta - 1) \right] Q^2 \right\} d\xi, \tag{37}$$

since the integral of the final term of (36)—namely, $\frac{1}{2}Q^2\xi^2 e^{-\xi} \big]_0^\infty$—must vanish in order to ensure the existence of I^*.

Comparison of (37) with (35) of 8-3(b), with $k = 1$, makes evident the convenience served by giving to the constant a that value for which $\alpha = -\frac{1}{4}$—or, according to the first of (33),

$$a = \frac{K}{\sqrt{-8mE}}. \tag{38}$$

Thus, according to the second of (33), the problem of determining the eigenvalues of E is reduced to that of determining the eigenvalues of β, with the correspondence

$$E = - \frac{mZ^2\epsilon^4}{2K^2\beta^2}, \tag{39}$$

as we find with the aid of (38).   (The fact of restricting the eigenvalues of E to negative values only by choosing $\alpha < 0$ is justified by physical considerations: The energy levels of an electron "bound" to a nucleus must be negative, since a *positive* amount of work is required to remove the electron to a position of rest infinitely far from the nucleus—in which state its energy is zero.) With the choice $\alpha = -\frac{1}{4}$ the integral (37) becomes

$$I^* = \frac{2\pi K^2 a}{m} \int_0^\infty e^{-\xi}\xi[\xi Q'^2 - (\beta - 1)Q^2]d\xi. \tag{40}$$

Reference to 8-3(b) reveals that for nonnegative integer values of $(\beta - 1)$ there exist extremizing functions Q for which the integral I^* exists; *i.e.*, we have for the eigenvalues of β

$$\beta = n = 1,2,3, \ldots . \tag{41}$$

The corresponding eigenfunctions are, according to 8-3(b),

$$Q = Q_n = C_n L_{n-1}^{(1)}(\xi) \qquad (n = 1,2,3, \ldots), \tag{42}$$

where $L_{n-1}^{(1)}(\xi)$ is the Laguerre polynomial given explicitly by

$$L_{n-1}^{(1)}(\xi) = \frac{e^\xi}{(n - 1)!\xi} \frac{d^{n-1}}{d\xi^{n-1}}(e^{-\xi}\xi^n) \qquad (n = 1,2,3, \ldots); \tag{43}$$

C_n is determined in accordance with the requirement of normalization

$$1 = \int_0^\infty \int_0^{2\pi} \int_0^\pi \psi_n^2 r^2 \sin\theta \, d\theta \, d\phi \, dr = 4\pi \int_0^\infty \psi_n^2 r^2 \, dr = 4\pi a^3 \int_0^\infty \psi_n^2 \xi^2 \, d\xi$$

$$= 4\pi a^3 \int_0^\infty e^{-\xi}Q_n^2 \xi^2 \, d\xi = 4\pi a^3 C_n^2 \int_0^\infty e^{-\xi}[L_{n-1}^{(1)}(\xi)]^2 \xi^2 \, d\xi. \tag{44}$$

With the aid of (43) we evaluate[1] the final integral and so obtain

$$C_n = \frac{1}{\sqrt{8\pi n^2 a^3}}. \tag{45}$$

We note, further, that the constant a depends upon E, according to (38), and therefore upon the index n, according to (39) and (41); that is, we have

$$a = a_n = \frac{K^2 n}{2mZ\epsilon^2}. \tag{46}$$

Finally, we have for the normalized eigenfunctions which depend only on r, according to (32), (35), (42), (45), and (46),

$$\psi_n = \frac{1}{\sqrt{8\pi n^2 a_n^3}} e^{-\frac{r}{2a_n}} L_{n-1}^{(1)}\left(\frac{r}{a_n}\right) \qquad \left(a_n = \frac{K^2 n}{2mZ\epsilon^2}\right), \tag{47}$$

[1] See end-chapter exercise 3.

for $n = 1, 2, 3, \ldots$. The corresponding energy levels are, according to (39) and (41),

$$E_n = -\frac{mZ^2\epsilon^4}{2K^2n^2}. \qquad (48)$$

In particular we have for the lowest energy state ($n = 1$)—the ground state—of the system consisting of a fixed nucleus and a single electron

$$\psi_1 = \frac{1}{\sqrt{8\pi a_1^3}}\, e^{-\frac{r}{2a_1}}L_0^{(1)}\left(\frac{r}{a_1}\right) = \frac{e^{-\frac{r}{2a_1}}}{\sqrt{8\pi a_1^3}} \qquad \left(a_1 = \frac{K^2}{2mZ\epsilon^2}\right) \qquad (49)$$

(as we find with the aid of (43), with $n = 1$) and

$$E_1 = -\frac{mZ^2\epsilon^4}{2K^2}. \qquad (50)$$

(b) Consideration of possible θ and ϕ dependence of the wave functions yields the same set of energy levels as the set given by (48). The ground state is nondegenerate; *i.e.*, there is but a single eigenfunction—namely, (49)—which corresponds to the lowest energy eigenvalue given by (50). The higher energy levels are all degenerate, with θ- and ϕ-dependent eigenfunctions, in addition to the spherically symmetric function (47), arising for each value of $n \geq 2$. (In the sections following—in particular, in 11-5(a)—we have occasion to use only the eigenfunction (49) for the ground state.)

The energy levels (48) are precisely those given by the old Bohr theory (1913) and are found to agree with the levels obtained by experiment upon hydrogen atoms ($Z = 1$) and singly ionized helium atoms ($Z = 2$) to within the refined corrections which are accounted for by considering the intrinsic angular momenta (spin) and magnetic moments of the electron and nucleus. (If, instead of assuming the nucleus to be at rest, we take into account its translational motion, the mass m must be replaced by the so-called reduced mass $\mu = [mM/(m + M)]$, where M is the nuclear mass, in all the results of (a) above. Since for hydrogen the ratio (m/M) is $(1/1837)$ and is about one-fourth as large for helium, the ratio $[(m + M)/M]$ of m to μ is quite close to unity; the difference between m and μ is sufficiently large, however, to be detectable in measurements of the energy levels of hydrogen and singly ionized helium atoms through the determination of spectral frequencies.[1])

(c) While the meaning of the eigenvalues of E was understood immediately on the first application of the Schrödinger equation, it was, for some time after, uncertain what physical interpretation should be assigned to the corresponding eigenfunctions. The interpretation of the wave func-

[1] See exercise 7 at the end of this chapter

tions whose eventual universal acceptance has been completely justified, because of theoretical consistency as well as through successful comparison with experiment, reads as follows:

We suppose $\psi = \psi(x,y,z)$ to be an eigenfunction, corresponding to a particular eigenvalue of E, of the Schrödinger equation involving a given potential-energy function $V = V(x,y,z)$. The quantity† ψ^2, a function of position, is interpreted as the position probability-density function of the particle whose potential energy is V; that is to say, if we seek to locate the given particle within a volume element $dx\,dy\,dz$ at the point (x,y,z), the probability of our finding it there is given by $\psi^2\,dx\,dy\,dz$. The probability of our locating the particle within a given region of space having extended dimensions is, accordingly, the volume integral of the function ψ^2 carried out over the given region. In particular the normalization

$$\iiint \psi^2\,dx\,dy\,dz = 1$$

takes on special significance:[1] The probability of locating the particle *somewhere* in space is unity; the particle is assumed to exist, that is.

We note two salient features which distinguish quantum mechanics from the classical mechanics studied in Chap. 6:

(i) In classical mechanics a conservative motion may have associated with it any value of the total energy greater than the absolute minimum of the potential-energy function. By continuously varying the initial conditions of a given problem, it is possible to obtain a continuous variation of the total energy associated with the problem. In quantum mechanics, on the other hand, we find problems in which the total energy is confined to sets of discrete values—the eigenvalues of the parameter E in the corresponding Schrödinger equation. In such cases the energy is said to be *quantized*.

(ii) The solution of a problem in classical mechanics consists of a detailed description of the motion of the particles of the system involved; *i.e.*, the solution gives the position and velocity of each particle for all instants of time once the initial positions and velocities are prescribed. In quantum mechanics, however, no such description is possible. All that one obtains in a solution to a problem is the relative *probability* of the existence of various position configurations[2] of a given system.

---

† In a more complete study of quantum mechanics than the present one the admissibility of complex eigenfunctions ψ is generally shown to be necessary. If ψ is complex, the quantity $|\psi|^2$ is employed as the position probability-density function inasmuch as ψ^2 is not restricted to real nonnegative values. If ψ is real, we of course have $|\psi|^2 = \psi^2$.

[1] The integral is carried out over *all* space.

[2] A more extensive development than the present one affords a velocity, or momentum, probability distribution as well.

The fundamental physical ideas underlying the above features which distinguish quantum mechanics from classical mechanics are embodied in the so-called *principle of indeterminacy,* or *uncertainty principle.* This principle takes into account the fact that the experimental determination of the position or velocity of a particle involves a disturbance of the particle's motion by the agency of measurement and that this disturbance is necessarily indeterminate in both magnitude and direction. The degree of indeterminacy is negligible when one deals with large-scale events; but when the objects of measurement are atomic or subatomic in size, the indeterminacy assumes full significance. Accordingly, since one cannot determine, by experiment, the precise location and velocity of an atomic or subatomic particle, it is in a certain sense meaningless to speak of its precise location or velocity; one should treat only of the *probability* distribution of its location or of its velocity. Quantum mechanics supplies only such information as is verifiable by experiment, and so avoids such "meaningless" concepts as orbits, position as a function of time, etc.[1]

11-4. Extension to Systems of Particles. Minimum Character of the Energy Eigenvalues

(a) Extension of the Schrödinger theory to a system of s particles may be carried out in the manner in which the Schrödinger equation for a single particle is derived in 11-1(a). In place of (1) of that section we consider the reduced Hamilton-Jacobi equation for a system of s particles, whereby the simple trio of terms

$$\frac{1}{2m}\left[\left(\frac{\partial S^*}{\partial x}\right)^2 + \left(\frac{\partial S^*}{\partial y}\right)^2 + \left(\frac{\partial S^*}{\partial z}\right)^2\right]$$

in (1) is replaced by the sum

$$\sum_{j=1}^{s} \frac{1}{2m_j}\left[\left(\frac{\partial S^*}{\partial x_j}\right)^2 + \left(\frac{\partial S^*}{\partial y_j}\right)^2 + \left(\frac{\partial S^*}{\partial z_j}\right)^2\right],$$

where x_j, y_j, z_j are the cartesian position coordinates and m_j is the mass of the jth particle. The potential-energy function V—which depends, in general, upon the $3s$ coordinate variables—describes the system of forces which influence the motion of the particles of the system.

[1] For amplification of this necessarily brief discussion of the principle of indeterminacy the reader is referred to the abundant literature on modern atomic mechanics. See, for example, Max Born, "Atomic Physics," pp. 85–90, Hafner Pub. Co., New York, 1936.

The procedure of 11-1(a) is followed[1] until we arrive at the $3s$-tuple integral carried out over the infinite range of each of the coordinate variables:

$$I^* = \iint \cdots \int \left\{ \sum_{j=1}^{s} \frac{K^2}{2m_j} \left[\left(\frac{\partial \psi}{\partial x_j}\right)^2 + \left(\frac{\partial \psi}{\partial y_j}\right)^2 + \left(\frac{\partial \psi}{\partial z_j}\right)^2 \right] \right.$$
$$\left. + (V - E)\psi^2 \right\} \prod_{j=1}^{s} dx_j \, dy_j \, dz_j. \quad (51)$$

The extremization of (51) leads to the Schrödinger equation for the given system of s particles—namely,

$$\sum_{j=1}^{s} \frac{K^2}{2m_j} \nabla_j^2 \psi + (E - V)\psi = 0, \quad (52)$$

where we write

$$\nabla_j^2 \psi = \frac{\partial^2 \psi}{\partial x_j^2} + \frac{\partial^2 \psi}{\partial y_j^2} + \frac{\partial^2 \psi}{\partial z_j^2} \quad (j = 1, 2, \ldots, s). \quad (53)$$

(The derivation of (52) is left for exercise 4(b) at the end of this chapter.)

As with the single-particle equation (4), the solution of (52) presents an eigenvalue-eigenfunction problem: Any value of E for which there is a solution ψ such that the integral (51) exists is an eigenvalue of E; the solution ψ is the corresponding eigenfunction.

(b) The physical interpretation of the eigenvalues of E in (52) is identical with the interpretation in the single-particle case: The eigenvalues of E are the physically realizable values of the total energy of the system under the influence of the potential energy V.

Similarly, the physical interpretation of the eigenfunctions of (52) as applied to a many-particle atomic problem is a direct generalization of the interpretation of the single-particle wave functions which is presented in 11-3(c): The quantity† $\psi^2 \prod_{j=1}^{s} dx_j \, dy_j \, dz_j$ is the probability of simultaneous location of the first particle of the system within the volume element $dx_1 \, dy_1 \, dz_1$ at the point (x_1, y_1, z_1), the second within the element $dx_2 \, dy_2 \, dz_2$ at (x_2, y_2, z_2), . . . , the sth within the element $dx_s \, dy_s \, dz_s$ at (x_s, y_s, z_s). Or if we wish to regard the position configuration of the entire system as described by the $3s$ coordinates of a single point in a space of $3s$ dimen-

---

[1] The details are left for end-chapter exercise 4(a).

† See footnote, p. 276: If ψ is complex, we must replace ψ^2 by $|\psi|^2$ as the position probability-density function.

sions, we have the equivalent statement that $\psi^2 \prod\limits_{j=1}^{s} dx_j \, dy_j \, dz_j$ is the prob-

ability of the system's location within the (3s-dimensional) volume ele-

ment $\prod\limits_{j=1}^{s} dx_j \, dy_j \, dz_j$ at the point $(x_1,y_1,z_1,x_2,y_2,z_2, \ldots ,x_s,y_s,z_s)$.   Since it

is assumed that the s particles of the given system are located *somewhere* in space—or, in the equivalent description, that the single point describing the position configuration of the system is located *somewhere* in the space of $3s$ dimensions—we must require the normalization

$$\iint \cdots \int \psi^2 \prod_{j=1}^{s} dx_j \, dy_j \, dz_j = 1. \tag{54}$$

(c) We consider briefly the special case in which the potential-energy function V which describes the forces influencing the motion of a system of s particles can be written as a sum of terms each of which involves the coordinates of only one particle.   That is, we deal with potential-energy functions of the special form

$$V = \sum_{j=1}^{s} V_j(x_j,y_j,z_j). \tag{55}$$

If V has the form (55), the many-particle Schrödinger equation (52) possesses solutions which are products of functions each of which involves the coordinates of only one particle—namely,

$$\psi = \prod_{j=1}^{s} \psi^{(j)}(x_j,y_j,z_j). \tag{56}$$

For if we substitute (56) together with (55) into (52), we obtain—on noting from (53) that $\nabla_j^2 \psi = (\psi/\psi^{(j)})\nabla_j^2 \psi^{(j)}$ and on dividing through by ψ—

$$\sum_{j=1}^{s} \left\{ \frac{K^2}{2m_j} \frac{\nabla_j^2 \psi^{(j)}}{\psi^{(j)}} - V_j \right\} + E = 0. \tag{57}$$

We transpose to the right-hand member a single term—the kth, say—of the sum over j in (57):

$$\sum_{j=1}^{s} \left\{ \frac{K^2}{2m_j} \frac{\nabla_j^2 \psi^{(j)}}{\psi^{(j)}} - V_j \right\} + E = - \left\{ \frac{K^2}{2m_k} \frac{\nabla_k^2 \psi^{(k)}}{\psi^{(k)}} - V_k \right\}. \tag{58}$$

(The prime of Σ' indicates omission of the term $j = k$.)   Since the right-hand member depends only on the independent variables x_k, y_k, z_k, and since the left is independent of these variables, we conclude that each member is a constant, which we denote by $E^{(k)}$. Since this result is independent of the choice of k, it must be true for all $j = 1, 2, \ldots, s$:

$$- \left\{ \frac{K^2}{2m_j} \frac{\nabla_j^2 \psi^{(j)}}{\psi^{(j)}} - V_j \right\} = E^{(j)}, \tag{59}$$

or

$$\frac{K^2}{2m_j} \nabla_j^2 \psi^{(j)} + (E^{(j)} - V_j)\psi^{(j)} = 0 \qquad (j = 1,2, \ldots, s). \tag{60}$$

Moreover, by substituting (59) into (57), we conclude that

$$E = \sum_{j=1}^{s} E^{(j)}. \tag{61}$$

We thus have the special result: If V exhibits the form (55), the many-particle Schrödinger equation (52) possesses solutions which may be written as a product of factors $\psi^{(j)}(x_j, y_j, z_j)$, where each $\psi^{(j)}$ is a solution of a single-particle Schrödinger equation (60), for $j = 1, 2, \ldots, s$. According to (61) the corresponding energy eigenvalue of the many-particle equation is the sum of the energy eigenvalues of the s single-particle equations (60). In particular if an eigenvalue of E in (52) is nondegenerate, such a product solution is *the* eigenfunction,[1] if (55) gives the form of V.

(d) It is shown in 11-1(b) that the single-particle Schrödinger eigen-value-eigenfunction problem is equivalent to a certain isoperimetric problem. In similar fashion it may[2] likewise be shown that the many-particle Schrödinger problem may be so characterized. Namely, the extremization of (51) of (a) above is equivalent to the extremization of

$$I = \iint \cdots \int \left\{ \sum_{j=1}^{s} \frac{K^2}{2m_j} \left[\left(\frac{\partial \psi}{\partial x_j} \right)^2 + \left(\frac{\partial \psi}{\partial y_j} \right)^2 + \left(\frac{\partial \psi}{\partial z_j} \right)^2 \right] \right.$$

$$\left. + V\psi^2 \right\} \prod_{j=1}^{s} dx_j \, dy_j \, dz_j \tag{62}$$

with respect to functions ψ which satisfy the normalization condition

$$\iint \cdots \int \psi^2 \prod_{j=1}^{s} dx_j \, dy_j \, dz_j = 1. \tag{63}$$

---

[1] Explicit justification of this final statement is left for end-chapter exercise 8.
[2] See end-chapter exercise 4(c).

Moreover, the successive eigenvalues of E, arranged in the ascending order

$$E_1 \leqq E_2 \leqq \cdots \leqq E_n \leqq E_{n+1} \leqq \cdots,$$

are the successive minima of (62) in the following sense: The nth eigenvalue E_n is the minimum of (62) with respect to sufficiently regular functions ψ which satisfy (63) along with the $(n-1)$ orthogonality conditions

$$\iint \cdots \int \psi \psi_k \prod_{j=1}^{s} dx_j\, dy_j\, dz_j = 0 \qquad (k = 1,2, \ldots, n-1),$$

where ψ_k is the eigenfunction of the problem corresponding to the eigenvalue E_k. Discussion of the proof of this assertion is reserved for exercise 6 at the end of this chapter.

Application of the minimum characterization of the Schrödinger energy eigenvalues is found in 11-5 below. We omit discussion of a maximum-minimum characterization such as that which appears in 9-11(a) in relation to the membrane eigenvalues.

11-5. Ritz Method : Ground State of the Helium Atom.   Hartree Model of the Many-electron Atom

(a) To illustrate the approximate solution of a quantum-mechanical problem through the direct minimization of the integral (62) of 11-4(d) we consider the problem of finding the lowest energy eigenvalue for the helium atom. As in the one-electron problem of 11-3(a), we suppose the nucleus to be in fixed position at the origin of coordinates; the two electron positions are described by the sets of cartesian coordinate variables (x_1,y_1,z_1) and (x_2,y_2,z_2), respectively. The potential energy is given by

$$V = V_1 + V_2 + V_{12}, \tag{64}$$

where

$$V_j = -\frac{2\epsilon^2}{r_j} \quad (j = 1,2), \qquad V_{12} = \frac{\epsilon^2}{r_{12}}, \tag{65}$$

with

$$r_j = \sqrt{x_j^2 + y_j^2 + z_j^2} \quad (j = 1,2) \tag{66}$$

and

$$r_{12} = \sqrt{(x_2 - x_1)^2 + (y_2 - y_1)^2 + (z_2 - z_1)^2}. \tag{67}$$

The term V_j of (64) represents the interaction between the nucleus and the jth electron $(j = 1,2)$; the term V_{12} represents the interaction between the two electrons. The quantity ϵ is the fundamental electronic charge, as introduced in 11-3(a).

In place of the Schrödinger problem defined by the relations (64) to

(67) we consider temporarily the problem in which (65) is replaced by

$$V_j = -\frac{Ze^2}{r_j} \quad (j = 1,2), \qquad V_{12} = 0; \tag{68}$$

i.e., we suppose no interaction between the electrons and leave unspecified the electric charge $(Z\epsilon)$ on the nucleus. In this problem the result of 11-4(c) is applicable, so that we may obtain a solution of the form

$$\psi = \psi^{(1)}(x_1,y_1,z_1)\psi^{(2)}(x_2,y_2,z_2), \tag{69}$$

where $\psi^{(j)}$ is the solution of a Schrödinger problem involving the coordinates (x_j,y_j,z_j) only, for each of $j = 1,2$. According to (60) of 11-4(c), $\psi^{(j)}$ is an eigenfunction of the equation

$$\frac{K^2}{2m}\nabla_j^2\psi^{(j)} + \left(E^{(j)} + \frac{Ze^2}{r_j}\right)\psi^{(j)} = 0 \qquad (j = 1,2), \tag{70}$$

with the aid of (64) and (68), and since $m_1 = m_2 = m$, the electronic mass.

Careful reference to the work of 11-3(a) reveals that each of the "separated" problems embodied in (70) is equivalent to the problem handled in 11-3(a)—with (x,y,z,r) replaced by (x_1,y_1,z_1,r_1) in one case and with (x,y,z,r) replaced by (x_2,y_2,z_2,r_2) in the other. The eigenfunction corresponding to the lowest energy eigenvalue of the problem is given, accordingly, by (69) and

$$\psi^{(j)} = \frac{e^{-(r_j/2a_1)}}{\sqrt{8\pi a_1^3}} \quad (j = 1,2), \qquad a_1 = \frac{K^2}{2mZ\epsilon^2}, \tag{71}$$

as we find from (49) of 11-3(a). Also, from (50) of 11-3(a), we have

$$E^{(j)} = E_1 = -\frac{mZ^2\epsilon^4}{2K^2} \quad (j = 1,2) \tag{72}$$

for the lowest eigenvalue of (70), and thus corresponding to the eigenfunction (71). We note, in passing, that the eigenfunction (71) is normalized—namely[1]

$$\iiint (\psi^{(j)})^2 \, dx_j \, dy_j \, dz_j = 1 \qquad (j = 1,2). \tag{73}$$

(b) Returning to the Schrödinger problem defined by (64) to (67), we substitute the product function (69) into the integral

[1] See end-chapter exercise 9.

$$I = \iiiiii \left\{ \frac{K^2}{2m} \sum_{j=1}^{2} \left[\left(\frac{\partial \psi}{\partial x_j} \right)^2 + \left(\frac{\partial \psi}{\partial y_j} \right)^2 + \left(\frac{\partial \psi}{\partial z_j} \right)^2 \right] \right.$$

$$\left. + (V_1 + V_2 + V_{12})\psi^2 \right\} \, dx_1 \, dy_1 \, dz_1 \, dx_2 \, dy_2 \, dz_2, \quad (74)$$

whose minimum with respect to functions ψ which satisfy the normalization

$$\iiiiii \psi^2 \, dx_1 \, dy_1 \, dz_1 \, dx_2 \, dy_2 \, dz_2 = 1 \tag{75}$$

is the lowest energy eigenvalue of the helium atom, according to (62) and (63) of 11-4(d). With the stated substitution we obtain I as a function of Z; subsequent minimization of I with respect to Z supplies an approximation from above to the actual minimum of I—to the lowest helium energy eigenvalue, that is. We proceed, in the paragraphs following, to the achievement of this approximation. (We note that (75) is fulfilled by the product function (69) by virtue of the normalization (73).)

We have, from (69), that $(\partial \psi / \partial x_1) = \psi^{(2)}(\partial \psi^{(1)} / \partial x_1)$, etc., so that we may write

$$\sum_{j=1}^{2} \left[\left(\frac{\partial \psi}{\partial x_j} \right)^2 + \left(\frac{\partial \psi}{\partial y_j} \right)^2 + \left(\frac{\partial \psi}{\partial z_j} \right)^2 \right]$$

$$= \sum_{j=1}^{2} (\psi^{(k)})^2 \left[\left(\frac{\partial \psi^{(j)}}{\partial x_j} \right)^2 + \left(\frac{\partial \psi^{(j)}}{\partial y_j} \right)^2 + \left(\frac{\partial \psi^{(j)}}{\partial z_j} \right)^2 \right],$$

with $k = (2/j)$. We therefore obtain, with the aid of the normalization (73)—with j replaced by k—and Green's formula (32) of 2-14(e),

$$\iiiiii \sum_{j=1}^{2} \left[\left(\frac{\partial \psi}{\partial x_j} \right)^2 + \left(\frac{\partial \psi}{\partial y_j} \right)^2 + \left(\frac{\partial \psi}{\partial z_j} \right)^2 \right] dx_1 \, dy_1 \, dz_1 \, dx_2 \, dy_2 \, dz_2$$

$$= \sum_{j=1}^{2} \iiint (\psi^{(k)})^2 \, dx_k \, dy_k \, dz_k \iiint \left[\left(\frac{\partial \psi^{(j)}}{\partial x_j} \right)^2 + \left(\frac{\partial \psi^{(j)}}{\partial y_j} \right)^2 + \left(\frac{\partial \psi^{(j)}}{\partial z_j} \right)^2 \right] dx_j \, dy_j \, dz_j$$

$$= - \sum_{j=1}^{2} \iiint \psi^{(j)} \nabla_j^2 \psi^{(j)} \, dx_j \, dy_j \, dz_j$$

$$= \frac{2m}{K^2} \sum_{j=1}^{2} \iiint \left(E_1 + \frac{Ze^2}{r_j} \right) (\psi^{(j)})^2 \, dx_j \, dy_j \, dz_j, \tag{76}$$

where the final form is obtained with the aid of (70) and partial use of (72). Further, we have use for the relation

$$\iiint\iiint V_j\psi^2 \, dx_1 \, dy_1 \, dz_1 \, dx_2 \, dy_2 \, dz_2$$

$$= \iiint (\psi^{(k)})^2 \, dx_k \, dy_k \, dz_k \iiint V_j(\psi^{(j)})^2 \, dx_j \, dy_j \, dz_j \quad \left(k = \frac{2}{j}\right)$$

$$= \iiint V_j(\psi^{(j)})^2 \, dx_j \, dy_j \, dz_j \qquad\qquad (j = 1,2), \quad (77)$$

as we find with the aid of (69) and (73).

Since the dependence of $\psi^{(1)}$ on the variables x_1, y_1, z_1 is identical with the dependence of $\psi^{(2)}$ on x_2, y_2, z_2, according to (71) and (66), the two terms of the final member of (76) are identical;[1] we may therefore replace j by 1 (or 2) and the summation sign by the factor 2 in the final member of (76). For the same reason, and because the dependence of V_1 on x_1, y_1, z_1 is identical with that of V_2 on x_2, y_2, z_2, according to (65) and (66), we may similarly replace the index j in the final member of (77) by 1 (or 2). Thus, on applying the results (76) and (77) to the substitution of (69) into (74), we obtain, with the definition

$$Q = \iiint\iiint V_{12}(\psi^{(1)})^2(\psi^{(2)})^2 \, dx_1 \, dy_1 \, dz_1 \, dx_2 \, dy_2 \, dz_2, \qquad (78)$$

the simplified expression

$$I = 2 \iiint (\psi^{(1)})^2 \left(E_1 + \frac{Z\epsilon^2}{r_1} + V_1\right) dx_1 \, dy_1 \, dz_1 + Q$$

$$= 2E_1 + 2(Z - 2)\epsilon^2 \iiint \frac{(\psi^{(1)})^2}{r_1} \, dx_1 \, dy_1 \, dz_1 + Q, \qquad (79)$$

as we find with the aid of (73) and (65).

We evaluate the middle term of the final member of (79) by introducing the spherical coordinates (r,θ,ϕ)—with omission of the superfluous subscript 1—as variables of integration. With (71) we obtain

$$\iiint \frac{(\psi^{(1)})^2}{r_1} \, dx_1 \, dy_1 \, dz_1 = \frac{1}{8\pi a_1^3} \int_0^\infty \int_0^{2\pi} \int_0^\pi \frac{e^{-(r/a_1)}}{r} \, r^2 \sin\theta \, d\theta \, d\phi \, dr$$

$$= \frac{1}{2a_1}. \qquad (80)$$

We evaluate the multiple integral (78) in the following manner: We first hold (x_2,y_2,z_2) fixed and introduce a change of variables from (x_1,y_1,z_1) to the cartesian set (x_1',y_1',z_1'), whose origin is the original origin of coordi-

---

[1] The integration is carried out over the infinite range of the variables involved in both cases.

nates but whose z_1' axis passes through the point (x_2, y_2, z_2); the orientation of the x_1' or y_1' axis is immaterial. The jacobian of this transformation is unity.[1] Moreover, we have $r_1^2 = x_1'^2 + y_1'^2 + z_1'^2$, so that we may introduce the transformation to spherical coordinates

$$x_1' = r_1 \sin \theta_1 \cos \phi_1, \qquad y_1' = r_1 \sin \theta_1 \sin \phi_1, \qquad z_1' = r_1 \cos \theta_1.$$

We thus have (see Fig. 11-1), according to (65) and (67),

$$V_{12} = \frac{\epsilon^2}{r_{12}} = \frac{\epsilon^2}{\sqrt{r_1^2 + r_2^2 - 2r_1 r_2 \cos \theta_1}},$$

with the aid of the law of cosines. Since $\psi^{(1)}$ is a function of r_1 alone,

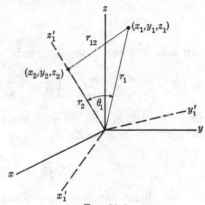

FIG. 11-1.

according to (71), it therefore follows that

$$\iiint (\psi^{(1)})^2 V_{12} \, dx_1 \, dy_1 \, dz_1 = \int_0^\infty \int_0^\pi \int_0^{2\pi} (\psi^{(1)})^2 V_{12} r_1^2 \sin \theta_1 \, d\phi_1 \, d\theta_1 \, dr_1$$

$$= 2\pi\epsilon^2 \int_0^\infty (\psi^{(1)})^2 r_1^2 \left\{ \int_0^\pi \frac{\sin \theta_1 \, d\theta_1}{\sqrt{r_1^2 + r_2^2 - 2r_1 r_2 \cos \theta_1}} \right\} dr_1$$

$$= 2\pi\epsilon^2 \int_0^\infty (\psi^{(1)})^2 r_1^2 \left\{ \frac{1}{r_1 r_2} \left[\sqrt{(r_1 + r_2)^2} - \sqrt{(r_1 - r_2)^2} \right] \right\} dr_1$$

$$= 2\pi\epsilon^2 \int_0^\infty (\psi^{(1)})^2 r_1^2 \begin{Bmatrix} 2/r_1 \\ 2/r_2 \end{Bmatrix} dr_1 \qquad \begin{Bmatrix} r_1 \geq r_2 \\ r_1 \leq r_2 \end{Bmatrix}$$

$$= 4\pi\epsilon^2 \left\{ \frac{1}{r_2} \int_0^{r_2} (\psi^{(1)})^2 r_1^2 \, dr_1 + \int_{r_2}^\infty (\psi^{(1)})^2 r_1 \, dr_1 \right\}$$

$$= \frac{\epsilon^2}{2a_1^3} \left\{ \frac{1}{r_2} \int_0^{r_2} e^{-(r_1/a_1)} r_1^2 \, dr_1 + \int_{r_2}^\infty e^{-(r_1/a_1)} r_1 \, dr_1 \right\}$$

$$= \frac{\epsilon^2}{2a_1} \left\{ 2 \left(\frac{a_1}{r_2} \right) - \left[2 \left(\frac{a_1}{r_2} \right) + 1 \right] e^{-(r_2/a_1)} \right\}.$$

[1] See end-chapter exercise 11 for the proof.

With this result the evaluation of (78) is completed by means of the transformation

$$x_2 = r_2 \sin \theta_2 \cos \phi_2, \qquad y_2 = r_2 \sin \theta_2 \sin \phi_2, \qquad z_2 = r_2 \cos \theta_2.$$

After integration over the angles ϕ_2 and θ_2—which results in affixing the factor 4π—we obtain

$$
\begin{aligned}
Q &= \frac{2\pi\epsilon^2}{a_1} \int_0^\infty (\psi^{(2)})^2 \left\{ 2\left(\frac{a_1}{r_2}\right) - \left[2\left(\frac{a_1}{r_2}\right) + 1\right] e^{-(r_2/a_1)} \right\} r_2^2 \, dr_2 \\
&= \frac{\epsilon^2}{4a_1^4} \int_0^\infty \left\{ 2\left(\frac{a_1}{r_2}\right) e^{-(r_2/a_1)} - \left[2\left(\frac{a_1}{r_2}\right) + 1\right] e^{-(2r_2/a_1)} \right\} r_2^2 \, dr_2 \\
&= \frac{5\epsilon^2}{16a_1}.
\end{aligned}
\tag{81}
$$

With the results (80) and (81) equation (79) reads

$$I = 2E_1 + (Z - 2) \frac{\epsilon^2}{a_1} + \frac{5\epsilon^2}{16a_1} = \left(Z^2 - \frac{27}{8} Z\right) \frac{m\epsilon^4}{K^2}, \tag{82}$$

according to (72) and (71). In accordance with the procedure outlined above we minimize (82) with respect to Z:

$$\frac{dI}{dZ} = \left(2Z - \frac{27}{8}\right) \frac{m\epsilon^4}{K^2} = 0, \qquad Z = \frac{27}{16} \qquad \left(\frac{d^2I}{dZ^2} > 0\right);$$

substituting into (82), we obtain for our approximation to the lowest energy eigenvalue of the helium atom

$$I = -\left(\frac{27}{16}\right)^2 \frac{m\epsilon^4}{K^2} = -2.85 \frac{m\epsilon^4}{K^2}. \tag{83}$$

This value is within 2 per cent of the ground-state energy E_1 of helium as determined by experiment. Although we have no *theoretical* criterion for the accuracy of the result (83), its derivation as an approximation *from above* validates the inequality $E_1 \leqq -2.85(m\epsilon^4/K^2)$ for the helium atom.

(c) The larger the number of electrons per atom, the more complicated is the problem of determining the energy eigenvalues and corresponding wave functions of the atom. We proceed to discuss one of the standard methods of approximation, the so-called Hartree method, which has been applied to many-electron atomic problems with great success.

We consider an atom with nucleus assumed at rest at the origin of coordinates, with s electrons associated with the cartesian coordinates (x_1,y_1,z_1), (x_2,y_2,z_2), . . . , (x_s,y_s,z_s), respectively. The total electric charge on the nucleus is $s\epsilon$, so that the atom as a whole is uncharged. The potential energy of the system is given by

$$V = \sum_{k=1}^{s} V_k + \tfrac{1}{2} \sum_{k=1}^{s} \sum_{j=1}^{s} V_{jk}, \tag{84}$$

where

$$V_k = -\frac{se^2}{r_k} \qquad (r_k = \sqrt{x_k^2 + y_k^2 + z_k^2}) \tag{85}$$

represents the interaction between the nucleus and the kth electron, and

$$V_{jk} = V_{kj} = \frac{\epsilon^2}{r_{jk}} \qquad (j \neq k),\ (V_{jj} = 0),$$

$$(r_{jk} = \sqrt{(x_j - x_k)^2 + (y_j - y_k)^2 + (z_j - z_k)^2}) \tag{86}$$

represents the interaction between the jth and the kth electron. (The factor $\tfrac{1}{2}$ appearing before the double sum in (84) takes care of the fact that each $V_{jk} = V_{kj} (j \neq k)$ appears twice, but must only be counted once.)

The Hartree method is based upon minimization of the integral (62) of 11-4(d) with respect to normalized functions ψ which exhibit the special form

$$\psi = \prod_{j=1}^{s} \psi^{(j)}(x_j, y_j, z_j). \tag{87}$$

The normalization (63) of 11-4(d) is fulfilled by requiring that each $\psi^{(j)}$ be normalized—namely, that

$$\iiint (\psi^{(j)})^2\, dx_j\, dy_j\, dz_j = 1 \qquad (j = 1,2, \ldots, s). \tag{88}$$

We substitute (87) into (62) of 11-4(d) and proceed to effect the minimization of this integral with respect to s sets of normalized functions $\psi^{(1)}, \psi^{(2)}, \ldots, \psi^{(s)}$. Upon substitution there occur several simplifications which are embodied in the results directly following:

From (87) we have that[1]

$$\iint \cdots \int \left[\left(\frac{\partial \psi}{\partial x_k}\right)^2 + \left(\frac{\partial \psi}{\partial y_k}\right)^2 + \left(\frac{\partial \psi}{\partial z_k}\right)^2 \right] \prod_{j=1}^{s} dx_j\, dy_j\, dz_j$$

$$= \iiint \left[\left(\frac{\partial \psi^{(k)}}{\partial x_k}\right)^2 + \left(\frac{\partial \psi^{(k)}}{\partial y_k}\right)^2 + \left(\frac{\partial \psi^{(k)}}{\partial z_k}\right)^2 \right] dx_k\, dy_k\, dz_k \prod_{j=1}^{s}{}' \iiint (\psi^{(j)})^2\, dx_j\, dy_j\, dz_j$$

$$= \iiint \left[\left(\frac{\partial \psi^{(k)}}{\partial x_k}\right)^2 + \left(\frac{\partial \psi^{(k)}}{\partial y_k}\right)^2 + \left(\frac{\partial \psi^{(k)}}{\partial z_k}\right)^2 \right] dx_k\, dy_k\, dz_k \qquad (k = 1,2,\ldots,s),$$

$$\tag{89}$$

because of (88). Further, from (85) and (87) we have

[1] We use Π' to indicate the absence of the factor for which $j = k$.

$$\iint \cdots \int V_k \psi^2 \prod_{j=1}^{s} dx_j \, dy_j \, dz_j$$

$$= \iiint V_k(\psi^{(k)})^2 \, dx_k \, dy_k \, dz_k \prod_{j=1}^{s}{}' \iiint (\psi^{(j)})^2 \, dx_j \, dy_j \, dz_j$$

$$= \iiint V_k(\psi^{(k)})^2 \, dx_k \, dy_k \, dz_k \qquad (k = 1,2, \ldots ,s), \qquad (90)$$

because of (88). Finally, we obtain in similar fashion, for each pair of values of j and k ($j \neq k$),

$$\iint \cdots \int V_{jk} \psi^2 \prod_{p=1}^{s} dx_p \, dy_p \, dz_p$$

$$= \iiint\!\!\iiint V_{jk}(\psi^{(j)})^2(\psi^{(k)})^2 \, dx_j \, dy_j \, dz_j \, dx_k \, dy_k \, dz_k, \qquad (91)$$

as a result of (87), (86), and (88).

Using the results (89), (90), and (91), we substitute (87) and (84) into (62) of 11-4(d) to obtain

$$I = \sum_{k=1}^{s} \iiint \left\{ \frac{K^2}{2m} \left[\left(\frac{\partial \psi^{(k)}}{\partial x_k} \right)^2 + \left(\frac{\partial \psi^{(k)}}{\partial y_k} \right)^2 + \left(\frac{\partial \psi^{(k)}}{\partial z_k} \right)^2 \right] \right.$$

$$\left. + (\psi^{(k)})^2 \left[V_k + \frac{1}{2} \sum_{j=1}^{s} \iiint V_{jk}(\psi^{(j)})^2 \, dx_j \, dy_j \, dz_j \right] \right\} dx_k \, dy_k \, dz_k \qquad (92)$$

for the quantity whose minimum we seek. (We set

$$m_1 = m_2 = \cdots = m_s = m,$$

the electronic mass.)

For the purpose of minimizing (92) we proceed in the following manner: We suppose all the functions $\psi^{(1)}$, $\psi^{(2)}$, $\ldots$, $\psi^{(s)}$ with the exception of one—say $\psi^{(i)}$—to be correctly determined for the minimization; we are thus left with the problem of choosing $\psi^{(i)}$ correctly for the minimization. To do this we need to consider only those terms of (92) which involve the particular $\psi^{(i)}$—*viz.*,

$$I_i = \iiint \left\{ \frac{K^2}{2m} \left[\left(\frac{\partial \psi^{(i)}}{\partial x_i} \right)^2 + \left(\frac{\partial \psi^{(i)}}{\partial y_i} \right)^2 + \left(\frac{\partial \psi^{(i)}}{\partial z_i} \right)^2 \right] \right.$$

$$\left. + (\psi^{(i)})^2 \left[V_i + \sum_{j=1}^{s} \iiint V_{ji}(\psi^{(j)})^2 \, dx_j \, dy_j \, dz_j \right] \right\} dx_i \, dy_i \, dz_i. \qquad (93)$$

(Omission of the factor $\frac{1}{2}$ before the summation over j in (93) arises from the fact that $(\psi^{(i)})^2$ appears *twice* in the double (j,k) sum in (92), once as coefficient of V_{ji} and once as coefficient of V_{ik}. Since j and k run through all values from 1 to s independently, and since $V_{ji} = V_{ij}$, the two terms in which $(\psi^{(i)})^2$ appears in (92) are lumped together in (93).)

Since $V_{ii} = 0$, according to the definition (86), the term $j = i$ does not appear in the sum over j in (93); thus since all the $\psi^{(j)}(j \neq i)$ are assumed determined, the triple integrals over x_j, y_j, z_j may be regarded as *known* functions of x_i, y_i, z_i within the integrand of I_i. In fact comparison of (93) with (5) of 11-1(b)—together with comparison of (88), for $j = i$, with (6) of 11-1(b)—shows that the extremization of I_i with respect to normalized functions $\psi^{(i)}$ is identical with the *single*-particle Schrödinger problem with the potential-energy function

$$V = V_i + \sum_{j=1}^{s} \iiint V_{ji}(\psi^{(j)})^2 \, dx_j \, dy_j \, dz_j.$$

Thus, according to (4) of 11-1(a), with obvious notational modification, we have

$$\frac{K^2}{2m} \nabla_i^2 \psi^{(i)} + \left[E^{(i)} - V_i - \sum_{j=1}^{s} \iiint V_{ji}(\psi^{(j)})^2 \, dx_j \, dy_j \, dz_j \right] \psi^{(i)} = 0$$
$$(\nabla_i^2 \psi^{(i)} = \psi_{x_i x_i}^{(i)} + \psi_{y_i y_i}^{(i)} + \psi_{z_i z_i}^{(i)}), \quad (i = 1, 2, \ldots, s) \quad (94)$$

for the differential equation which must be satisfied by the function $\psi^{(i)}$ which minimizes I_i. (Since the result (94) holds for *any* value of i from 1 to s, the designation $i = 1, 2, \ldots, s$ is affixed.)

(d) From the viewpoint of obtaining a solution for the functions $\psi^{(i)}$ in precise analytical form, the s equations (94), a system of nonlinear integro-differential equations, are of little use. They do, however, lend themselves to a procedure of numerical solution which, although formidably laborious, has yielded results of high accuracy and of wide utility. The procedure, roughly, is the following: First, one makes a simple reasonable assumption as to the potential energy represented by the sum over j in (94) and then solves the resulting Schrödinger equation for each $\psi^{(i)}(i = 1, 2, \ldots, s)$. These solutions are then inserted into their proper positions in the sum over j in (94) and a new solution of the resulting linear differential equations for the various $\psi^{(i)}$ is effected. The process of solution and substitution is reiterated until the successive sets of solutions for the $\psi^{(i)}$ differ sufficiently little from one another. (Treatment of such simplifications as the assumption of spherical symmetry, restrictions according to the so-called *exclusion principle*, and other details of the Hartree method are necessarily omitted from the discussion here.)

The physical interpretation of the sum over j in (94) is of some interest. According to (86) a representative term of this sum is

$$\iiint \frac{\epsilon^2}{r_{ij}}(\psi^{(j)})^2 \, dx_j \, dy_j \, dz_j$$

—the potential energy of a point charge of magnitude ϵ at (x_i, y_i, z_i) as the result of its interaction with a continuously distributed charge of magnitude density $\epsilon(\psi^{(j)})^2$ which has the same sign ($+$ or $-$) as that of the point charge. If, then, we regard $\psi^{(i)}$ as the wave function associated with the ith electron, we may split its potential energy, according to (94), into the following s independent parts:

(1) nuclear influence represented by the term V_i,

(2,3, . . . , s) the influence of $(s-1)$ continuous distributions of charge of densities[1] $-\epsilon(\psi^{(j)})^2$ for $j = 1, 2, \ldots, s$ ($j \neq i$), each of which is associated with one of the remaining $(s-1)$ electrons of the atom.

But if $\psi^{(j)}$ is the wave function associated with the jth electron, $(\psi^{(j)})^2$ is the probability-density function of the jth electron's position in space, according to 11-3(c). Thus, in its electrical influence upon each of the remaining $(s-1)$ electrons of the atom, the jth electron behaves as if it were a continuous distribution of charge of total value[2] $-\epsilon$ whose density at each point is proportional to the probability density of locating the jth electron there.

The foregoing model of the many-electron atom—the so-called *Hartree model*—is surely oversimplified, for it is based upon the restricted form (87) for the electronic wave functions. It does, however, appear sufficiently accurate to yield results in excellent agreement with a wide variety of experiments.

EXERCISES

1. Show that each of the real and imaginary parts of any solution of the wave equation (7) of 11-2(a) is also a solution of the same equation (7).

2. In a given direction of space, superimposed plane waves whose frequencies are confined to a narrow range are propagated with velocities which depend slightly upon the wave frequencies. It can be shown that any measurement of the velocity of such a group of waves yields the so-called *group velocity* U, given through the formula

$$\frac{1}{U} = 2\pi \frac{d(1/\lambda)}{d\omega}.$$

Show that the "matter waves" discussed in 11-2(d,e) are such that the group velocity is identical with v, the classical velocity of the particle with which a given wave is associated. HINT: Use (25) and (26).

[1] The charge on the electron, of magnitude ϵ, is negative.

[2] Integrated over all space, that is (see (88) of (c) above).

3. Use (43) of 11-3(*a*) to evaluate the final integral of (44). HINT: Substitute (43) for one factor of the integrand and integrate by parts $(n - 1)$ times. Take note that the factor $[\xi L_{n-1}^{(1)}(\xi)]$ is a polynomial of degree n, so that $(n - 1)$ differentiations "destroy" all but two of its terms.

4. (*a*) Carry through the procedure of 11-1(*a*) in order to derive (51) of 11-4(*a*).

(*b*) Use 9-1(*b*) to derive (52) from (51).

(*c*) Use 9-1(*c*) to prove the statement in the opening paragraph of 11-4(*d*).

5. (*a*) Prove the orthogonality relation

$$\iiint \psi_j \psi_k \, dx \, dy \, dz = 0 \qquad (95)$$

for the Schrödinger eigenfunctions if $E_j \neq E_k$. HINT: Use (4) with $\psi = \psi_j$, $E = E_j$, then with $\psi = \psi_k$, $E = E_k$; compare 9-6(*b*).

Further, extend the validity of (95) to include the case $E_j = E_k$ $(j \neq k)$ by means of the argument (Schmidt orthogonalization process) of 9-6(*c*).

(*b*) Extend the results of part (*a*) to include the eigenfunctions of the many-particle Schrödinger equation (52). HINT: In the absence of a many-dimensional Green's theorem use direct integration by parts.

6. (*a*) On the basis of an expansion theorem completely analogous to the theorem given in 9-6(*d*) give a formal proof of the asserted minimum characterization of the Schrödinger eigenvalues given in 11-4(*d*). HINT: Compare 9-9(*b*). Note in particular the condition which must be imposed on the "expanded" functions at infinity.

(*b*) The validity of the expansion theorem mentioned in part (*a*) may not obtain if the potential-energy function V is not sufficiently well behaved. Inasmuch as one has very often to deal with potential-energy functions in quantum mechanics which exhibit singularities (for example, (29) of 11-3(*a*) at $r = 0$), the question of the expansion of arbitrary functions is an exceedingly difficult one—more so, for example, than the corresponding question as related to vibrating-membrane eigenfunctions. For the problems considered in the foregoing chapter, however, there is no question as to the validity of the minimum property of the Schrödinger eigenvalues.

7. (*a*) With the nucleus regarded as free (and not fixed, as in 11-3) the integral whose extremization results in the Schrödinger equation for the hydrogen atom is, according to (51),

$$I^* = \iiint\!\!\!\iint \left\{ \frac{K^2}{2m} |\nabla_1 \psi|^2 + \frac{K^2}{2M} |\nabla_2 \psi|^2 + (V - E)\psi^2 \right\} \prod_{i=1}^{2} dx_i \, dy_i \, dz_i, \qquad (06)$$

where

$$|\nabla_i \psi|^2 = \left(\frac{\partial \psi}{\partial x_i} \right)^2 + \left(\frac{\partial \psi}{\partial y_i} \right)^2 + \left(\frac{\partial \psi}{\partial z_i} \right)^2 \qquad (i = 1,2);$$

(x_1, y_1, z_1) and (x_2, y_2, z_2) are respectively the coordinates of electron and nucleus; m is the electronic mass, and M is the mass of the nucleus; and V depends only on the distance

$$r = \sqrt{(x_1 - x_2)^2 + (y_1 - y_2)^2 + (z_1 - z_2)^2}$$

between the electron and the nucleus. Effect the transformation

$$x = x_1 - x_2, \qquad y = y_1 - y_2, \qquad z = z_1 - z_2 \qquad (97)$$

$$X = \frac{x_1 m + x_2 M}{m + M}, \qquad Y = \frac{y_1 m + y_2 M}{m + M}, \qquad Z = \frac{z_1 m + z_2 M}{m + M} \qquad (98)$$

to bring (96) into the form

$$I = \iiint\!\!\!\iint \left\{ \frac{K^2}{2(m+M)} |\nabla_c\psi|^2 + \frac{K^2}{2\mu} |\nabla\psi|^2 + (V - E)\psi^2 \right\} dx\, dy\, dz\, dX\, dY\, dZ, \quad (99)$$

where

$$|\nabla_c\psi|^2 = \psi_X^2 + \psi_Y^2 + \psi_Z^2, \qquad |\nabla\psi|^2 = \psi_x^2 + \psi_y^2 + \psi_z^2,$$

and

$$\mu = \frac{mM}{m+M}. \tag{100}$$

HINT: First show that the absolute value of the jacobian of the transformation is unity. Then derive $\psi_{x_1} = \psi_x + [m/(m+M)]\psi_x$, $\psi_{X_2} = -\psi_x + [M/(m+M)]\psi_X$, etc.

(b) Carry out the extremization of (99) to derive the Schrödinger equation of the problem. HINT: Use 9-1(b).

Show that an eigenfunction ψ of this equation may be written in the form

$$\psi = \psi^{(c)}(X,Y,Z)\psi^{(e)}(x,y,z),$$

where $\psi^{(c)}$ satisfies the equation

$$\frac{K^2}{2(m+M)}(\psi_{XX}^{(c)} + \psi_{YY}^{(c)} + \psi_{ZZ}^{(c)}) + E^{(c)}\psi^{(c)} = 0 \tag{101}$$

—the Schrödinger equation for a *free* (zero-potential-energy) particle—and $\psi^{(e)}$ satisfies the Schrödinger equation for the hydrogen atom with stationary nucleus (11-3), but with m replaced by μ. HINT: Use the fact that V depends only on x, y, z; compare 11-4(c).

(The variables X, Y, Z, defined by (98) are the coordinates of the *center of mass* of the atom. The result (101) may thus be interpreted that the atom *as a whole* is to be considered, *in its translational motion*, as a free particle located at its center of mass. The final result of the preceding paragraph justifies the parenthetic remark made at the conclusion of 11-3(b). What do the variables x, y, z, defined in (97), represent?)

8. Give explicit justification for the final statement made in 11-4(c).

9. Verify directly the fact that $\psi^{(j)}$ of (71) is normalized. (This is of course a very simple special consequence of exercise 3 above.)

10. Suppose that a single (approximate) eigenfunction

$$\psi = \prod_{i=1}^{s} \psi^{(i)}(x_i, y_i, z_i) \tag{102}$$

has been obtained for a given s-electron atom by means of the Hartree method (11-5(c)); each $\psi^{(i)}$ is properly normalized. Let, further, $E^{(1)}$, $E^{(2)}$, . . . $E^{(s)}$ be the corresponding set of eigenvalues obtained by solving the equations (94). Show that the total (approximate) energy of the atom which corresponds to (102) is given by

$$E = \sum_{i=1}^{s} E^{(i)} - \tfrac{1}{2} \sum_{k=1}^{s} \sum_{j=1}^{s} \iiint\!\!\!\iint V_{jk}(\psi^{(j)}\psi^{(k)})^2 \, dx_j\, dy_j\, dz_j\, dx_k\, dy_k\, dz_k. \tag{103}$$

HINT: Compare (92), whose minimum is E, with (93), whose minimum is $E^{(i)}$.

What is the physical significance of the appearance of the double sum in (103)?

11. (a) Show that a linear homogeneous transformation

$$x_i = \sum_{j=1}^{3} b_{ij} x_j' \qquad (i = 1,2,3) \qquad (104)$$

from a cartesian system (x_1,x_2,x_3) is a pure rotation of axes to a second such cartesian system (x_1',x_2',x_3') if and only if the six relations

$$\sum_{i=1}^{3} b_{ij} b_{ik} = \delta_{jk} \qquad (j,k = 1,2,3, \text{ independently}) \qquad (105)$$

hold. HINT: The necessary and sufficient condition that the transformation be a pure rotation is that

$$\sum_{i=1}^{3} x_i^2 = \sum_{i=1}^{3} x_i'^2 \qquad (106)$$

for all values of the x_i and x_i'. Derive (105) directly from (104) and (106).

(b) Use (105) to show that the absolute value of the jacobian of the transformation (104) is unity. HINT: Use the rule for multiplying determinants (2-8(c)) to form the square of the jacobian.

CHAPTER 12

ELECTROSTATICS

12-1. Laplace's Equation. Capacity of a Condenser

(a) To say that there exists an electrostatic field in a given region of space is equivalent to asserting the existence of a vector whose cartesian components E_x, E_y, E_z are, in general, functions of the position coordinates x, y, z (but *not* of the time t) such that a point charge Q located at (x,y,z) experiences a force whose cartesian components are QE_x, QE_y, QE_z. The vector (E_x,E_y,E_z) is called the *electrostatic intensity*. The electrostatic field is conservative in the sense of 6-1; that is, there exists a function $\phi(x,y,z)$ with continuous second partial derivatives from which the components of the electrostatic intensity are derivable as

$$E_x = -\frac{\partial \phi}{\partial x}, \qquad E_y = -\frac{\partial \phi}{\partial y}, \qquad E_z = -\frac{\partial \phi}{\partial z}. \tag{1}$$

The function ϕ, which is actually the potential energy of a unit charge ($Q = 1$), is called the electrostatic potential function—or, simply, the *potential*—of the field. The component of the electrostatic intensity in any given direction is the negative of the derivative of the potential taken with respect to that direction.

For the sake of simplicity we may define a metallic conductor—or, briefly, a *conductor*—as a body in which the electrostatic potential has the same value at all points; in particular the surface of a conductor in an electrostatic field is characterized by a constant potential. (From (1) it thus follows that the electrostatic intensity is everywhere zero in the interior of a conductor and has a zero component in every direction tangential to the surface of a conductor.)

We consider the three-dimensional region R which is exterior to a given number of isolated fixed conductors and interior to a single closed conducting surface; the region R is unoccupied. Owing to the assumed presence of an electric charge on at least one of the conducting surfaces the region R constitutes an electrostatic field. The potential energy per unit volume associated with such an electrostatic field is given[1] by the

[1] A demonstration of this result is far beyond our present scope. See, for example, Max Abraham and Richard Becker, "The Classical Theory of Electricity and Magnetism," pp. 81–84, Blackie & Son, Ltd., Glasgow, 1932.

expression $(1/8\pi)(E_x^2 + E_y^2 + E_z^2)$. Integrating this quantity over the region R occupied by the field, we obtain, with the aid of (1), the total potential energy of the field—namely,

$$W = \frac{1}{8\pi} \iiint\limits_R \left[\left(\frac{\partial \phi}{\partial x}\right)^2 + \left(\frac{\partial \phi}{\partial y}\right)^2 + \left(\frac{\partial \phi}{\partial z}\right)^2 \right] dx\, dy\, dz. \qquad (2)$$

This field potential energy—not to be confused with the potential function $\phi = \phi(x,y,z)$—represents the amount of mechanical work which would be required in order to bring the electric charges which give rise to the electrostatic field from infinitely great mutual distances to their actual distributions on the conductor surfaces.

The principle which characterizes a system in stable equilibrium as possessing a minimum of potential energy consistent with its constraints applies to an electrostatic field as well as to a mechanical system. We may thus expect to derive the differential equation satisfied by the potential function by rendering the integral (2) a minimum with respect to continuously differentiable functions ϕ which possess a prescribed constant value on each of the conducting surfaces which constitute the boundary B of R. The boundary condition for the functions ϕ eligible for the minimization of (2) springs from the definition above of a conductor; thus a different constant value is in general assumed on each isolated conductor.

To extremize (2) we may employ the general Euler-Lagrange equation (9) of 9-1(a), with $w = \phi$ and[1]

$$f = \psi_x^2 + \psi_y^2 + \psi_z^2.$$

We thus obtain for the extremizing function ϕ—namely, the actual potential function of the electrostatic field—the partial differential equation

$$\phi_{xx} + \phi_{yy} + \phi_{zz} = 0, \quad \text{or} \quad \nabla^2 \phi = 0. \qquad (3)$$

The equation (3)—so-called *Laplace's equation*—finds applicability not only in electrostatic theory but also in the studies of classical (Newtonian) gravitation, hydrodynamics, heat flow, and other physical phenomena.

In exercise 1(a) at the end of this chapter a proof that the ϕ which extremizes (2) is actually a *minimizing* function is called for. Further, in exercise 2(b) it is proved that the solution of (3) under the given boundary conditions is uniquely determined. (Adequate hints are provided in each case.) The question of the *existence* of the minimum of

---

[1] From this point forward we employ subscripts to denote partial differentiation, as in preceding chapters—but *not* in (1) above!

(2) is discussed in 12-4 below; we make the generally valid assumption of its existence.

(b) The problem of finding the solution of Laplace's equation (3) in a given region R, with ϕ required to assume specific values on the boundary surface B of R, is called the *Dirichlet problem* for R. The solution of the Dirichlet problem in closed analytical form has been accomplished in several cases; these are discussed adequately in the literature on potential theory.[1] We direct our attention, rather, to the possibility of effecting approximate solutions through the direct minimization of the integral (2) for cases in which a solution in closed form cannot be achieved. One general method of such approximation is completely analogous to the Ritz method as applied in 7-6, 9-13, 10-10, 11-5, etc.: A class of functions ϕ is defined by the various sets of values of a finite number of parameters borne by a single analytical expression which assumes the required values on B for all values of the parameters. The parameter-laden expression is substituted for ϕ in the integrand of (2), and the minimum of W with respect to the parameters is effected. The minimizing values of the parameters thus define that function of the given class which gives the "best"—in the sense of rendering W the smallest—approximation to the actual potential.

The method described in the preceding paragraph is in general quite laborious in its execution. Justification for the amount of labor required can of course lie only in the degree of urgency attached to the (approximate) solution of any given problem. For the purpose of illustration we carry out the method for a problem of nontypical simplicity—one in which the procedure leads us directly to the known precise solution:

We choose for the region R the exterior of a given sphere of radius b; the outer boundary of R may be considered to be "a sphere of infinite radius concentric with the given sphere." We set up a system of spherical coordinates[2] (r, θ, ψ) with origin at the center of the sphere. The class of functions with respect to which we choose to minimize (2) is defined by the single parameter p in

$$\phi = \phi_1 \left(\frac{r}{b}\right)^p \qquad (p < 0), \qquad (4)$$

where ϕ_1 is the potential on the sphere and the "potential at infinity" is taken to be zero.

In spherical coordinates (2) reads, on substitution of (4) with the aid of (36) of 11-3(a),

---

[1] See, for example, Kellogg.

[2] Because of the use of ϕ for the potential function, we substitute the symbol ψ for the usual ϕ as third spherical coordinate.

$$W = \frac{p^2\phi_1^2}{8\pi b^{2p}} \int_b^\infty \int_0^\pi \int_0^{2\pi} r^{2p-2}r^2 \sin\theta \, d\psi \, d\theta \, dr \qquad (p < 0)$$

$$= -\frac{p^2\phi_1^2 b}{2(2p+1)} = -\frac{1}{8} b\phi_1^2 \left(2p - 1 + \frac{1}{2p+1}\right).$$

To minimize W we form

$$\frac{dW}{dp} = -\frac{1}{4} b\phi_1^2 \left(1 - \frac{1}{(2p+1)^2}\right) = 0,$$

whence $p = 0, -1$. Since $(d^2W/dp^2) > 0$ for $p = -1$, W is a minimum for this value of the parameter, and the best approximation to the potential, as supplied by (4), is

$$\phi = \phi_1\left(\frac{b}{r}\right). \tag{5}$$

(We must reject the solution $p = 0$ in advance because of the necessary requirement $p < 0$.) It happens that (5) actually satisfies Laplace's equation (see end-chapter exercise $3(a)$) as well as the given boundary conditions and is thus the precise solution of the problem.

(c) We devote the remainder of this section to the consideration of regions R of the type which lie exterior to a single given closed conducting surface B_1 and interior to a second given closed conducting surface B_2; the two conductors are then said to constitute a *condenser*. The essential quantity associated with a condenser is its *capacity*, which is defined by the formula

$$C = \frac{1}{4\pi(\phi_2 - \phi_1)^2} \iiint_R (\phi_x^2 + \phi_y^2 + \phi_z^2)dx \, dy \, dz, \tag{6}$$

where ϕ_1 is the constant potential on B_1, ϕ_2 the constant potential on B_2, and $\phi = \phi(x,y,z)$ the potential in the intervening region R.

In view of the minimizing character of the potential function enunciated in (a) above, and since the integral in (6) coincides with that in (2), an equivalent definition of the capacity of a condenser is the *minimum* of (6) with respect to continuously differentiable functions which satisfy $\phi = \phi_1$ on B_1, $\phi = \phi_2$ on B_2. The merit of this minimum definition lies, of course, in its usefulness for the approximation of the capacity of a given condenser along the lines sketched in (b) above. Moreover, there exists a method of approximation which is far more elegant and, at least in some cases, simpler in its application than the direct method of (b) in which the integral of (6) is minimized with respect to a finite set of parameters. We proceed to develop this method.

We suppose that B_1 and B_2 are members of the one-parameter family of closed surfaces $u(x,y,z) = A$ $(u_1 \leqq A \leqq u_2)$, with $u(x,y,z) = u_1$ on B_1 and $u(x,y,z) = u_2$ on B_2. Further, we restrict the continuously differentiable function $u(x,y,z)$ to be such that through each point of R there passes one and only one surface $u(x,y,z) = A$ lying entirely within R, with the values of the parameter A so ordered that the surface $u = A_1$ is everywhere interior to the region bounded by the surface $u = A_2$ whenever $A_1 < A_2$. (See Fig. 12-1 for a plane section of R. Discussion of the existence of the required function $u(x,y,z)$ when B_1 and B_2 are given

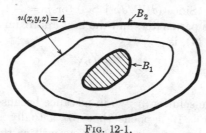

FIG. 12-1.

is reserved for exercise 5(d) at the end of this chapter.)   We seek to minimize (6) with respect to functions which exhibit the special form

$$\phi = G(u), \qquad \text{with } G(u_1) = \phi_1, G(u_2) = \phi_2. \tag{7}$$

From (7) it follows that

$$\phi_x = G'(u)u_x, \qquad \phi_y = G'(u)u_y, \qquad \phi_z = G'(u)u_z,$$

where the prime (') indicates differentiation with respect to the argument u.   Thus, on substitution of (7), the integral of (6) (which we seek to minimize through proper choice of the function G) becomes

$$I = \iiint_R (\phi_x^2 + \phi_y^2 + \phi_z^2)dx\,dy\,dz = \iiint_R [G'(u)]^2(u_x^2 + u_y^2 + u_z^2)dx\,dy\,dz. \tag{8}$$

We proceed to reduce (8) to a simple integral over the variable u.

We introduce the two continuously differentiable functions $v(x,y,z)$ and $w(x,y,z)$ such that through every point of R there passes one and only one surface $v(x,y,z) = $ constant and one and only one surface $w(x,y,z) = $ constant for some pair of ranges of values $v_1 \leqq v \leqq v_2$ and $w_1 \leqq w \leqq w_2$. Thus we have in (u,v,w) a coordinate system related to the cartesian system by the transformation equations

$$u = u(x,y,z), \qquad v = v(x,y,z), \qquad w = w(x,y,z); \tag{9}$$

the inverse transformation is obtained by solving the system (9) for x,y,z:

$$x = x(u,v,w), \qquad y = y(u,v,w), \qquad z = z(u,v,w). \tag{10}$$

The numbers v_1, v_2, w_1, w_2 are so chosen that to every (u,v,w), with $u_1 \leqq u \leqq u_2$, $v_1 \leqq v \leqq v_2$, $w_1 \leqq w \leqq w_2$, there corresponds a single point of R. (An example of a suitable assignment of v and w is given in (d) below, where the method under development is applied to a specific problem.)

Using (10), we transform (8) to read

$$I = \int_{u_1}^{u_2} \int_{v_1}^{v_2} \int_{w_1}^{w_2} [G'(u)]^2 (u_x^2 + u_y^2 + u_z^2) \left| \frac{\partial(x,y,z)}{\partial(u,v,w)} \right| dw \, dv \, du, \qquad (11)$$

according to the rule given in 2-8(f); the jacobian $[\partial(x,y,z)/\partial(u,v,w)]$ and $(u_x^2 + u_y^2 + u_z^2)$ are supposed expressed in terms of[1] (u,v,w). Since the factor $[G'(u)]^2$ is independent of v and w, we may define

$$H(u) = \int_{v_1}^{v_2} \int_{w_1}^{w_2} (u_x^2 + u_y^2 + u_z^2) \left| \frac{\partial(x,y,z)}{\partial(u,v,w)} \right| dw \, dv \qquad (12)$$

and so rewrite (11) as the simple integral

$$I = \int_{u_1}^{u_2} [G'(u)]^2 H(u) du. \qquad (13)$$

Thus the problem of minimizing I with respect to functions ϕ having the form (7) is reduced to the problem of minimizing the simple integral (13) with respect to functions G of a single variable. To accomplish this we may apply the result (26) of 3-4(a)—first integral of the Euler-Lagrange equation (25) of 3-3(b) in the event the integrand f is explicitly independent of the dependent variable G—to the integrand $f = G'^2 H$ of (13). We thus obtain

$$G'H = c_1, \qquad \text{or} \quad G(u) = c_1 \int_{u_1}^{u} \frac{du}{H(u)} + c_2, \qquad (14)$$

where c_1 and c_2 are constants. (In exercises 2 and 3, Chap. 3, it is shown that the extremizing function given by (14) actually *minimizes* (13).) To evaluate c_1 and c_2 we employ the second and third relations of (7) and so obtain from (14)

$$c_2 = \phi_1, \qquad c_1 = \frac{\phi_2 - \phi_1}{\int_{u_1}^{u_2} du/H(u)}. \qquad (15)$$

Thus, according to the first of (14) and the second of (15), we have

$$HG'^2 = \frac{c_1^2}{H} = \frac{(\phi_2 - \phi_1)^2}{H \left[\int_{u_1}^{u_2} du/H(u) \right]^2},$$

---

[1] See end-chapter exercise 6(b).

so that (6) becomes, with the aid of (13) and (8), the *approximate* capacity

$$C = \frac{1}{4\pi \int_{u_1}^{u_2} du/H(u)} \geqq C_0, \qquad (16)$$

where C_0 is the *actual* capacity (the minimum of (6) with respect to ϕ) of the condenser under consideration.

Under exercise 6(c) at the end of this chapter it is shown that the function $H(u)$ given by (12) is independent of the choice of functions v and w in (9) and (10). Thus the possibility of improving the approximation given in (16) through a "better" choice of $v(x,y,z)$ and $w(x,y,z)$ does not exist.

The only possible way to decrease the difference $(C - C_0)$—if it does not already vanish—is by means of a more suitable function $u(x,y,z)$ in terms of which the surfaces B_1 and B_2 are represented. In fact if the family of surfaces $u(x,y,z) = $ constant happens to be identical with the family of surfaces $\phi_0(x,y,z) = $ constant (where ϕ_0 is the *actual* potential in R), it is directly seen that the approximation (16) is perfect; that is, $C = C_0$. For with such a choice of u there is a one-to-one[1] functional relationship between ϕ_0 and u; thus the function $G(u)$ is some function $F(\phi_0)$; the minimum of (6) with respect to $\phi = G(u) = F(\phi_0)$ is clearly achieved for $F = \phi_0$. From this fact it follows that the method of this subsection may be used to solve the following condenser problem: Given any one-parameter representation of the equipotential surfaces (surfaces on which the potential is constant), find the potential function itself.

The main limitations of the method of the preceding paragraphs lie, first, in the possible difficulty of finding a sufficiently simple function $u(x,y,z)$; and, second, once the choice of $u(x,y,z)$ is made, the possible failure of the integral (12) for $H(u)$ to be evaluable in explicit form. In the specific example of (d) below, however, neither of these difficulties prevails.

In 12-2 below we consider a second method for approximating the capacity of a condenser—a method which results in a *lower* bound for the capacity. We are thus enabled to estimate the accuracy of any approximation from above achieved by the methods of this section.

(d) The remainder of this section is devoted to applying the method of (c) above to the approximation of the capacity of the condenser formed by the similar ellipsoids of revolution described by the surfaces $u = u_1$ and $u = u_2$ ($0 < u_1 < u_2$), where

$$u = \sqrt{x^2 + y^2 + \alpha^2 z^2} \qquad (\alpha > 0). \qquad (17)$$

---

[1] See end-chapter exercise 5(d) for a fuller discussion of this point.

(If $\alpha < 1$, we deal with prolate spheroids; if $\alpha > 1$, we have oblate spheroids.) In the special case $\alpha = 1$, the surfaces $u = u_1$ and $u = u_2$ are concentric spheres with their centers at the origin of coordinates and u is the distance from the origin. In this case we should almost certainly choose for the functions v, w (see (9) and (10) of (c) above) the spherical coordinates usually denoted by θ, ϕ. That is, (10) would read

$$x = u \sin v \cos w, \qquad y = u \sin v \sin w, \qquad z = u \cos v \qquad (\alpha = 1). \qquad (18)$$

Since passing from $\alpha = 1$ to $\alpha \neq 1$ in (17) means replacing z by αz, we should expect to obtain a suitable choice of the functions v and w by replacing z by αz in (18)—namely,

$$x = u \sin v \cos w, \qquad y = u \sin v \sin w, \qquad z = \frac{u}{\alpha} \cos v. \qquad (19)$$

We note, first, that (17) is satisfied by (19). Further, it is clear that each point of R is associated, through (19), with one and only one triple of values (u, v, w), with

$$u_1 \leqq u \leqq u_2, \qquad 0 \leqq v \leqq \pi, \qquad 0 \leqq w < 2\pi. \qquad (20)$$

(Proof of this fact is called for in exercise 9 at the end of this chapter.)
From (17) we compute directly

$$u_x^2 + u_y^2 + u_z^2 = \frac{x^2 + y^2 + \alpha^4 z^2}{x^2 + y^2 + \alpha^2 z^2} = \sin^2 v + \alpha^2 \cos^2 v, \qquad (21)$$

according to (19). From (19) we compute the jacobian

$$\frac{\partial(x,y,z)}{\partial(u,v,w)} = \begin{vmatrix} \sin v \cos w & \sin v \sin w & \frac{1}{\alpha} \cos v \\ u \cos v \cos w & u \cos v \sin w & -\frac{u}{\alpha} \sin v \\ -u \sin v \sin w & u \sin v \cos w & 0 \end{vmatrix}$$

$$= \frac{u^2}{\alpha} \sin v. \qquad (22)$$

With (21) and (22) the equation (12) of (c) above becomes, with the aid of (20),

$$H(u) = \frac{u^2}{\alpha} \int_0^\pi \int_0^{2\pi} (\sin^2 v + \alpha^2 \cos^2 v) \sin v \, dw \, dv = \frac{4\pi}{3\alpha} (2 + \alpha^2) u^2, \qquad (23)$$

from which we compute

$$\int_{u_1}^u \frac{du}{H(u)} = \frac{3\alpha}{4\pi(2 + \alpha^2)} \left(\frac{1}{u_1} - \frac{1}{u} \right). \qquad (24)$$

Substituting (24) into (14) and (15) of (c) above, we obtain for the approximate potential function

$$\phi = G(u) = -\frac{\phi_2 - \phi_1}{u_2 - u_1}\left(\frac{u_1 u_2}{u}\right) + \frac{u_2\phi_2 - u_1\phi_1}{u_2 - u_1}. \tag{25}$$

Substituting (24), with $u = u_2$, into (16) above, we obtain for the approximate capacity of the condenser under consideration

$$C = \frac{2 + \alpha^2}{3\alpha}\left(\frac{u_1 u_2}{u_2 - u_1}\right) \geqq C_0, \tag{26}$$

where C_0 is the precise capacity. We note, on comparison with the results of end-chapter exercise 8, that both (25) and (26) give precise results in the case $\alpha = 1$—for the condenser consisting of concentric spheres, that is, according to (17). We should therefore expect the method to give its most reliable results in this problem when α is in the neighborhood of unity; this fact is borne out in 12-2 below, where a lower bound for the capacity of the ellipsoidal condenser is derived.

(e) The capacity of a *single* conducting surface is defined as the limit of the capacity of the condenser, of which the given surface is the inner conductor, as the outer conductor recedes to infinity in all directions. In the case of the ellipsoid $u = u_1$ of (d) above we obtain the approximation C' to its capacity C_0' by letting $u_2 \to \infty$ in (26)—namely,

$$C' = \frac{2 + \alpha^2}{3\alpha}\, u_1 \geqq C_0'. \tag{27}$$

If we write the equation of the ellipsoid $u = u_1$ in the familiar form

$$\frac{x^2 + y^2}{b^2} + \frac{z^2}{a^2} = 1, \tag{28}$$

we have, according to (17), $u_1 = b$ and $\alpha = (b/a)$, so that (27) reads

$$C' = \frac{2a^2 + b^2}{3a} = \frac{a}{1 - \frac{1}{3}[1 - (b^2/a^2)]} \geqq C_0'. \tag{29}$$

In exercise 10 we compare (29) with the formula giving C_0' precisely for the ellipsoid (28) in the case $a > b$.

12-2. Approximation of the Capacity from Below (Relaxed Boundary Conditions)

(a) The minimum of the expression (6) of 12-1(c) with respect to continuously differentiable functions ϕ for which $\phi = \phi_1$ on B_1 and $\phi = \phi_2$

on B_2 is by definition the capacity of the condenser whose inner (B_1) and outer (B_2) conducting surfaces bound the region R. We proceed to demonstrate that the capacity may be defined equivalently as the *maximum* of (6) with respect to functions which (i) satisfy Laplace's equation and (ii) satisfy a "relaxed" boundary condition expressed in (34) below.

We suppose that ϕ is the *actual* potential in the region R bounded by B_1, on which the constant potential is ϕ_1, and B_2, on which the constant potential is ϕ_2. That is, we have

$$\nabla^2\phi = 0 \text{ in } R, \qquad \phi = \phi_k \text{ on } B_k \ (k = 1,2). \tag{30}$$

Thus the actual capacity C_0 of the condenser under attention is

$$C_0 = \frac{1}{4\pi(\phi_2 - \phi_1)^2} \iiint\limits_R (\phi_x^2 + \phi_y^2 + \phi_z^2)dx\, dy\, dz. \tag{31}$$

We write

$$\phi = \psi + Q, \tag{32}$$

where the function $\psi = \psi(x,y,z)$ satisfies Laplace's equation—namely,

$$\nabla^2\psi = 0 \qquad \text{in } R \tag{33}$$

—and the relaxed boundary condition

$$\iint\limits_B (\phi - \psi)\frac{\partial\psi}{\partial n}\, dS = 0. \tag{34}$$

(The function Q merely represents the difference between ϕ and ψ as defined.) The surface integral which appears in (34) is carried out over the boundary B of R—that is, over the *two* surfaces B_1 and B_2. The derivative $(\partial\psi/\partial n)$ is computed with respect to the normal to B directed *outward* from R.

(The boundary condition (34) is called "relaxed" in that it is less stringent than the boundary conditions imposed upon the functions ϕ eligible for the minimization of (6). That is, any function ψ which satisfies the latter conditions—namely, $\psi = \phi_1$ on B_1, $\psi = \phi_2$ on B_2—clearly satisfies (34), because of (30); on the other hand we see below that there exist functions ψ which do not satisfy these conditions but for which (34) holds.)

We substitute (32) into (31):

$$4\pi(\phi_2 - \phi_1)^2 C_0$$
$$= \iiint\limits_R (\psi_x^2 + \psi_y^2 + \psi_z^2)dx\, dy\, dz + \iiint\limits_R (Q_x^2 + Q_y^2 + Q_z^2)dx\, dy\, dz$$
$$+ 2\iiint\limits_R (\psi_x Q_x + \psi_y Q_y + \psi_z Q_z)dx\, dy\, dz. \tag{35}$$

According to Green's theorem (30) of 2-14 we have

$$\iiint_R (\psi_x Q_x + \psi_y Q_y + \psi_z Q_z)dx\,dy\,dz$$

$$= \iint_B Q\,\frac{\partial\psi}{\partial n}\,dS - \iiint_R Q\nabla^2\psi\,dx\,dy\,dz = 0, \quad (36)$$

because of (34), with (32), and (33). Since the second integral on the right of (35) cannot be negative, it follows from (35) and (36) that

$$C_0 \geqq \frac{1}{4\pi(\phi_2 - \phi_1)^2} \iiint_R (\psi_x^2 + \psi_y^2 + \psi_z^2)dx\,dy\,dz. \quad (37)$$

Since the equality sign holds in (37) if $\psi = \phi$ and since $\psi = \phi$ satisfies both (33) and (34), because of (30), we are justified in defining the capacity C_0 as the maximum of the right-hand member of (37) with respect to functions ψ which satisfy both (33) and (34).

The difference between the two members of (37) is, according to (35) and (36), proportional to the positive quantity

$$I = \iiint_R (Q_x^2 + Q_y^2 + Q_z^2)dx\,dy\,dz, \quad (38)$$

where Q is given by (32). The smaller the value of (38), therefore, the better is the approximation to C_0 which we achieve through the right-hand member of (37). We proceed to show that, if we write ψ as a linear combination

$$\psi = \sum_{i=1}^{N} a_i U_i \quad [\nabla^2 U_i = 0\ (i = 1,2,\ \ldots\ ,N)] \quad (39)$$

of N given functions $U_i(x,y,z)$ which individually satisfy Laplace's equation, the set of values of $a_1, a_2, \ldots, a_N$ for which (38) is a minimum is a set for which ψ satisfies the relaxed boundary condition (34).

With (32) and (39) we substitute

$$Q = \phi - \sum_{i=1}^{N} a_i U_i. \quad (40)$$

For (38) to be a minimum with respect to $a_1, a_2, \ldots, a_N$ we must have $(\partial I/\partial a_k) = 0$ for $k = 1, 2, \ldots, N$. (Since $I \geqq 0$, the minimum surely exists inasmuch as I is a continuous function of the a_k.) From (38) we have

$$\frac{\partial I}{\partial a_k} = 2 \iiint_R \left(Q_x \frac{\partial Q_x}{\partial a_k} + Q_y \frac{\partial Q_y}{\partial a_k} + Q_z \frac{\partial Q_z}{\partial a_k} \right) dx\, dy\, dz$$

$$= -2 \iiint_R \left(\frac{\partial Q}{\partial x} \frac{\partial U_k}{\partial x} + \frac{\partial Q}{\partial y} \frac{\partial U_k}{\partial y} + \frac{\partial Q}{\partial z} \frac{\partial U_k}{\partial z} \right) dx\, dy\, dz, \quad (41)$$

as we find by performing the requisite differentiations in (40). With the aid of Green's theorem (30) of 2-14, equation (41) becomes

$$\frac{1}{2} \frac{\partial I}{\partial a_k} = \iiint_R Q \nabla^2 U_k \, dx\, dy\, dz - \iint_B Q \frac{\partial U_k}{\partial n} \, dS$$

$$= - \iint_B \left(\phi - \sum_{i=1}^{N} a_i U_i \right) \frac{\partial U_k}{\partial n} \, dS,$$

because of the bracketed portion of (39) and (40). The required vanishing of all the $(\partial I / \partial a_k)$ for the minimum of I thus gives the set of N linear inhomogeneous equations

$$\iint_B \phi \frac{\partial U_k}{\partial n} \, dS - \sum_{i=1}^{N} a_i \iint_B U_i \frac{\partial U_k}{\partial n} \, dS = 0 \qquad (k = 1, 2, \ldots, N) \quad (42)$$

for the best choice of the a_i. That this choice also renders (34) satisfied by (39) is shown by multiplying the kth equation of (42) by a_k, for $k = 1, 2, \ldots, N$, and by adding the resulting N equations to obtain

$$\iint_B \left(\phi - \sum_{i=1}^{N} a_i U_i \right) \frac{\partial}{\partial n} \left(\sum_{k=1}^{N} a_k U_k \right) dS = 0;$$

this is identical with (34), through (39).

Thus, in order to achieve a lower bound for the capacity C_0 by means of (37) we form the linear combination (39) where $U_1, U_2, \ldots, U_N$ are given functions known to satisfy Laplace's equation and where $a_1, a_2, \ldots, a_N$ are obtained through solution of the system (42). Moreover, we can expect, in general, to have an improved approximation to C_0 in the right-hand member of (37) by adding more terms to the linear combination (39). For since (42) ensures that the difference between the members of (37) is a minimum with respect to the class of functions defined by (39), with $U_1, U_2, \ldots, U_N$ given, the widening of this class cannot increase the minimum achieved.

In applying the foregoing method to a given condenser, we note that each surface integral of (42) is the sum of two surface integrals, one carried out over B_1, with $\phi = \phi_1$, and the other carried out over B_2, with $\phi = \phi_2$.

(b) We apply the method of (a) above to the condenser of 12-1(d), consisting of concentric coaxial similar ellipsoids of revolution described by the equations $u = u_1$ and $u = u_2$ ($0 < u_1 < u_2$), with

$$u = \sqrt{x^2 + y^2 + \alpha^2 z^2} \qquad (\alpha > 0). \tag{43}$$

In the case $\alpha = 1$—whereby the condenser consists of concentric spherical surfaces—the precise potential is given by

$$\phi = \frac{a_1}{\sqrt{x^2 + y^2 + z^2}} + a_2,$$

where a_1 and a_2 are constants selected so as to fit the boundary conditions (see end-chapter exercise 8). We may thus expect to achieve a good approximation to the capacity C_0, at least in the neighborhood of $\alpha = 1$, by the choice

$$\psi = a_1 U_1, \qquad U_1 = \frac{1}{\sqrt{x^2 + y^2 + z^2}} \tag{44}$$

for the linear combination (39). With (44) the result (42) becomes a single equation which—when solved for a_1 and with the surface integral over B split into its component parts—reads, since $\phi = \phi_1$ on B_1 and $\phi = \phi_2$ on B_2,

$$a_1 = \frac{\phi_1 \iint\limits_{B_1} (\partial U_1/\partial n)dS + \phi_2 \iint\limits_{B_2} (\partial U_1/\partial n)dS}{\iint\limits_{B_1} U_1(\partial U_1/\partial n)dS + \iint\limits_{B_2} U_1(\partial U_1/\partial n)dS} = \frac{\displaystyle\sum_{k=1}^{2} \phi_k J_k}{\displaystyle\sum_{k=1}^{2} L_k}, \tag{45}$$

where

$$J_k = \iint\limits_{B_k} \frac{\partial U_1}{\partial n}\, dS, \qquad L_k = \iint\limits_{B_k} U_1 \frac{\partial U_1}{\partial n}\, dS \qquad (k = 1,2). \tag{46}$$

To evaluate the integrals (46) we use as coordinates on the surface B_k the variables v, w introduced in (19) of 12-1(d); thus, for points of the surface $u = u_k$, we have

$$x = u_k \sin v \cos w, \qquad y = u_k \sin v \sin w, \qquad z = \frac{u_k}{\alpha} \cos v \qquad (k = 1,2),$$

$$(0 \leqq v \leqq \pi, 0 \leqq w < 2\pi). \tag{47}$$

According to 2-11(b) the element of surface area on B_k is given by

$$dS = \sqrt{(x_v^2 + y_v^2 + z_v^2)(x_w^2 + y_w^2 + z_w^2) - (x_v x_w + y_v y_w + z_v z_w)^2} \, dv \, dw$$
$$= \frac{u_k^2}{\alpha} \sqrt{\sin^2 v + \alpha^2 \cos^2 v} \, \sin v \, dv \, dw \qquad (k = 1,2), \quad (48)$$

as we compute with the aid of (47). To express the normal derivative $(\partial U_1/\partial n)$ in terms of v and w we first note that the *outward* (from R) normal direction is in the direction of *decreasing* u on B_1 and in the direction of *increasing* u on B_2, according to (43). Thus, in applying the result (19) of 2-11(c) we write

$$\frac{\partial U_1}{\partial n} = (-1)^k \frac{(U_1)_x u_x + (U_1)_y u_y + (U_1)_z u_z}{\sqrt{u_x^2 + u_y^2 + u_z^2}}$$
$$= -(-1)^k \frac{x^2 + y^2 + \alpha^2 z^2}{\sqrt{x^2 + y^2 + \alpha^4 z^2}\,(x^2 + y^2 + z^2)^{\frac{3}{2}}}$$
$$= \frac{(-1)^{k+1}\alpha^3}{u_k^2 \sqrt{\sin^2 v + \alpha^2 \cos^2 v}\,(\alpha^2 \sin^2 v + \cos^2 v)^{\frac{3}{2}}}, \quad (49)$$

as we find with the use of (43) and (44), then with the aid of (47). With the results (48) and (49), and with (47) applied to (44), the integrals (46) become

$$J_k = (-1)^{k+1}\alpha^2 \int_0^\pi \int_0^{2\pi} \frac{\sin v \, dw \, dv}{(\alpha^2 \sin^2 v + \cos^2 v)^{\frac{3}{2}}} \qquad (k = 1,2),$$
$$L_k = (-1)^{k+1}\frac{\alpha^3}{u_k} \int_0^\pi \int_0^{2\pi} \frac{\sin v \, dw \, dv}{(\alpha^2 \sin^2 v + \cos^2 v)^2} \qquad (k = 1,2).$$

Evaluation of J_k and L_k is accomplished by elementary methods. In both cases integration over w merely yields the factor 2π; the substitution $\xi = \cos v$ reduces both integrands to standard algebraic forms. We obtain, finally,

$$J_k = 4\pi(-1)^{k+1}, \qquad L_k = \frac{2\pi(-1)^{k+1}}{u_k} F(\alpha), \quad (50)$$

$$F(\alpha) = \begin{cases} \dfrac{\cos^{-1}\alpha}{\sqrt{1 - \alpha^2}} + \alpha & (0 < \alpha < 1), \\[2mm] 2 & (\alpha = 1), \\[2mm] \dfrac{\log(\alpha + \sqrt{\alpha^2 - 1})}{\sqrt{\alpha^2 - 1}} + \alpha & (\alpha > 1). \end{cases} \quad (51)$$

With (50) equation (45) becomes

$$a_1 = \frac{2(\phi_1 - \phi_2)}{F(\alpha)\left(\dfrac{1}{u_1} - \dfrac{1}{u_2}\right)}. \quad (52)$$

Substituting (44) into the right-hand member of (37), we obtain

$$C_0 \geqq \frac{a_1^2}{4\pi(\phi_2 - \phi_1)^2} \iiint\limits_{R} [(U_1)_x^2 + (U_1)_y^2 + (U_1)_z^2] dx\, dy\, dz,$$

or, on applying Green's theorem (30) of 2-14 plus the fact that $\nabla^2 U_1 = 0$,

$$C_0 \geqq \frac{a_1^2}{4\pi(\phi_2 - \phi_1)^2} \iint\limits_{B} U_1 \frac{\partial U_1}{\partial n}\, dS = \frac{a_1^2}{4\pi(\phi_2 - \phi_1)^2} (L_1 + L_2), \quad (53)$$

according to the second definition of (46). With (50) and (52) the inequality (53)—our final result—reads

$$C_0 \geqq \frac{2}{F(\alpha)} \left(\frac{u_1 u_2}{u_2 - u_1} \right), \tag{54}$$

where $F(\alpha)$ is given by (51).

Combining (54) with the result (26) of 12-1(d) for the condenser under consideration, we obtain the double inequality

$$\frac{2 + \alpha^2}{3\alpha} \left(\frac{u_1 u_2}{u_2 - u_1} \right) \geqq C_0 \geqq \frac{2}{F(\alpha)} \left(\frac{u_1 u_2}{u_2 - u_1} \right). \tag{55}$$

For $\alpha = 1$ the upper and lower bounds are equal, according to (51), to the precise expression for the concentric spherical condenser.

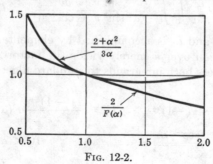

Fig. 12-2.

Figure 12-2 exhibits the behavior of the numerical coefficients of $[u_1 u_2/(u_2 - u_1)]$ in the upper and lower approximations which appear in (55), for the range $0.5 \leqq \alpha \leqq 2.0$.

12-3. Remarks on Problems in Two Dimensions

(a) An important class of problems in electrostatics is characterized by the fact that the potential function ϕ is independent of one of the cartesian coordinates—which, for the sake of definiteness, we designate as z. In geometric terms the conducting surfaces which bound the electrostatic field are cylindrical with generators parallel to the z axis. Thus every plane $z = $ constant contains the same cross-sectional configuration of these surfaces, so that we may confine our attention to a single such plane—say $z = 0$. Thus, in describing the conducting surfaces, we speak of them as curves—their cross sections—in the xy plane. A right-circular cylinder of radius a whose axis is the z axis, for example, is thus spoken of as the *circle* $x^2 + y^2 = a^2$ in the xy plane. The region R bounded by two cylindrical surfaces thus becomes the domain D in the xy plane bounded by two curves—the cross sections of the surfaces; etc.

The physical realization of a situation in which ϕ is independent of z is at best approximate; the mathematical results achieved are applicable to cases in which the fields are bounded by cylindrical surfaces cut off by parallel planes separated by distances which are great compared with the cross-sectional dimensions of the field, and only at points between, but distant from, these planes.

Two-dimensional problems in potential theory are most easily handled by the methods of the theory of functions of a complex variable. Nevertheless, the techniques introduced in the two preceding sections of this chapter are also of some use, particularly for the approximation of solutions which are unattainable in precise form. In this section we indicate the lines along which these techniques may be applied to two-dimensional problems. Many of the details, as well as applications, are left for the exercises at the end of this chapter.

(b) The *capacity per unit length* of a cylindrical condenser consisting of the inner curve C_1 and the outer curve C_2 which bound the domain D may be defined as the minimum of

$$c = \frac{1}{4\pi(\phi_2 - \phi_1)^2} \iint\limits_{D} (\phi_x^2 + \phi_y^2)\,dx\,dy \qquad (56)$$

with respect to continuously differentiable functions ϕ for which $\phi = \phi_1$ on C_1 and $\phi = \phi_2$ on C_2. In end-chapter exercise 1(b) it is proved that the minimizing ϕ satisfies the two-dimensional Laplace equation

$$\phi_{xx} + \phi_{yy} = 0, \qquad \text{or} \qquad \nabla^2\phi = 0. \qquad (57)$$

The techniques for approximating (from above) the minimum of (56)—and thus approximating the solution of (57)—are essentially identical with those discussed in 12-1(b,c,d) above; in general the details of application to two-dimensional problems are somewhat simpler.

Analogously to 12-2(a) we may also define the capacity per unit length of the condenser of the preceding paragraph as the *maximum* of (56) with respect to functions ϕ which satisfy (57) and the "relaxed" boundary condition

$$\int_{C_1} (\phi_1 - \phi) \frac{\partial \phi}{\partial n} \, ds + \int_{C_2} (\phi_2 - \phi) \frac{\partial \phi}{\partial n} \, ds = 0. \tag{58}$$

The proof is called for in end-chapter exercise 13(a).

(c) We may develop the *minimum* definition of the capacity per unit length of (b) above along the line of 12-1(c). With the details left for end-chapter exercise 14(a) we arrive at the result

$$\frac{1}{4\pi \int_{u_1}^{u_2} du/h(u)} \geqq c_0, \tag{59}$$

where c_0 is the actual capacity per unit length of the condenser bounded by the curves C_1 and C_2 which are members of the family of closed curves $u(x,y) = a$, with $u = u_1$ on C_1 and $u = u_2$ on C_2. The function $u(x,y)$ is such that $u(x,y) = a_1$ lies entirely within the domain bounded by $u(x,y) = a_2$ whenever $u_1 \leqq a_1 < a_2 \leqq u_2$. The function $h(u)$ is defined by the integral

$$h(u) = \int_{v_1}^{v_2} (u_x^2 + u_y^2) \left| \frac{\partial(x,y)}{\partial(u,v)} \right| \, dv, \tag{60}$$

where v is a continuously differentiable function of position in D which varies monotonically from v_1 to v_2 as any curve $u = $ constant is traversed exactly once ($v_1 < v_2$).

12-4. The Existence of Minima of the Dirichlet Integral

In the foregoing sections of this chapter we have occasion to consider the so-called Dirichlet integral

$$I = \iiint_R (\phi_x^2 + \phi_y^2 + \phi_z^2) dx \, dy \, dz, \tag{61}$$

whose minima with respect to certain classes of functions ϕ characterize the solutions of problems in free-space electrostatics. It is assumed throughout that the required minima actually exist. Similarly, much of the work of the preceding chapters is intimately connected with the tacitly assumed existence of certain minima of integrals very closely related to (61), in one, two, and three dimensions. Although it would carry us far beyond the scope of our study to go into the question of the existence of the stated minima with any degree of thoroughness, certain facts concerning this question should be recognized.

Clearly, the integral (61) possesses a lower bound: I can never be negative. It therefore seems plausible to suppose that there exists one from among the functions eligible for the minimization which bestows upon I a value less than (or equal to) any value which any other eligible function gives to I. The plausibility of this supposition is strengthened by the well-known theorem for continuous functions: In a closed region a continuous function actually assumes its minimum value for some set of values of the independent variables. A further argument supporting the existence of a minimum of I with respect to a given class of functions ϕ rests upon physical considerations: "Since the solution of the corresponding physical problem must exist, so also must the solution of the minimum problem."

The foregoing line of argument, widely referred to as Dirichlet's principle, dominated mathematical thinking around the middle of the nineteenth century and was actually responsible for a large body of significant discoveries. With his enormous critical faculty brought to bear upon the question, however, Weierstrass found Dirichlet's principle unreliable and in 1870 produced an example which conclusively demonstrated the principle to be false *in the form in which it was understood at that date.* As a natural result of this discovery, reliance upon Dirichlet's principle was abandoned, and many of its consequences—in particular, many important theorems on the existence of solutions of boundary-value problems related to partial differential equations—were accordingly viewed with serious doubt.

The effect of Weierstrass's negative discovery had significant positive aspects as well. With the collapse of one of their stanchest pillars mathematicians labored hard, and in large measure successfully, to provide substitute foundations for the consequences of the discredited principle of Dirichlet. But the discredit was only temporary; in 1899 Hilbert established an unassailable basis for Dirichlet's principle—*under proper conditions satisfied by the region R and by the functions ϕ admitted to eligibility for the minimization.* We do not discuss here what these conditions are;[1] what is significant to the scope of our study is the fact that their nonsatisfaction is associated with problems of distinctly "pathological" cast —problems whose main interest lies in their contradiction of Dirichlet's principle *as it was understood prior to* 1899.

EXERCISES

1. (*a*) In (2) of 12-1(*a*) write $\phi = \phi_0 + \psi$, where ψ is continuously differentiable, and

$$\nabla^2\phi_0 = 0 \text{ in } R, \qquad \psi = 0 \text{ on } B, \tag{62}$$

[1] For full discussion the reader is directed, for example, to Kellogg, Chap. 11, and Courant (1).

and so prove that $W \geqq W_0$, where W_0 is the value of W when $\psi = 0$ ($\phi = \phi_0$). Thus, since ϕ exhibits the same boundary values as does ϕ_0, we have the required proof that the function which extremizes (2)—and thus, according to (3), satisfies Laplace's equation—also renders (2) a minimum with respect to functions ϕ which assume the requisite values on B. HINT: Transform the integral which involves both ϕ_0 and ψ according to Green's theorem (30) of 2-14(c). Show that this integral vanishes because of (62).

(b) Use the method of part (a) to demonstrate the analogous two-dimensional result stated at the opening of 12-3(b).

2. (a) Use Green's theorem (30) of 2-14 to show that a function ψ which satisfies Laplace's equation in R and

$$\iint\limits_{B} \psi \, \frac{\partial \psi}{\partial n} \, dS = 0,$$

where B is the boundary surface of R, is necessarily a constant. If, further, $\psi = 0$ anywhere on B, it thus follows that $\psi = 0$ identically in R.

(b) Use part (a) to prove that $\nabla^2\phi = 0$ in R, with ϕ prescribed everywhere on B, is sufficient to determine ϕ uniquely. HINT: Assume the two solutions $\phi = \phi_1$ and $\phi = \phi_2$, so that, with $\psi = \phi_2 - \phi_1$, we have $\nabla^2\psi = 0$ in R, $\psi = 0$ on B.

3. (a) Prove that the function ϕ given in (5) actually satisfies $\nabla^2\phi = 0$. HINT: Either use $r = (x^2 + y^2 + z^2)^{\frac{1}{2}}$ with the cartesian form of Laplace's equation or, more directly, use the polar form of the laplacian found in (29) of 9-2(c).

(b) Use the definitions given at the opening of 12-1(c) and 12-1(e) to compute the capacity of the sphere considered in 12-1(b). ANSWER: b.

4. (a) Use Green's theorem (31) of 2-14 to prove that

$$\iint\limits_{B'} \frac{\partial \phi}{\partial n} \, dS = 0, \tag{63}$$

if B' is the boundary surface of any region R' in which $\nabla^2\phi = 0$ everywhere.

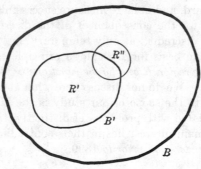

FIG. 12-3.

(b) Let $\nabla^2\phi = 0$ in R; ϕ is not identically constant in R. Use (63) to prove that there is no subregion R' of R, which has a finite (nonzero) volume less than that of R, in which ϕ is constant throughout. PROOF: Let the closed surface B', which lies entirely in R, be the boundary of the *largest* connected subregion R' in which ϕ assumes the constant value ϕ'_1. B' may coincide with part, *but not all*, of the boundary B of R since ϕ is by hypothesis not constant throughout R. It is thus possible to construct a sphere Q which lies entirely within R and whose center is at some point of B', with radius so small as to exclude from Q every point of R exterior to R' at which $\phi = \phi_1'$. Let the portion of Q exterior to R' be denoted by R''. Because ϕ is continuous, its value in R'' is either everywhere greater than ϕ_1' or everywhere less than ϕ_1'; for the sake of definiteness we suppose $\phi > \phi_1'$ in R''. Thus, as we traverse any *sufficiently short* straight line segment from any point of B' (within Q) into R'', the function ϕ always increases. We may therefore choose the radius of Q so small that $(\partial\phi/\partial n) > 0$ at every point of its surface B^* which is exterior to R'. Since $(\partial\phi/\partial n) = 0$ at every point of B^* within R', it thus follows that

$$\iint_{B^*} \frac{\partial\phi}{\partial n}\, dS > 0, \tag{64}$$

in contradiction to (63), since $\nabla^2\phi = 0$ in Q. (If $\phi < \phi_1'$ in R'', the inequality in (64) is reversed.)

(c) Use (63) to prove that ϕ, not identically constant, can assume its maximum and minimum values only on the boundary B of any region R in which $\nabla^2\phi = 0$ throughout. PROOF: Suppose that ϕ assumes its maximum at any point P of R not on B. Construct, with P as center, a spherical surface B' in R having radius so small that $(\partial\phi/\partial n) \leqq 0$ everywhere on B'. Because of part (b) we cannot have $(\partial\phi/\partial n) = 0$ identically on B' for *all* sufficiently small radius. Thus (63) is contradicted. (The argument against a minimum at P is similar.)

(d) Use part (c) to prove that, if $\nabla^2\phi = 0$ throughout R and $\phi = $ constant on the complete boundary B of R, then $\phi = $ constant throughout R. Also, use 2(b) to prove this result.

5. Use exercise 4 to prove the following sequence of facts concerning the potential ϕ in the region R bounded by the conducting surfaces of a condenser as described in the opening paragraph of 12-1(c). That is, R is the region exterior to B_1, on which $\phi = \phi_1$, and interior to B_2, on which $\phi = \phi_2$; $\phi_1 < \phi_2$, and $\nabla^2\phi = 0$ in R.

(a) If $\phi = \phi_1'$ at any interior point P_1 of R, then $\phi_1 < \phi_1' < \phi_2$. Through any such point P_1 there passes a closed surface B_1' on which ϕ has the constant value ϕ_1'. B_1' is everywhere interior to B_2, and B_1 is everywhere interior to B_1'. HINT: On every straight line segment joining B_1 to B_2 the continuous function ϕ must assume the value ϕ_1' at least once since $\phi_1 < \phi_1' < \phi_2$. Etc.

(b) For a given value of ϕ_1' there is but a single closed surface $\phi = \phi_1'$ in R ($\phi_1 < \phi_1' < \phi_2$). HINT: Use exercise 4(b,d) to show, first, that there can be no surface $\phi = $ constant in whose interior B_1 does not lie. From 4(d), also conclude that the existence of two surfaces $\phi = \phi_1'$, both of which enclose regions which contain B_1, necessitates a finite subregion of R throughout which $\phi = \phi_1'$, which is ruled out by 4(b) and $\phi_1 < \phi_2$.

(c) From parts (a) and (b) conclude that there exists a one-parameter family of equipotential (constant-ϕ) surfaces, closed in R, with the following property: The surface $\phi = \phi_1'$ lies everywhere interior to the surface $\phi = \phi_2'$ if $\phi_1' < \phi_2'$. (These surfaces cannot have a point in common since ϕ is single-valued.)

(d) Let $u = F(\phi)$, where $F'(\phi) > 0$ in R and $F(\phi_1) = u_1$, $F(\phi_2) = u_2$. Show that u satisfies the requirement laid down in 12-1(c) for the function $u(x,y,z)$: Namely, $u = u_k$ on B_k ($k = 1,2$); through each point of R there passes one and only one surface of the family $u(x,y,z) = A$ lying entirely within R, such that $u = A_1$ is everywhere interior to the region bounded by $u = A_2$ whenever $A_1 < A_2$.

Thus the existence of the required function u follows directly from the existence of the potential function ϕ; the latter we tacitly assume throughout (see 12-4). It should be kept in mind, however, that it is not always a simple matter to determine the function u in spite of the highly arbitrary character of the function $F(\phi)$; prior to solution of the problem in which the use of u is required, the potential ϕ is not available to us!

6. (a) Given the differentiable function $u = u(x,y,z)$ with x, y, z expressed in terms of the *independent* variables u, v, w through the differentiable relationships

$$x = x(u,v,w), \qquad y = y(u,v,w), \qquad z = z(u,v,w), \tag{65}$$

derive the set of equations

$$\begin{aligned}
1 &= u_x x_u + u_y y_u + u_z z_u, \\
0 &= u_x x_v + u_y y_v + u_z z_v, \\
0 &= u_x x_w + u_y y_w + u_z z_w.
\end{aligned} \tag{66}$$

(b) Use (66) to obtain the result

$$u_x^2 + u_y^2 + u_z^2 = \frac{\left[\dfrac{\partial(y,z)}{\partial(v,w)}\right]^2 + \left[\dfrac{\partial(z,x)}{\partial(v,w)}\right]^2 + \left[\dfrac{\partial(x,y)}{\partial(v,w)}\right]^2}{\left[\dfrac{\partial(x,y,z)}{\partial(u,v,w)}\right]^2}, \tag{67}$$

with the denominator assumed not to vanish. Thus, through (65), equation (67) provides the explicit expression of $(u_x^2 + u_y^2 + u_z^2)$ in terms of (u,v,w), as called for in (11) and (12).

(c) On the basis of (67) fill in the details of the following proof that the function $H(u)$ defined by (12) is independent of the particular choice of the functions v, w consistent with the requirements enunciated in conjunction with equation (9):

Let $v^* = v^*(x,y,z)$ and $w^* = w^*(x,y,z)$ be a *second* choice of the functions v and w. We thus form, according to (12),

$$\begin{aligned}
H^*(u) &= \int_{v_1^*}^{v_2^*} \int_{w_1^*}^{w_2^*} (u_x^2 + u_y^2 + u_z^2) \left|\frac{\partial(x,y,z)}{\partial(u,v^*,w^*)}\right| dw^*\, dv^* \\
&= \int_{v_1^*}^{v_2^*} \int_{w_1^*}^{w_2^*} \frac{\left[\dfrac{\partial(y,z)}{\partial(v^*,w^*)}\right]^2 + \left[\dfrac{\partial(z,x)}{\partial(v^*,w^*)}\right]^2 + \left[\dfrac{\partial(x,y)}{\partial(v^*,w^*)}\right]^2}{\left|\dfrac{\partial(x,y,z)}{\partial(u,v^*,w^*)}\right|}\, dw^*\, dv^*.
\end{aligned} \tag{68}$$

If $v = v(x,y,z)$ and $w = w(x,y,z)$ are the *original* choice of the required functions, express (68) as an integral over the variables v and w by means of the transformation whose jacobian is $[\partial(v^*,w^*)/\partial(v,w)]$. With the aid of the relations (see 2-8(f))

$$\frac{\partial(y,z)}{\partial(v^*,w^*)} \frac{\partial(v^*,w^*)}{\partial(v,w)} = \frac{\partial(y,z)}{\partial(v,w)}, \qquad \text{etc.,} \qquad \frac{\partial(u,v^*,w^*)}{\partial(u,v,w)} = \frac{\partial(v^*,w^*)}{\partial(v,w)}, \tag{69}$$

and

$$\frac{\partial(x,y,z)}{\partial(u,v^*,w^*)} \frac{\partial(u,v^*,w^*)}{\partial(u,v,w)} = \frac{\partial(x,y,z)}{\partial(u,v,w)},$$

together with (67), thus show that $H^*(u) = H(u)$ as given by (12).

(d) Prove the final relation of (69).

7. (a) Show that the integral (8) may be written

$$I = \int_{u_1}^{u_2} \iint_A [G'(u)]^2(u_x^2 + u_y^2 + u_z^2)dS \frac{du}{\left|\frac{\partial u}{\partial n}\right|} = \int_{u_1}^{u_2} [G'(u)]^2 \left(\iint_A \left|\frac{\partial u}{\partial n}\right| dS \right) du, \quad (70)$$

where the surface A is a representative surface $u = A$ in R ($u_1 \leqq A \leqq u_2$). The partial derivative $(\partial u/\partial n)$ represents the rate of change of u along the normal to the surfaces $u = $ constant. HINT: For the first form of (70) choose the volume element $dS\, dn$ and use the transformation $du = |(\partial u/\partial n)|dn$; for the second form use the fact that the magnitude $\sqrt{u_x^2 + u_y^2 + u_z^2}$ of the gradient of u (see 2-12(a)) is the rate of change of u along the normal to a surface $u = $ constant. The use of the absolute value of $(\partial u/\partial n)$ makes immaterial the choice between inward or outward normal.

(b) Using (70), show that (12) becomes

$$H(u) = \iint_A \left|\frac{\partial u}{\partial n}\right| dS. \quad (71)$$

(c) With (12) written in the form (71) the method and results of 12-1(c) become applicable to cases in which it is no simple matter to write an expression of the function u explicitly in terms of x, y, z. Thus, if it is most convenient to define u in *geometric* terms, the integral (71) may lend itself to ready evaluation while (12) may not.

For example, consider the case of the condenser formed by the concentric parallel cubes of edge a_1 and a_2 ($a_1 < a_2$). We define the family of surfaces $u = $ constant as the aggregate of cubes concentric with, and parallel to, the boundary cubes and having edge lengths between a_1 and a_2. On any given member of this family, having edge a, we assign the constant value $u = \frac{1}{2}a$.

Show that, with this geometric assignment of u, we have $|(\partial u/\partial n)| = 1$, so that evaluation of (71) gives $H(u) = 6a^2 = 24u^2$. Hence, with the aid of (16), show that the capacity C_0 of our cubical condenser satisfies the inequality

$$C_0 \leqq \frac{6u_1u_2}{\pi(u_2 - u_1)} = \frac{3a_1a_2}{\pi(a_2 - a_1)} = 0.95 \frac{a_1a_2}{u_2 - u_1}. \quad (72)$$

(d) Use the above method to obtain an upper bound for the capacity C_0' of the *single* conductor consisting of the cube of side a_1, but employ for the surfaces $u = $ constant the parallel surfaces of the cube. (A given "parallel surface" B' of the cubical surface B_1 is a closed surface exterior to B_1 such that the distance from any point of B' to B_1 along an internal normal to B' is the same at all points of B'. Thus B' consists of six squares, twelve quarter-circle cylinders, and eight octants of a sphere; these twenty-six parts are joined smoothly. It is clear that any line normal to one of the parallel surfaces of a given cube is normal to all.) HINT: If we associate with a given parallel surface a value of u equal to the perpendicular distance from one of its plane portions to the center of the given cube, the constant normal distance to B_1 is $(u - \frac{1}{2}a_1)$ and $|(\partial u/\partial n)| = 1$, so that $H(u)$, according to (71), is the area of the parallel surface:

$$H(u) = 6a_1^2 + 6\pi a_1(u - \tfrac{1}{2}a_1) + 4\pi(u - \tfrac{1}{2}a_1)^2.$$

ANSWER: $C_0' \leqq 0.7105a_1$. Compare with (72), with $a_2 \to \infty$.

8. Given that the equipotential surfaces in a condenser consisting of two concentric spheres are themselves spheres concentric with the conductors, derive the precise expression for the potential according to the method outlined in the antepenultimate paragraph of 12-1(c).

9. Show that the transformation (19) of 12-1(d) is such that each point of the region R between the concentric ellipsoids of revolution $u = u_1$ and $u = u_2$ is associated with one and only one triple of values (u,v,w), with $u_1 \leq u \leq u_2$, $0 \leq v \leq \pi$, $0 \leq w < 2\pi$.

10. The precise capacity C_0' of the single conducting prolate spheroidal surface given by (28), with $b < a$, is given by (see Abraham and Becker (referred to in 12-1(a)), p. 64)

$$C_0' = \frac{\sqrt{a^2 - b^2}}{\log\left[(a + \sqrt{a^2 - b^2})/b\right]}. \tag{73}$$

Expand the reciprocal of C_0' as a power series in $[1 - (b/a)^2]$, and compare the result with the similar expansion of $(1/C')$ as given by (29). Thus prove directly that $C' \geq C_0'$, and show that the accuracy of approximation of C_0' by C' is the better the smaller the value of $[1 - (b/a)^2]$.

11. (a) Show that the inequality (37) of 12-2(a) may be rewritten in the form

$$4\pi(\phi_2 - \phi_1)^2 C_0 \geq \iint_B \psi \frac{\partial \psi}{\partial n} dS = \iint_B \phi \frac{\partial \psi}{\partial n} dS, \tag{74}$$

with the aid of Green's theorem (30) of 2-14 and (34). HINT: Use the fact that $\nabla^2 \psi = 0$ in R.

(b) Use Green's theorem (31) of 2-14 to prove that

$$\iint_B \frac{\partial \psi}{\partial n} dS = \iint_{B_1} \frac{\partial \psi}{\partial n} dS + \iint_{B_2} \frac{\partial \psi}{\partial n} dS = 0 \tag{75}$$

if $\nabla^2 \psi = 0$ in the region R bounded by the condenser surfaces B_1 and B_2.

(c) Since $\phi = \phi_1$ on B_1 and $\phi = \phi_2$ on B_2 ($\phi_2 > \phi_1$), rewrite (74) as

$$4\pi(\phi_2 - \phi_1)C_0 \geq \iint_{B_2} \frac{\partial \psi}{\partial n} dS, \tag{76}$$

with the aid of (75).

(d) In case we let $\psi = a_1 U_1$ ($\nabla^2 U_1 = 0$), use (42) and (75) to rewrite (76) in the form

$$C_0 \geq \frac{J_2^2}{4\pi(L_1 + L_2)}, \tag{77}$$

where

$$J_2 = \iint_{B_2} \frac{\partial U_1}{\partial n} dS, \qquad L_k = \iint_{B_k} U_1 \frac{\partial U_1}{\partial n} dS \qquad (k = 1,2). \tag{78}$$

12. (a) Use (77) and (78) of exercise 11 to obtain a lower bound for the capacity C_0 of the cubical condenser described in exercise 7(c). HINT: Place the origin of coordinates at the center of the condenser, with axes parallel to the cube edges, and make the choice of U_1 given by (44). By symmetry argument thus show that any integral of (78) is equal to six times the integral carried out over one face of the cube. On the face $x = \frac{1}{2}a_1$ of B_1, $(\partial U_1/\partial n) = -(\partial U_1/\partial x)$; on $x = \frac{1}{2}a_2$ of B_2, $(\partial U_1/\partial n) = (\partial U_1/\partial x)$. Thus evaluate

$$J_2 = -12a_2 \int_0^{\frac{1}{2}a_2} \int_0^{\frac{1}{2}a_2} \frac{dy\,dz}{(\frac{1}{4}a_2^2 + y^2 + z^2)^{\frac{3}{2}}} = -4\pi, \tag{79}$$

$$L_k = 12a_k(-1)^{k+1} \int_0^{\frac{1}{2}a_k} \int_0^{\frac{1}{2}a_k} \frac{dy\,dz}{(\frac{1}{4}a_k^2 + y^2 + z^2)^2} = \frac{24\sqrt{2}}{a_k}(-1)^{k+1}\cot^{-1}\sqrt{2}$$
$$(k = 1,2).$$

ANSWER: $C_0 \geq \dfrac{\pi}{6\sqrt{2}\cot^{-1}\sqrt{2}}\dfrac{a_1 a_2}{a_2 - a_1} = 0.60\dfrac{a_1 a_2}{a_2 - a_1}$; compare with the upper-bound result (72) and—for $a_2 \to \infty$—with the upper bound for the single cube obtained in exercise 7(d).

(b) Prove that it is not mere coincidence that the result (79) is identical with the value of J_2 found in (50). HINT: Note (75).

13. (a) Prove that the capacity per unit length c_0 of a cylindrical condenser, as described in 12-3(a), is given by the maximum of (56) with respect to functions ϕ which satisfy both (57) and (58). HINT: Compare 12-2(a).

(b) Develop the principle embodied in part (a) along the lines of 12-2(a) and exercise 11 above to derive the result

$$c_0 \geq \frac{J_2^2}{4\pi(L_1 + L_2)}, \tag{80}$$

where here

$$J_2 = \int_{C_2} \frac{\partial U}{\partial n}\,ds, \qquad L_k = \int_{C_k} U\frac{\partial U}{\partial n}\,ds \qquad (k = 1,2), \tag{81}$$

$$U = U(x,y), \qquad \nabla^2 U = 0. \tag{82}$$

(c) Use (80) to obtain a lower bound to the capacity per unit length c_0 of a cylindrical condenser whose trace in the xy plane consists of concentric parallel squares of sides a_1 and a_2, respectively ($a_1 < a_2$). For the function U satisfying (82) use $\log \sqrt{x^2 + y^2}$, where the origin of coordinates is the common center of the squares. HINT: Do not try to evaluate L_1 or L_2 individually as given in (81); the required sum $(L_1 + L_2)$, however, is readily evaluable. (Use polar coordinates for the integration.) ANSWER:

$$c_0 \geq \frac{1}{2\log(a_2/a_1)}. \tag{83}$$

(d) Show that result (83) applies also to a cylindrical condenser whose trace in the xy plane consists of the pair of concentric coaxial similar ellipses

$$\frac{x^2}{a_1^2} + \frac{y^2}{\alpha^2 a_1^2} = 1, \qquad \frac{x^2}{a_2^2} + \frac{y^2}{\alpha^2 a_2^2} = 1 \qquad (a_2 > a_1 > 0). \tag{84}$$

(e) Prove that the result (83) applies to any cylindrical condenser whose trace in the xy plane is described, in plane polar coordinates, by the two equations $r = a_1 g(\theta)$ and $r = a_2 g(\theta)$, with $a_2 > a_1 > 0$. Show that the equality sign holds in (83) if $g(\theta)$ is a constant.

14. (a) Prove the inequality (59), where $h(u)$ is given by (60). HINT: Compare 12-1(c).

(b) Show that (60) may be written also in the form

$$h(u) = \int_C \left| \frac{\partial u}{\partial n} \right| ds, \tag{85}$$

where C is any curve $u = $ constant as described in 12-3(c). HINT: Compare exercise 7(a,b) above.

(c) Employ (59) together with (60) to obtain an upper bound for the capacity per unit length c_0 of the cylindrical condenser described by (84). ANSWER:

$$c_0 \leqq \frac{(\alpha^2 + 1)}{4\alpha \log (a_2/a_1)}.$$

Compare with the lower bound given by (83).

(d) Employ (59) together with (85) to obtain an upper bound for the capacity per unit length c_0 for the concentric-square cylindrical condenser described in exercise 13(c) above. HINT: Compare the treatment given the cubical condenser in exercise 7(c). ANSWER: $c_0 \leqq 2/[\pi \log (a_2/a_1)]$. Compare with the lower bound given by (83).

BIBLIOGRAPHY

BLISS, GILBERT A., (1) "Calculus of Variations," Mathematical Association of America, 1944.

———, (2) "Lectures on the Calculus of Variations," University of Chicago Press, Chicago, 1947.

BOLZA, OSKAR, "Lectures on the Calculus of Variations," Hafner Pub. Co., New York, 1946. Dover reprint.

COURANT, R., (1) "Dirichlet's Principle," Interscience Publishers, Inc., New York, 1950.

———, (2) Über die Eigenwerte bei den Differentialgleichungen der mathematischen Physik, *Mathematische Zeitschrift*, Bd. 7, pp. 1–57, 1920.

——— and D. HILBERT, "Methoden der mathematischen Physik," Vol. I, Chaps. 4,6, Julius Springer, Berlin, 1931.

FOX, CHARLES, "An Introduction to the Calculus of Variations," Oxford University Press, New York, 1950.

FRANKLIN, PHILIP, "Methods of Advanced Calculus," McGraw-Hill Book Company, Inc., New York, 1944.

GOURSAT, ÉDOUARD, "Mathematical Analysis" (translated by E. R. Hedrick), Vol. I, Ginn & Company, Boston, 1904. Dover reprint.

INCE, E. L., "Ordinary Differential Equations," Dover Publications, New York, 1944.

JACKSON, DUNHAM, "Fourier Series and Orthogonal Polynomials," Mathematical Association of America, Oberlin, 1941.

KELLOGG, OLIVER D., "Foundations of Potential Theory," Frederick Ungar Publishing Co., New York, 1940. Dover reprint.

LOVE, A. E. H., "Mathematical Theory of Elasticity," Dover Publications, New York, 1944.

PÓLYA, G. and G. SZEGÖ, "Isoperimetric Inequalities in Mathematical Physics," Princeton University Press, Princeton, 1951.

RAYLEIGH, LORD J. W. S., "Theory of Sound," Dover Publications, New York, 1945.

SCHIFF, L. I., "Quantum Mechanics," McGraw-Hill Book Company, Inc., New York, 1949.

SOKOLNIKOFF, IVAN S., "Mathematical Theory of Elasticity," McGraw-Hill Book Company, Inc., New York, 1946.

WHITTAKER, E. T., "Analytical Dynamics," Dover Publications, New York, 1944.

INDEX

A CATALOGUE OF SELECTED DOVER BOOKS
IN ALL FIELDS OF INTEREST

A CATALOGUE OF SELECTED DOVER BOOKS
IN ALL FIELDS OF INTEREST

THE NOTEBOOKS OF LEONARDO DA VINCI, edited by J.P. Richter. Extracts from manuscripts reveal great genius; on painting, sculpture, anatomy, sciences, geography, etc. Both Italian and English. 186 ms. pages reproduced, plus 500 additional drawings, including studies for Last Supper, Sforza monument, etc. 860pp. 7⅞ x 10¾.   USO 22572-0, 22573-9 Pa., Two vol. set $12.00

ART NOUVEAU DESIGNS IN COLOR, Alphonse Mucha, Maurice Verneuil, Georges Auriol. Full-color reproduction of Combinaisons ornementales (c. 1900) by Art Nouveau masters. Floral, animal, geometric, interlacings, swashes — borders, frames, spots — all incredibly beautiful. 60 plates, hundreds of designs. 9⅜ x 8¹/₁₆.   22885-1 Pa. $4.00

GRAPHIC WORKS OF ODILON REDON. All great fantastic lithographs, etchings, engravings, drawings, 209 in all. Monsters, Huysmans, still life work, etc. Introduction by Alfred Werner. 209pp. 9⅛ x 12¼.   21996-8 Pa. $6.00

EXOTIC FLORAL PATTERNS IN COLOR, E.-A. Seguy. Incredibly beautiful full-color pochoir work by great French designer of 20's. Complete Bouquets et frondaisons, Suggestions pour étoffes. Richness must be seen to be believed. 40 plates containing 120 patterns. 80pp. 9⅜ x 12¼.   23041-4 Pa. $6.00

SELECTED ETCHINGS OF JAMES A. McN. WHISTLER, James A. McN. Whistler. 149 outstanding etchings by the great American artist, including selections from the Thames set and two Venice sets, the complete French set, and many individual prints. Introduction and explanatory note on each print by Maria Naylor. 157pp. 9⅜ x 12¼.   23194-1 Pa. $5.00

VISUAL ILLUSIONS: THEIR CAUSES, CHARACTERISTICS, AND APPLICATIONS, Matthew Luckiesh. Thorough description, discussion; shape and size, color, motion; natural illusion. Uses in art and industry. 100 illustrations. 252pp.
                                                        21530-X Pa. $2.50

TEN BOOKS ON ARCHITECTURE, Vitruvius. The most important book ever written on architecture. Early Roman aesthetics, technology, classical orders, site selection, all other aspects. Stands behind everything since. Morgan translation. 331pp.
                                                        20645-9 Pa. $3.50

THE CODEX NUTTALL, A PICTURE MANUSCRIPT FROM ANCIENT MEXICO, as first edited by Zelia Nuttall. Only inexpensive edition, in full color, of a pre-Columbian Mexican (Mixtec) book. 88 color plates show kings, gods, heroes, temples, sacrifices. New explanatory, historical introduction by Arthur G. Miller. 96pp. 11³/₈ x 8½.   23168-2 Pa. $7.50

EAST O' THE SUN AND WEST O' THE MOON, George W. Dasent. Considered the best of all translations of these Norwegian folk tales, this collection has been enjoyed by generations of children (and folklorists too). Includes True and Untrue, Why the Sea is Salt, East O' the Sun and West O' the Moon, Why the Bear is Stumpy-Tailed, Boots and the Troll, The Cock and the Hen, Rich Peter the Pedlar, and 52 more. The only edition with all 59 tales. 77 illustrations by Erik Werenskiold and Theodor Kittelsen. xv + 418pp. 22521-6 Paperbound $4.00

GOOPS AND HOW TO BE THEM, Gelett Burgess. Classic of tongue-in-cheek humor, masquerading as etiquette book. 87 verses, twice as many cartoons, show mischievous Goops as they demonstrate to children virtues of table manners, neatness, courtesy, etc. Favorite for generations. viii + 88pp. 6½ x 9¼.
22233-0 Paperbound $2.00

ALICE'S ADVENTURES UNDER GROUND, Lewis Carroll. The first version, quite different from the final *Alice in Wonderland*, printed out by Carroll himself with his own illustrations. Complete facsimile of the "million dollar" manuscript Carroll gave to Alice Liddell in 1864. Introduction by Martin Gardner. viii + 96pp. Title and dedication pages in color. 21482-6 Paperbound $1.50

THE BROWNIES, THEIR BOOK, Palmer Cox. Small as mice, cunning as foxes, exuberant and full of mischief, the Brownies go to the zoo, toy shop, seashore, circus, etc., in 24 verse adventures and 266 illustrations. Long a favorite, since their first appearance in St. Nicholas Magazine. xi + 144pp. 6⅝ x 9¼.
21265-3 Paperbound $2.50

SONGS OF CHILDHOOD, Walter De La Mare. Published (under the pseudonym Walter Ramal) when De La Mare was only 29, this charming collection has long been a favorite children's book. A facsimile of the first edition in paper, the 47 poems capture the simplicity of the nursery rhyme and the ballad, including such lyrics as I Met Eve, Tartary, The Silver Penny. vii + 106pp. (USO) 21972-0 Paperbound $2.00

THE COMPLETE NONSENSE OF EDWARD LEAR, Edward Lear. The finest 19th-century humorist-cartoonist in full: all nonsense limericks, zany alphabets, Owl and Pussycat, songs, nonsense botany, and more than 500 illustrations by Lear himself. Edited by Holbrook Jackson. xxix + 287pp. (USO) 20167-8 Paperbound $3.00

BILLY WHISKERS: THE AUTOBIOGRAPHY OF A GOAT, Frances Trego Montgomery. A favorite of children since the early 20th century, here are the escapades of that rambunctious, irresistible and mischievous goat—Billy Whiskers. Much in the spirit of *Peck's Bad Boy*, this is a book that children never tire of reading or hearing. All the original familiar illustrations by W. H. Fry are included: 6 color plates, 18 black and white drawings. 159pp. 22345-0 Paperbound $2.75

MOTHER GOOSE MELODIES. Faithful republication of the fabulously rare Munroe and Francis "copyright 1833" Boston edition—the most important Mother Goose collection, usually referred to as the "original." Familiar rhymes plus many rare ones, with wonderful old woodcut illustrations. Edited by E. F. Bleiler. 128pp. 4½ x 6⅜. 22577-1 Paperbound $1.50

MANUAL OF THE TREES OF NORTH AMERICA, Charles S. Sargent. The basic survey of every native tree and tree-like shrub, 717 species in all. Extremely full descriptions, information on habitat, growth, locales, economics, etc. Necessary to every serious tree lover. Over 100 finding keys. 783 illustrations. Total of 986pp.
20277-1, 20278-X Pa., Two vol. set $9.00

BIRDS OF THE NEW YORK AREA, John Bull. Indispensable guide to more than 400 species within a hundred-mile radius of Manhattan. Information on range, status, breeding, migration, distribution trends, etc. Foreword by Roger Tory Peterson. 17 drawings; maps. 540pp.
23222-0 Pa. $6.00

THE SEA-BEACH AT EBB-TIDE, Augusta Foote Arnold. Identify hundreds of marine plants and animals: algae, seaweeds, squids, crabs, corals, etc. Descriptions cover food, life cycle, size, shape, habitat. Over 600 drawings. 490pp.
21949-6 Pa. $5.00

THE MOTH BOOK, William J. Holland. Identify more than 2,000 moths of North America. General information, precise species descriptions. 623 illustrations plus 48 color plates show almost all species, full size. 1968 edition. Still the basic book. Total of 551pp. 6½ x 9¼.
21948-8 Pa. $6.00

AN INTRODUCTION TO THE REPTILES AND AMPHIBIANS OF THE UNITED STATES, Percy A. Morris. All lizards, crocodiles, turtles, snakes, toads, frogs; life history, identification, habits, suitability as pets, etc. Non-technical, but sound and broad. 130 photos. 253pp.
22982-3 Pa. $3.00

OLD NEW YORK IN EARLY PHOTOGRAPHS, edited by Mary Black. Your only chance to see New York City as it was 1853-1906, through 196 wonderful photographs from N.Y. Historical Society. Great Blizzard, Lincoln's funeral procession, great buildings. 228pp. 9 x 12.
22907-6 Pa. $6.00

THE AMERICAN REVOLUTION, A PICTURE SOURCEBOOK, John Grafton. Wonderful Bicentennial picture source, with 411 illustrations (contemporary and 19th century) showing battles, personalities, maps, events, flags, posters, soldier's life, ships, etc. all captioned and explained. A wonderful browsing book, supplement to other historical reading. 160pp. 9 x 12.
23226-3 Pa. $4.00

PERSONAL NARRATIVE OF A PILGRIMAGE TO AL-MADINAH AND MECCAH, Richard Burton. Great travel classic by remarkably colorful personality. Burton, disguised as a Moroccan, visited sacred shrines of Islam, narrowly escaping death. Wonderful observations of Islamic life, customs, personalities. 47 illustrations. Total of 959pp.
21217-3, 21218-1 Pa., Two vol. set $10.00

INCIDENTS OF TRAVEL IN CENTRAL AMERICA, CHIAPAS, AND YUCATAN, John L. Stephens. Almost single-handed discovery of Maya culture; exploration of ruined cities, monuments, temples; customs of Indians. 115 drawings. 892pp.
22404-X, 22405-8 Pa., Two vol. set $8.00

DECORATIVE ALPHABETS AND INITIALS, edited by Alexander Nesbitt. 91 complete alphabets (medieval to modern), 3924 decorative initials, including Victorian novelty and Art Nouveau. 192pp. 7¾ x 10¾. 20544-4 Pa. $4.00

CALLIGRAPHY, Arthur Baker. Over 100 original alphabets from the hand of our greatest living calligrapher: simple, bold, fine-line, richly ornamented, etc. —all strikingly original and different, a fusion of many influences and styles. 155pp. 11³/8 x 8¼. 22895-9 Pa. $4.50

MONOGRAMS AND ALPHABETIC DEVICES, edited by Hayward and Blanche Cirker. Over 2500 combinations, names, crests in very varied styles: script engraving, ornate Victorian, simple Roman, and many others. 226pp. 8⅛ x 11. 22330-2 Pa. $5.00

THE BOOK OF SIGNS, Rudolf Koch. Famed German type designer renders 493 symbols: religious, alchemical, imperial, runes, property marks, etc. Timeless. 104pp. 6⅛ x 9¼. 20162-7 Pa. $1.75

200 DECORATIVE TITLE PAGES, edited by Alexander Nesbitt. 1478 to late 1920's. Baskerville, Dürer, Beardsley, W. Morris, Pyle, many others in most varied techniques. For posters, programs, other uses. 222pp. 8⅜ x 11¼. 21264-5 Pa. $5.00

DICTIONARY OF AMERICAN PORTRAITS, edited by Hayward and Blanche Cirker. 4000 important Americans, earliest times to 1905, mostly in clear line. Politicians, writers, soldiers, scientists, inventors, industrialists, Indians, Blacks, women, outlaws, etc. Identificatory information. 756pp. 9¼ x 12¾. 21823-6 Clothbd. $30.00

ART FORMS IN NATURE, Ernst Haeckel. Multitude of strangely beautiful natural forms: Radiolaria, Foraminifera, jellyfishes, fungi, turtles, bats, etc. All 100 plates of the 19th century evolutionist's Kunstformen der Natur (1904). 100pp. 9⅜ x 12¼. 22987-4 Pa. $4.00

DECOUPAGE: THE BIG PICTURE SOURCEBOOK, Eleanor Rawlings. Make hundreds of beautiful objects, over 550 florals, animals, letters, shells, period costumes, frames, etc. selected by foremost practitioner. Printed on one side of page. 8 color plates. Instructions. 176pp. 9³/16 x 12¼. 23182-8 Pa. $5.00

AMERICAN FOLK DECORATION, Jean Lipman, Eve Meulendyke. Thorough coverage of all aspects of wood, tin, leather, paper, cloth decoration — scapes, humans, trees, flowers, geometrics — and how to make them. Full instructions. 233 illustrations, 5 in color. 163pp. 8⅜ x 11¼. 22217-9 Pa. $3.95

WHITTLING AND WOODCARVING, E.J. Tangerman. Best book on market; clear, full. If you can cut a potato, you can carve toys, puzzles, chains, caricatures, masks, patterns, frames, decorate surfaces, etc. Also covers serious wood sculpture. Over 200 photos. 293pp. 20965-2 Pa. $3.00

JEWISH GREETING CARDS, Ed Sibbett, Jr. 16 cards to cut and color. Three say "Happy Chanukah," one "Happy New Year," others have no message, show stars of David, Torahs, wine cups, other traditional themes. 16 envelopes. 8¼ x 11.
23225-5 Pa. $2.00

AUBREY BEARDSLEY GREETING CARD BOOK, Aubrey Beardsley. Edited by Theodore Menten. 16 elegant yet inexpensive greeting cards let you combine your own sentiments with subtle Art Nouveau lines. 16 different Aubrey Beardsley designs that you can color or not, as you wish. 16 envelopes. 64pp. 8¼ x 11.
23173-9 Pa. $2.00

RECREATIONS IN THE THEORY OF NUMBERS, Albert Beiler. Number theory, an inexhaustible source of puzzles, recreations, for beginners and advanced. Divisors, perfect numbers. scales of notation, etc. 349pp.
21096-0 Pa. $4.00

AMUSEMENTS IN MATHEMATICS, Henry E. Dudeney. One of largest puzzle collections, based on algebra, arithmetic, permutations, probability, plane figure dissection, properties of numbers, by one of world's foremost puzzlists. Solutions. 450 illustrations. 258pp.
20473-1 Pa. $3.00

MATHEMATICS, MAGIC AND MYSTERY, Martin Gardner. Puzzle editor for Scientific American explains math behind: card tricks, stage mind reading, coin and match tricks, counting out games, geometric dissections. Probability, sets, theory of numbers, clearly explained. Plus more than 400 tricks, guaranteed to work. 135 illustrations. 176pp.
20335-2 Pa. $2.00

BEST MATHEMATICAL PUZZLES OF SAM LOYD, edited by Martin Gardner. Bizarre, original, whimsical puzzles by America's greatest puzzler. From fabulously rare Cyclopedia, including famous 14-15 puzzles, the Horse of a Different Color, 115 more. Elementary math. 150 illustrations. 167pp.
20498-7 Pa. $2.50

MATHEMATICAL PUZZLES FOR BEGINNERS AND ENTHUSIASTS, Geoffrey Mott-Smith. 189 puzzles from easy to difficult involving arithmetic, logic, algebra, properties of digits, probability. Explanation of math behind puzzles. 135 illustrations. 248pp.
20198-8 Pa. $2.75

BIG BOOK OF MAZES AND LABYRINTHS, Walter Shepherd. Classical, solid, and ripple mazes; short path and avoidance labyrinths; more — 50 mazes and labyrinths in all. 12 other figures. Full solutions. 112pp. 8⅛ x 11.
22951-3 Pa. $2.00

COIN GAMES AND PUZZLES, Maxey Brooke. 60 puzzles, games and stunts — from Japan, Korea, Africa and the ancient world, by Dudeney and the other great puzzlers, as well as Maxey Brooke's own creations. Full solutions. 67 illustrations. 94pp.
22893-2 Pa. $1.50

HAND SHADOWS TO BE THROWN UPON THE WALL, Henry Bursill. Wonderful Victorian novelty tells how to make flying birds, dog, goose, deer, and 14 others. 32pp. 6½ x 9¼.
21779-5 Pa. $1.25

MODERN CHESS STRATEGY, Ludek Pachman. The use of the queen, the active king, exchanges, pawn play, the center, weak squares, etc. Section on rook alone worth price of the book. Stress on the moderns. Often considered the most important book on strategy. 314pp. 20290-9 Pa. $3.50

CHESS STRATEGY, Edward Lasker. One of half-dozen great theoretical works in chess, shows principles of action above and beyond moves. Acclaimed by Capablanca, Keres, etc. 282pp. USO 20528-2 Pa. $3.00

CHESS PRAXIS, THE PRAXIS OF MY SYSTEM, Aron Nimzovich. Founder of hypermodern chess explains his profound, influential theories that have dominated much of 20th century chess. 109 illustrative games. 369pp. 20296-8 Pa. $3.50

HOW TO PLAY THE CHESS OPENINGS, Eugene Znosko-Borovsky. Clear, profound examinations of just what each opening is intended to do and how opponent can counter. Many sample games, questions and answers. 147pp. 22795-2 Pa. $2.00

THE ART OF CHESS COMBINATION, Eugene Znosko Borovsky. Modern explanation of principles, varieties, techniques and ideas behind them, illustrated with many examples from great players. 212pp. 20583-5 Pa. $2.50

COMBINATIONS: THE HEART OF CHESS, Irving Chernev. Step-by-step explanation of intricacies of combinative play. 356 combinations by Tarrasch, Botvinnik, Keres, Steinitz, Anderssen, Morphy, Marshall, Capablanca, others, all annotated. 245 pp. 21744-2 Pa. $3.00

HOW TO PLAY CHESS ENDINGS, Eugene Znosko-Borovsky. Thorough instruction manual by fine teacher analyzes each piece individually; many common endgame situations. Examines games by Steinitz, Alekhine, Lasker, others. Emphasis on understanding. 288pp. 21170-3 Pa. $2.75

MORPHY'S GAMES OF CHESS, Philip W. Sergeant. Romantic history, 54 games of greatest player of all time against Anderssen, Bird, Paulsen, Harrwitz; 52 games at odds; 52 blindfold, 100 consultation, informal, other games. Analyses by Anderssen, Steinitz, Morphy himself. 352pp 20386-7 Pa. $4.00

500 MASTER GAMES OF CHESS, S. Tartakower, J. du Mont. Vast collection of great chess games from 1798-1938, with much material nowhere else readily available. Fully annotated, arranged by opening for easier study. 665pp. 23208-5 Pa. $6.00

THE SOVIET SCHOOL OF CHESS, Alexander Kotov and M. Yudovich. Authoritative work on modern Russian chess. History, conceptual background. 128 fully annotated games (most unavailable elsewhere) by Botvinnik, Keres, Smyslov, Tal, Petrosian, Spassky, more. 390pp. 20026-4 Pa. $3.95

WONDERS AND CURIOSITIES OF CHESS, Irving Chernev. A lifetime's accumulation of such wonders and curiosities as the longest won game, shortest game, chess problem with mate in 1220 moves, and much more unusual material —356 items in all, over 160 complete games. 146 diagrams. 203pp. 23007-4 Pa. $3.50

VISUAL ILLUSIONS: THEIR CAUSES, CHARACTERISTICS, AND APPLICATIONS, Matthew Luckiesh. Thorough description and discussion of optical illusion, geometric and perspective, particularly; size and shape distortions, illusions of color, of motion; natural illusions; use of illusion in art and magic, industry, etc. Most useful today with op art, also for classical art. Scores of effects illustrated. Introduction by William H. Ittleson. 100 illustrations. xxi + 252pp.

21530-X Paperbound $2.50

A HANDBOOK OF ANATOMY FOR ART STUDENTS, Arthur Thomson. Thorough, virtually exhaustive coverage of skeletal structure, musculature, etc. Full text, supplemented by anatomical diagrams and drawings and by photographs of undraped figures. Unique in its comparison of male and female forms, pointing out differences of contour, texture, form. 211 figures, 40 drawings, 86 photographs. xx + 459pp. 5⅜ x 8⅜.

21163-0 Paperbound $5.00

150 MASTERPIECES OF DRAWING, Selected by Anthony Toney. Full page reproductions of drawings from the early 16th to the end of the 18th century, all beautifully reproduced: Rembrandt, Michelangelo, Dürer, Fragonard, Urs, Graf, Wouwerman, many others. First-rate browsing book, model book for artists. xviii + 150pp. 8⅜ x 11¼.

21032-4 Paperbound $4.00

THE LATER WORK OF AUBREY BEARDSLEY, Aubrey Beardsley. Exotic, erotic, ironic masterpieces in full maturity: Comedy Ballet, Venus and Tannhauser, Pierrot, Lysistrata, Rape of the Lock, Savoy material, Ali Baba, Volpone, etc. This material revolutionized the art world, and is still powerful, fresh, brilliant. With *The Early Work,* all Beardsley's finest work. 174 plates, 2 in color. xiv + 176pp. 8⅛ x 11.

21817-1 Paperbound $4.00

DRAWINGS OF REMBRANDT, Rembrandt van Rijn. Complete reproduction of fabulously rare edition by Lippmann and Hofstede de Groot, completely reedited, updated, improved by Prof. Seymour Slive, Fogg Museum. Portraits, Biblical sketches, landscapes, Oriental types, nudes, episodes from classical mythology—All Rembrandt's fertile genius. Also selection of drawings by his pupils and followers. "Stunning volumes," *Saturday Review.* 550 illustrations. lxxviii + 552pp. 9⅛ x 12¼.

21485-0, 21486-9 Two volumes, Paperbound $12.00

THE DISASTERS OF WAR, Francisco Goya. One of the masterpieces of Western civilization—83 etchings that record Goya's shattering, bitter reaction to the Napoleonic war that swept through Spain after the insurrection of 1808 and to war in general. Reprint of the first edition, with three additional plates from Boston's Museum of Fine Arts. All plates facsimile size. Introduction by Philip Hofer, Fogg Museum. v + 97pp. 9⅜ x 8¼.

21872-4 Paperbound $3.00

GRAPHIC WORKS OF ODILON REDON. Largest collection of Redon's graphic works ever assembled: 172 lithographs, 28 etchings and engravings, 9 drawings. These include some of his most famous works. All the plates from *Odilon Redon: oeuvre graphique complet,* plus additional plates. New introduction and caption translations by Alfred Werner. 209 illustrations. xxvii + 209pp. 9⅛ x 12¼.

21966-8 Paperbound $6.00

THE RED FAIRY BOOK, Andrew Lang. Lang's color fairy books have long been children's favorites. This volume includes Rapunzel, Jack and the Bean-stalk and 35 other stories, familiar and unfamiliar. 4 plates, 93 illustrations x + 367pp.
21673-X Paperbound **$3.00**

THE BLUE FAIRY BOOK, Andrew Lang. Lang's tales come from all countries and all times. Here are 37 tales from Grimm, the Arabian Nights, Greek Mythology, and other fascinating sources. 8 plates, 130 illustrations. xi + 390pp.
21437-0 Paperbound **$3.50**

HOUSEHOLD STORIES BY THE BROTHERS GRIMM. Classic English-language edition of the well-known tales — Rumpelstiltskin, Snow White, Hansel and Gretel, The Twelve Brothers, Faithful John, Rapunzel, Tom Thumb (52 stories in all). Translated into simple, straightforward English by Lucy Crane. Ornamented with headpieces, vignettes, elaborate decorative initials and a dozen full-page illustrations by Walter Crane. x + 269pp.
21080-4 Paperbound **$3.00**

THE MERRY ADVENTURES OF ROBIN HOOD, Howard Pyle. The finest modern versions of the traditional ballads and tales about the great English outlaw. Howard Pyle's complete prose version, with every word, every illustration of the first edition. Do not confuse this facsimile of the original (1883) with modern editions that change text or illustrations. 23 plates plus many page decorations. xxii + 296pp.
22043-5 Paperbound **$4.00**

THE STORY OF KING ARTHUR AND HIS KNIGHTS, Howard Pyle. The finest children's version of the life of King Arthur; brilliantly retold by Pyle, with 48 of his most imaginative illustrations. xviii + 313pp. 6⅛ x 9¼.
21445-1 Paperbound **$3.50**

THE WONDERFUL WIZARD OF OZ, L. Frank Baum. America's finest children's book in facsimile of first edition with all Denslow illustrations in full color. The edition a child should have. Introduction by Martin Gardner. 23 color plates, scores of drawings. iv + 267pp.
20691-2 Paperbound **$3.00**

THE MARVELOUS LAND OF OZ, L. Frank Baum. The second Oz book, every bit as imaginative as the Wizard. The hero is a boy named Tip, but the Scarecrow and the Tin Woodman are back, as is the Oz magic. 16 color plates, 120 drawings by John R. Neill. 287pp.
20692-0 Paperbound **$3.00**

THE MAGICAL MONARCH OF MO, L. Frank Baum. Remarkable adventures in a land even stranger than Oz. The best of Baum's books not in the Oz series. 15 color plates and dozens of drawings by Frank Verbeck. xviii + 237pp.
21892-9 Paperbound **$2.95**

THE BAD CHILD'S BOOK OF BEASTS, MORE BEASTS FOR WORSE CHILDREN, A MORAL ALPHABET, Hilaire Belloc. Three complete humor classics in one volume. Be kind to the frog, and do not call him names . . . and 28 other whimsical animals. Familiar favorites and some not so well known. Illustrated by Basil Blackwell. 156pp.
(USO) 20749-8 Paperbound **$2.00**

How to Solve Chess Problems, Kenneth S. Howard. Practical suggestions on problem solving for very beginners. 58 two-move problems, 46 3-movers, 8 4-movers for practice, plus hints. 171pp. 20748-X Pa. $2.00

A Guide to Fairy Chess, Anthony Dickins. 3-D chess, 4-D chess, chess on a cylindrical board, reflecting pieces that bounce off edges, cooperative chess, retrograde chess, maximummers, much more. Most based on work of great Dawson. Full handbook, 100 problems. 66pp. 7⅞ x 10¾. 22687-5 Pa. $2.00

Win at Backgammon, Millard Hopper. Best opening moves, running game, blocking game, back game, tables of odds, etc. Hopper makes the game clear enough for anyone to play, and win. 43 diagrams. 111pp. 22894-0 Pa. $1.50

Bidding a Bridge Hand, Terence Reese. Master player "thinks out loud" the binding of 75 hands that defy point count systems. Organized by bidding problem—no-fit situations, overbidding, underbidding, cueing your defense, etc. 254pp. EBE 22830-4 Pa. $3.00

The Precision Bidding System in Bridge, C.C. Wei, edited by Alan Truscott. Inventor of precision bidding presents average hands and hands from actual play, including games from 1969 Bermuda Bowl where system emerged. 114 exercises. 116pp. 21171-1 Pa. $1.75

Learn Magic, Henry Hay. 20 simple, easy-to-follow lessons on magic for the new magician: illusions, card tricks, silks, sleights of hand, coin manipulations, escapes, and more —all with a minimum amount of equipment. Final chapter explains the great stage illusions. 92 illustrations. 285pp. 21238-6 Pa. $2.95

The New Magician's Manual, Walter B. Gibson. Step-by-step instructions and clear illustrations guide the novice in mastering 36 tricks; much equipment supplied on 16 pages of cut-out materials. 36 additional tricks. 64 illustrations. 159pp. 6⅝ x 10. 23113-5 Pa. $3.00

Professional Magic for Amateurs, Walter B. Gibson. 50 easy, effective tricks used by professionals —cards, string, tumblers, handkerchiefs, mental magic, etc. 63 illustrations. 223pp. 23012-0 Pa. $2.50

Card Manipulations, Jean Hugard. Very rich collection of manipulations; has taught thousands of fine magicians tricks that are really workable, eye-catching. Easily followed, serious work. Over 200 illustrations. 163pp. 20539-8 Pa. $2.00

Abbott's Encyclopedia of Rope Tricks for Magicians, Stewart James. Complete reference book for amateur and professional magicians containing more than 150 tricks involving knots, penetrations, cut and restored rope, etc. 510 illustrations. Reprint of 3rd edition. 400pp. 23206-9 Pa. $3.50

The Secrets of Houdini, J.C. Cannell. Classic study of Houdini's incredible magic, exposing closely-kept professional secrets and revealing, in general terms, the whole art of stage magic. 67 illustrations. 279pp. 22913-0 Pa. $2.50

EGYPTIAN MAGIC, E.A. Wallis Budge. Foremost Egyptologist, curator at British Museum, on charms, curses, amulets, doll magic, transformations, control of demons, deific appearances, feats of great magicians. Many texts cited. 19 illustrations. 234pp.   USO 22681-6 Pa. $2.50

THE LEYDEN PAPYRUS: AN EGYPTIAN MAGICAL BOOK, edited by F. Ll. Griffith, Herbert Thompson. Egyptian sorcerer's manual contains scores of spells: sex magic of various sorts, occult information, evoking visions, removing evil magic, etc. Transliteration faces translation. 207pp.   22994-7 Pa. $2.50

THE MALLEUS MALEFICARUM OF KRAMER AND SPRENGER, translated, edited by Montague Summers. Full text of most important witchhunter's "Bible," used by both Catholics and Protestants. Theory of witches, manifestations, remedies, etc. Indispensable to serious student. 278pp. 6⅝ x 10.   USO 22802-9 Pa. $3.95

LOST CONTINENTS, L. Sprague de Camp. Great science-fiction author, finest, fullest study: Atlantis, Lemuria, Mu, Hyperborea, etc. Lost Tribes, Irish in pre-Columbian America, root races; in history, literature, art, occultism. Necessary to everyone concerned with theme. 17 illustrations. 348pp.   22668-9 Pa. $3.50

THE COMPLETE BOOKS OF CHARLES FORT, Charles Fort. Book of the Damned, Lo!, Wild Talents, New Lands. Greatest compilation of data: celestial appearances, flying saucers, falls of frogs, strange disappearances, inexplicable data not recognized by science. Inexhaustible, painstakingly documented. Do not confuse with modern charlatanry. Introduction by Damon Knight. Total of 1126pp.
                                                           23094-5 Clothbd. $15.00

FADS AND FALLACIES IN THE NAME OF SCIENCE, Martin Gardner. Fair, witty appraisal of cranks and quacks of science: Atlantis, Lemuria, flat earth, Velikovsky, orgone energy, Bridey Murphy, medical fads, etc. 373pp.   20394-8 Pa. $3.50

HOAXES, Curtis D. MacDougall. Unbelievably rich account of great hoaxes: Locke's moon hoax, Shakespearean forgeries, Loch Ness monster, Disumbrationist school of art, dozens more; also psychology of hoaxing. 54 illustrations. 338pp.   20465-0 Pa. $3.50

THE GENTLE ART OF MAKING ENEMIES, James A.M. Whistler. Greatest wit of his day deflates Wilde, Ruskin, Swinburne; strikes back at inane critics, exhibitions. Highly readable classic of impressionist revolution by great painter. Introduction by Alfred Werner. 334pp.   21875-9 Pa. $4.00

THE BOOK OF TEA, Kakuzo Okakura. Minor classic of the Orient: entertaining, charming explanation, interpretation of traditional Japanese culture in terms of tea ceremony. Edited by E.F. Bleiler. Total of 94pp.   20070-1 Pa. $1.25

Prices subject to change without notice.
Available at your book dealer or write for free catalogue to Dept. GI, Dover Publications, Inc., 180 Varick St., N.Y., N.Y. 10014. Dover publishes more than 150 books each year on science, elementary and advanced mathematics, biology, music, art, literary history, social sciences and other areas.